Introduction to Hazardous Waste Incineration

Introduction to Hazardous Waste Incineration

LOUIS THEODORE

JOSEPH REYNOLDS

Manhattan College
Riverdale, New York

A WILEY-INTERSCIENCE PUBLICATION

JOHN WILEY & SONS

NEW YORK   CHICHESTER   BRISBANE   TORONTO   SINGAPORE

Library of Congress Cataloging in Publication Data:

Theodore, Louis.
  Introduction to hazardous waste incineration.

  "A Wiley-Interscience publication."
  Includes bibliographies and index.
  1. Incineration. 2. Hazardous wastes.
I. Reynolds, Joseph P. II. Title.

TD796.T48 1987 628.4'457 87-8300
ISBN 0-471-84976-6

Printed in the United States of America

10 9 8 7 6 5 4 3

To
the QUEEN—one more time

and

To
BARBARA

Preface

The engineering profession has recently expanded its responsibilities to society to include the management of hazardous wastes, with particular emphasis on control by incineration. Increasing numbers of engineers, technicians, and maintenance personnel are being confronted with problems in this most important area. Since the problem of hazardous wastes is a relatively new concern, the environmental engineer of today and tomorrow must develop a proficiency and an improved understanding of the incineration of hazardous wastes in order to cope with these challenges. Although this is not the first professional book to treat this particular subject, it is the only book dealing with the technical and engineering aspects of hazardous waste incineration that may be used as a text.

This text–reference book is intended primarily for practicing engineers and engineering students. It is assumed that the reader has already taken basic courses in physics and chemistry, and has a minimum background in mathematics through calculus. The authors' aim is to offer the reader the fundamentals of incineration with appropriate practical application to the incineration of hazardous wastes, and to provide an introduction to the specialized literature in this and related areas. The readers are encouraged, through the bibliography, to continue their own development beyond the scope of this book.

As is usually the case in preparing a book, the problem of what to include and what to omit has been particularly difficult. However, every attempt has been made to offer engineering course material to individuals with a technical background at a level that should enable them to better cope with some of the complex problems encountered in waste incineration today.

The book is divided into four parts: an overview of the hazardous waste field, incineration principles, equipment, and facility design. In Part I, the general subject of hazardous waste incineration is examined. A separate chapter is devoted to other options for controlling hazardous wastes. Standards and regulations are also reviewed. Part II covers the broad subject of incineration principles. This section includes chapters on basic concepts, stoichiometric calculations, and thermochemical considerations. Much of this material is original and appears in print for the first time. Part III is highlighted by individual chapters devoted to equipment that may be found in a hazardous waste incineration facility. These include the incinerator, quencher, waste heat boiler, and air pollution control equipment. The last chapter examines some of the ancillary equipment. Part IV contains three chapters devoted to design principles and their application to a hazardous waste incineration facility. The book concludes with two extensive illustrative examples detailing the design of waste facilities. A short writeup on International System of units (SI/metric) is provided in the Appendix.

No engineering tool is complete without information on how to use it. By the same token, no engineering text is complete without its illustrative examples, which serve the important purpose of demonstrating the use of the procedures, equations, tables, graphs, and so on, presented in the text. There are many such examples as well as practice problems at the end of each of the technical chapters.

During the preparation of this book, the authors were ably assisted in many ways by a number of graduate students in Manhattan College's Chemical Engineering Masters Program. These students contributed much time and energy researching and writing parts of various sections of this book. We gratefully acknowledge their invaluable assistance; the names of these students are listed under the titles of the chapters to which they contributed (*contributing author* for those who played a major role in writing a part or several parts of the chapter and *partial contributor* for those who made a finite—though less than a major—contribution to the chapter). Several undergraduate students also gave generously of their time and energy performing numerous calculations (many of which required use of the computer) and helping with the preparation of the manuscript. These students are Antonella Caruso, Ann Corsetti, Mary Hewitt, Michael Kelly, Douglas Matthews, Nicola Peill, and Romona Pezzella. The authors are particularly indebted to Ui Young Choi for her technical contributions to the text. Our sincere gratitude is due to Mrs. Ann Kaptanis for her dedicated and patient efforts in typing the manuscript.

Somehow the editor usually escapes acknowledgment. We were particularly fortunate to have Jim Smith of John Wiley & Sons serve as our editor. He had the vision early on to realize the present need and timeliness for a project of this nature. He tolerated our mistakes and ignored our idiosyncrasies. His patient handling of the numerous problems that arose is particularly appreciated. It would have been difficult to complete the manuscript in its present form without his able direction.

LOUIS THEODORE
JOSEPH REYNOLDS

Riverdale, New York
July 1987

Contents

ix

Introduction to Hazardous Waste Incineration

I

THE HAZARDOUS WASTE PROBLEM

Part I serves the purpose of placing the subject of hazardous waste incineration in historical perspective and in perspective with respect to the field of hazardous waste management. Three chapters are included in this part.

Chapter 1 presents some of the historical background of the hazardous waste problem which has been compounded over the last half-century and, as a result, now requires enormous efforts to solve. This chapter also covers the topic of where the hazardous waste problem stands today. Hazardous waste management is a rather broad field with many different solutions and techniques for the disposal of hazardous waste material, of which incineration is only one option.

Chapter 2, entitled *Other Options*, describes many of these methods. Some of them are alternatives to incineration and are sometimes used instead of incineration to handle a particular waste stream. Others are techniques that are used to process wastes that cannot, because of the physical or chemical properties of the waste, be treated by incineration.

The legal aspects of the hazardous waste problem (i.e., the standards and regulations imposed on industry by federal, state, and local governments), which has a significant impact on the methods industry uses to handle particular wastes, are covered in Chapter 3. The possible future of incineration is also discussed—a future determined just as much by legal standards as by social, economic, and technological considerations.

1

Hazardous Waste Incineration Overview

Contributing author: Kent Murphy

1.1 INTRODUCTION

Hazardous wastes are legally defined as those wastes that may cause adverse or chronic effects on human health or the environment when not properly controlled. Hazardous wastes are generated either because processes have converted harmless materials into hazardous substances or because natural materials that are hazardous to begin with have been concentrated and released into the environment. These substances may be ignitable, reactive, corrosive, radioactive, infectious, or toxic. They may exist as solids, liquids, sludges, powders, or slurries. About 90% of them are liquid or semiliquid. Some of these wastes are nondegradable and may persist in nature indefinitely.[1]

The hazardous waste issue is a by-product of this nation's economic growth over the past 30 to 40 years. As industry has expanded, society has developed a greater reliance on a broader range of chemicals and chemical by-products in plastics, automobiles, and other major manufacturing sectors. When chemicals are used to develop new products, wastes are almost always produced. As the economy has grown and as industries have diversified, the result has been the advent of many new waste streams.

Table 1.1.1 shows the type of wastes generated by various industries as well as the usual methods of disposing of these wastes.[2] Some industries produce waste streams that are typical of the industry, for example, textile, pulp and paper, leather tanning, and plating wastes. For a given industry, the characteristics of these waste streams do not vary much from facility to facility. Other industries have highly variable waste streams; an example is the chemical industry where over a thousand different products may be produced at a single facility.

It is estimated that 264 million metric tons of hazardous waste were generated in 1981, a quantity equal to more than 70 billion gal. These are enormous numbers, although large portions of the total were actually mixtures of hazardous and non-hazardous wastes, such as wastes mixed with industrial process liquids. Since 1981,

3

TABLE 1.1.1. Hazardous Material Use, Production, and Waste by Industry

Industry	Properties of Products Used[a]					Properties of Products Made[a]					Toxic Materials Used[b]					Toxic Materials Produced[b]					Disposal Methods[c]				Toxic Wastes Produced[b]				
	C	E	F	R	P	C	E	F	R	P	S	H	G	O	I	S	H	G	O	I	DW	SEA	CON	AEC	S	H	G	O	I
Mining	x	x		x											x											x	x		x
Food		x	x							x			x												x			x	
Textiles	x	x	x						x		x	x															x		
Paper and allied products	x	x				x					x	x								x		x							x
Alkalis and chlorine	x					x		x			x	x	x	x	x			x				x			x		x		x
Cyclic intermediates	x	x	x			x	x	x			x	x	x	x	x	x	x	x	x	x	x		x		x	x	x	x	x
Organic chemicals	x	x	x			x	x	x			x	x	x	x	x	x	x	x	x	x	x	x	x		x	x	x	x	x
Inorganic chemicals	x	x				x		x			x	x	x	x	x		x		x	x	x	x			x	x	x	x	x
Plastic materials	x	x				x		x			x	x	x	x	x		x	x	x	x	x	x	x		x	x	x	x	x
Drugs	x			x		x			x		x	x		x	x	x	x		x	x			x		x		x	x	x
Soaps and cleaners	x										x	x		x	x				x	x					x			x	x
Paints and allied products	x							x			x	x	x	x	x				x	x		x	x		x	x	x	x	x
Agricultural chemicals	x							x			x	x	x	x	x	x		x	x	x	x	x	x		x	x	x	x	x
Explosives	x							x			x	x		x	x				x	x	x		x		x		x	x	x
Petroleum and coal products	x	x				x		x			x	x	x	x	x	x		x	x	x	x	x	x		x	x	x	x	x
Leather tanning	x										x		x	x	x					x					x	x		x	x
Asbestos products												x		x	x					x						x			x
Blast furnaces and steel	x											x		x	x								x			x	x		x
Nonferrous metals	x											x		x	x											x			x
Metal services	x										x	x		x	x										x	x	x	x	x
Department of Defense	x	x	x			x	x	x			x	x	x	x			x				x	x	x		x	x	x	x	x
Atomic Energy Commission	x		x					x	x		x	x	x	x							x			x	x	x	x	x	x
National Aeronautics and Space Administration	x	x	x	x		x	x	x			x	x	x	x							x				x	x	x	x	x
Consumers	x	x	x	x		x		x			x	x	x	x	x								x		x	x	x	x	x

[a] Hazard code: C = corrosive, E = explosive, F = flammable, R = radioactive, P = pathogenic.
[b] Toxicity code: S = toxic solvents, H = toxic metals, G = toxic gases, O = toxic organics, I = toxic inorganics.
[c] Disposal code: DW = deep well, SEA = deep sea dump, CON = contract disposal, AEC = radioactive disposal.

the amount of hazardous waste generated annually has increased. About 14,000 installations generate regulated quantities of hazardous waste, while 4800 companies manage these wastes using various treatment, storage, and disposal (TSD) techniques. The overwhelming fraction of hazardous waste generated in this country, 96%, is managed on-site. Of this, about 14.7 billion gal are disposed of in or on the land each year and about 500 million gal are incinerated.[3]

Hazardous waste generators and management facilities are concentrated in the manufacturing industries. An estimated 85% of all generators and 72% of all treatment, storage, and disposal facilities are associated with industrial manufacturing operations. Of the total quantity of wastes generated, manufacturers account for 92%; the chemical industry alone generates about 68% of the total.[2]

1.2 HISTORICAL BACKGROUND

Wastes that are now identified as *hazardous* have been around for a long time. The present day problems with hazardous wastes seem to be fairly recent, however. Some reasons for this are the following:

- *Expansion.* Urban centers are more concentrated than they were years ago; the population has increased; industry in general, and the chemical industry in particular, have greatly expanded.
- *Reaction of the Environment.* It takes time for the land to react to abuse, and there are fewer places to leave wastes around without causing problems.
- *Public Awareness.* Above all, society has become much more aware of the health and environmental dangers associated with hazardous wastes.

To fully understand the reasons why there is a hazardous waste problem today, it is necessary to examine the past. The U.S. government has generally been inclined to leave industry alone and avoid over regulation. This national attitude of freedom has been clearly reflected in a reluctance to regulate and enforce health and environmental issues, either on a state or federal level. Prior to World War II, there were no environmental regulatory programs. What federal and state activities there were can best be described as *advisory* and mildly *persuasive*. No authority existed to assure that good waste management was practiced. Rivers handled the liquid wastes, and the open burning dump took care of the solids. With the end of World War II and the economic prosperity that followed, the amounts and types of wastes produced by society expanded exponentially. At first, waste streams were handled in the same manner as before the war when lesser amounts of waste were produced. The same technologies were used to treat the waste and the same institutional structures were employed to oversee their management. For the most part, these wastes ended up in the streams and skies; the balance went into the land.

Substantive water and air pollution control legislation was not passed until as late as the 1950s and 1960s. (The history of solid and hazardous waste legislation is covered in detail in Chapter 3.) These landmark efforts by the federal government led to the development of federal and state regulatory programs to require adequate treatment of wastewater and waste air streams before they were discharged into the ambient environment. However, as wastewater treatment plants and air pollution control systems were built, the residual wastes produced by such efforts were diverted to the land. At that time, there was no federal legislation to protect the land because few people considered it necessary. Some protection was provided by a myriad of controls emanating from local governments. Local control, however, was normally directed at regulating the use of land, and not preserving its quality. Consequently, there were almost no restrictions on diverting concentrated waste streams resulting from air and water cleanup to the land.

The technology for land disposal of wastes was not adequate for the types of wastes received. The use of a sanitary landfill was not established until after World War II, and even then the ability of the landfill to handle liquids, sludges, and persistent chemicals in hazardous wastes was not adequate. Pits, ponds, and lagoons utilized for these wastes were also inadequate. More sophisticated technologies, such as incineration, were not considered because the less expensive option of using the land remained.[4]

Technology is available today to render hazardous wastes sufficiently innocuous so that they can be disposed of in ways that will not create unreasonable risks to public health and the environment. There are many methods for handling process waste streams. With the exception of incineration, these are covered in Chapter 2. Of these, seven are quite commonly used, although not all are suitable for the disposal of *hazardous* wastes. Brief descriptions are given in the following list.

1. *Incineration.* This involves oxidative conversion of combustible materials to harmless gases suitable for atmospheric release. Undesired gaseous products such as HCl, SO_2, NO_x, must be removed prior to release. (Note: the symbol NO_x represents various oxides of nitrogen.) Heats of combustion may sometimes be recovered, and occasionally additional combustible material must be added to insure adequate combustion. Residual solids, such as ash, are landfilled.

2. *Dispersal into Contiguous Water Bodies.* Historically this procedure has been widely employed, but EPA (the Environmental Protection Agency) action has now curtailed it in many industries. Usually the waste solid is temporarily stored in a lagoon, wherein it is treated to minimize environmental impact. For example, acids and bases are neutralized and suspended solids are permitted to settle. In the case of river discharge, attempts are made to release waste during periods of high flow.

3. *Ocean Dumping.* This practice involves transporting the material out to sea and then releasing it. Usually the material is conveyed by barge. It may be released directly or within containers. Costs depend on distance of transport, and whether the material can be loaded in-plant or must be intermediately transported.

4. *Lagooning.* Lagoons may be employed for either temporary or permanent storage. Permanent storage occurs when the solid material tends not to settle. For example, phosphate slimes have undesirable properties if water cover is not maintained. Other examples include red mud (*blows*) from alumina and phossy water (*burns*) from phosphorus. *Phossy water* is water that contains particles of phosphorus metal. The metal bursts into flames when exposed to oxygen.

5. *Sanitary Landfill.* This pertains to the burial of nonhazardous solid wastes under controls sufficient to preclude degradation of the surrounding environment.

6. *Chemical Landfill.* This is the extension of sanitary landfill to enable the acceptance of hazardous waste.

7. *Subsurface Injection.* In this case, the solid material is slurried and then

pumped into underground cavities. The EPA opposes such disposal practice for hazardous waste.

The EPA is currently in the process of further examining alternative technologies for treating hazardous wastes.[5] Included in the list are biological treatment processes, dechlorination, carbon adsorption, recovery–reuse–recycle processes, and solidification and stabilization processes.

Today, in spite of its cost, incineration is a popular method for hazardous waste disposal; one reason for this fact is that it has a proven track record. The following is a brief history of incineration systems in the United States.

Incineration of chemicals now considered hazardous is a technology that is < 30 years old. After World War II, the chemical industries in the United States began to grow rapidly and with that growth came a commensurate growth in chemical wastes to be disposed of. Solid materials, whether municipal wastes or by-products of process operations, were often burned in open dumps. In the early 1950s, the widespread installation of municipal incinerators for the treating of these solid wastes occurred, but most liquid process wastes were released directly to sewers or nearby streams and rivers.

In specific industries, there were some early attempts at waste incineration before 1960, but such attempts were fairly crude. By banning waste discharge to sewers and streams, the *Federal Water Quality Act* promulgated in 1966 provided a driving force for the incineration of combustible waste materials. During the late 1960s and early 1970s, industrial incineration became an important disposal method for processing wastes from the chemical and allied industries. Most of the early systems, however, were not designed to meet current air pollution control regulations and would not be permitted to operate today.

Dow Chemical Company began a large on-site incineration facility at Midland, Michigan in the late 1960s, utilizing a rotary kiln as the basis of the incineration process. (Various types of incinerators are discussed in Chapter 7.) Liquid injection incinerators for liquid process wastes were first introduced by companies such as John Zink in Tulsa, Oklahoma, Thermal Research and Engineering Corporation (now Trane Thermal Company) in Conshohocken, Pennsylvania, and other combustion equipment manufacturers who anticipated liquid process incineration as a new market. Liquid injection incinerators of this period were often vertically fired and discharged hot gases from the combustion chamber directly to the atmosphere. In most cases, the environmental criterion was opacity, which sometimes permitted high particulate discharge and often concentrations of acid gases such as hydrogen chloride and sulfur dioxide.

After the *Clean Air Amendments* of 1970, certain limitations were placed on the types of emissions that could be discharged to the atmosphere. Incinerator manufacturers had to add particulate removal devices such as absorbers, high energy scrubbers, electrostatic precipitators, or baghouses. The capital costs required for a complete system thereby increased substantially over that for the incinerator itself.

During the 1970s more sophisticated liquid injection systems were developed and refined to handle a greater variety of process wastes. For example, new incinerator designs were needed to handle some wastes containing inorganic salts that attacked refractories and for the destruction of waste gases from vinyl chloride monomer plants. During the same period, rotary kiln technology was the major system employed for the destruction of process wastes in solid or sludge form and which could not be atomized in a liquid injection system. The rotary kiln was equipped with the same type of air pollution control devices that were needed in the liquid injection system. In 1974, about the time of the oil crisis, the first heat recovery system was employed with a liquid injection incinerator. In 1975, additional waste heat boilers were installed, some of these even on flue gases with corrosive atmospheres. (Refer to Chapter 8 for a discussion of waste heat boilers.)

In 1979 and 1980, the *Resource Conservation and Recovery Act* (RCRA), passed in 1976, began to exert its influence through regulatory promulgation and enforcement. Temporary *Part A* permits were issued for incineration facilities under RCRA authorization and EPA began changing the rules for incineration. First, detailed design requirements were proposed, followed by destruction and removal efficiency standards (DRE). These were later changed to the current performance standards. The new design criteria stressed the well-known combustion parameters of retention time, adequate operating temperature, turbulence, and sufficient oxygen.

Federal regulation of chemical waste incineration under RCRA quite naturally brought about significant progress in source testing. Under RCRA, EPA proposed the first regulation for hazardous waste incinerators in May, 1980 and promulgated an interim final regulation in July, 1982. The regulation set performance standards for three process parameters: particulate emissions [0.08 gr/dscf (grains per dry standard cubic foot) of exhaust, corrected to 50% excess air], HCl emissions (< 4 lb/h or, if a scrubber is employed, better than 99% removal efficiency), and destruction and removal efficiency for organic materials (a minimum of 99.99%). The *trial burn* for compliance demonstration was established. Over a 3-yr period following proposal of the regulation, earlier analytical methods were upgraded and refined. In 1983, a national program sponsored by EPA for source testing of hazardous waste incinerators spurred the development of more sensitive and reliable field sampling methods such as the volatile organic sampling train (VOST).

Today, permitting of hazardous waste incinerators is a lengthy and complex process, using testing methods that are continually being refined. After all general facility standards have been met and a detailed trial burn plan has been approved by the appropriate state or EPA region, performance, and operating limitations for the incinerator are established through the trial burn.[6] More extensive details on the current permitting process for hazardous waste incinerators are given in Chapter 3.

1.3   TODAY'S PROBLEM

A decade has passed since hazardous waste control became a public concern. The key starting point for this concern was the passing of the RCRA in 1976. In the ensuing years, EPA gradually issued regulations based on the RCRA; states became actively involved in managing hazardous waste programs, and another law, the *Comprehensive Environmental Response, Compensation and Liability Act* (CERCLA), better known as *Superfund* came into being in 1980. During the same period the Chemical Process Industries (CPI) began tightening up production operations to generate fewer or less harmful wastes. Also, an emerging waste-management service industry became a big time business; current estimates are that this represents a $3 to 5 billion/yr market of its own.

From many perspectives, however, this does not mean that the hazardous wastes problem is under control. Despite the efforts of many companies, and sometimes state or local governments, to build new waste management facilities, local resistance to their siting is still extremely intense. The lack of available sites has caused many problems in disposing of wastes and, especially where wastes must be transported long distances, has created other risks, for example, the possibility of truck or rail accidents. The NIMBY (*not in my back yard*) syndrome remains the overriding response of most citizens to the siting of treatment facilities. Similar problems abound with most alternative waste treatment technologies, such as landfills. Regulatory, industrial, or environmental structures have prevented many such technologies from being widely applied, even though these methods may be safer or cheaper in the long term. Legal, economic, and societal hurdles have also hampered the Superfund program, the intent of which has been the cleaning up of derelict dump sites by using a fund generated from taxes on chemical feedstocks.[7]

The difficulty of identifying hazardous wastes and assessing the resulting health and environmental risks is another current problem. Some questions that need answers are

1. How do hazardous wastes move through the environment?
2. To how much of the wastes are human populations actually exposed?
3. Based on the answer to Question (2), what degree of health hazard do these amounts represent?
4. After their final treatment and disposal, what is the ultimate fate of pollutants?

Finding the answers to these and more questions like these will require time and money. The present list of hazardous wastes defined by RCRA is primarily based upon short-term studies. Though the list is quite extensive, it is certain to become much longer.

Sampling and analytical methods for identifying hazardous wastes are generally costly, complex, time consuming, and were developed for specific media. The proper identification of hazardous wastes present at a site and their concentration is the first step in assessing the potential risks posed. Problems arise, however, when dealing with the complex and heterogeneous waste mixtures and conditions normally found at a hazardous waste dump site. As a result, it is difficult to obtain representative samples. Standardized and validated sampling methods are therefore needed to insure reproducible results. If representative samples are not obtained, the true extent of the risks posed to human health and the environment by a particular site cannot be measured.

In addition to high costs for chemical analysis, there are other analytical and quality assurance/quality control (QA/QC) problems, particularly with nonaqueous samples. Some limitations include: difficulties in preserving samples at the concentration levels found on-site; problems in preparing samples for analysis; and complexities in identifying interference effects, where the presence of some compounds may be hidden from analysis by the presence of others. Biological tests may provide quick screening tools to reduce chemical analysis costs. Biological tests use animals, plants, microorganisms, and cells to evaluate the toxicity of sample materials. Biological analysis may also overcome some of the difficulties posed by mixtures and the synergistic–antagonistic reactions of codisposed wastes, since these tests study reactions to the whole waste stream and, unlike chemical tests, do not identify individual compounds in that stream. A synergistic–antagonistic reaction involves two or more compounds reacting with one another to create a greater or lesser hazardous waste stream.[8]

Adequate treatment and disposal capacity is critical to the management of hazardous wastes. The development of environmentally sound disposal facilities is essential to the successful implementation of the hazardous waste regulatory program currently mandated by law. This capacity does not exist. There is a shortage of suitable disposal sites and this problem will become even more acute as additional wastes are determined to be hazardous, existing sites are closed because they do not meet environmental criteria, and wastes that are being disposed of on private property are taken to off-site facilities. Although not all states lack appropriate capacity, other problems prevail. In New York, for example, where adequate capacity does exist, the disposal facility is often not within a reasonable distance of the waste generator. This increases the cost and energy associated with long transportation hauls. There are also legislative barriers to aggravate the capacity problem. The state of Connecticut, for example, passed a law allowing local governments to prohibit, through zoning, land usage for hazardous waste disposal. No disposal site can be built, established, or altered without the approval of both the local zoning planning board and the Connecticut Department of Environmental Protection. This law almost guarantees that the department will not get approval from local governing bodies for a site to be used for regional purposes. Advocates of the law contend that the smaller towns should not be forced to serve as dumping grounds for wastes from other communities.

There are techniques to reduce waste volumes, but these are not expected to alleviate the capacity problem at the present time. Such techniques include the restriction of hazardous chemicals used in operations, substitution of less hazardous materials, and better quality control to reduce production spoilage. The more hazardous or toxic waste streams can be isolated from mixtures in which they occur, and wastes can be concentrated by dewatering, resulting in reduced waste volumes and disposal costs. Many wastes also contain valuable basic materials, which makes material recovery logical from both resource conservation and environmental viewpoints. However, these techniques have not gained general acceptance or wide use because at present they are usually more expensive than land disposal.

1.4   SOCIOECONOMIC CONCERNS

Socioeconomic and environmental considerations, together with public acceptance, dictate the location of a proposed site for a hazardous waste incinerator. Unfortunately, the controlling factors seem to be the socioenvironmental concerns rather than economic or technical factors. *Economic* concerns include installation and operating costs of the incinerator, costs of transporting and storing the hazardous wastes, and disposal costs of the residue. If the incineration system is land based, the costs include adaptation of the site, which involves access roads, water and sewage problems, topography modifications, soil problems, aesthetic treatment, and compensation for possible losses by neighboring taxpayers from air pollution damage to and devaluation of their property. *Social* and *environmental* concerns mainly involve potential effects upon public health. The adverse consequences of living within an environment that receives low levels of various air contaminants on a daily basis is not known. However, the inhalation of high levels of many specific metallic or salt particulates, polynuclear organic compounds, and acid gases is known both to have serious respiratory effects and to contribute to development of chronic illnesses such as cancer. These contaminants can be emitted continuously from a hazardous waste incinerator if not properly removed from the stack gas.

Siting the Incinerator

The siting of a proposed incinerator, has been one of the main social problems. A large segment of the public is apparently unconvinced that hazardous wastes can be disposed of safely. Furthermore, the siting process is not always perceived by the public as being fair, open, and equitable. Those who live near the selected site feel that they carry too much of the cost, including health concerns, lower property values, and the stigma of living near a site. The challenge to the industries involved and the federal and state governments is to convince the public that the site is needed, that reasonable steps have been taken to protect public health, that the solution is fair, and that appropriate incentives have been developed for community acceptance.

Despite the ever-increasing quantities of hazardous wastes, there are at present only a limited number of commercially available hazardous waste incinerators in the United States. During the past several years, disposal of oil and hazardous substances resulting from cleanup of accidental spills and abandoned waste sites has become increasingly difficult. As described previously, greater public awareness of dangers associated with hazardous waste disposal has led to local opposition to proposed land disposal sites. The justifications for new incineration facilities are obvious and critical but finding a site is extremely difficult.

Under today's regulations, the private sector would not under normal circumstances be able to obtain a site for a hazardous waste incineration facility. For this reason, some public role in siting has become necessary. This does not guarantee a site, however. As an example, no commercial hazardous waste incinerators are currently operating in New York State. To meet the obvious need, New York State proposed to site and finance a high technology hazardous waste disposal facility in 1980 with an approved budget of $60 million. The facility would have consisted of the necessary equipment and structures for transporting, receiving, and storing wastes; a laboratory for analytical and testing services; various processing units; an incinerator; secure landfills; waste lagoons and ponds; and storage tanks and wastes piles. After almost two aggressive years of searching, a rural site of 600 acres near Sterling in central New York State was finally chosen. The selected site met with such strong public opposition that the plan for the high technology hazardous waste disposal facility was suspended.

In a more positive example, Dow Chemical Company in Midland, Michigan has treated and disposed of virtually all of its wastes on-site for the past 30 yr in an effort to avoid involving other parties and to control its liability. In addition to an intensive program of waste prevention and recovery, Dow Chemical has stressed thermal destruction rather than land disposal. Less than 1% by volume of Dow's chemical wastes is deposited in the ground. This approach has proven to be cost effective and has not invoked credible local criticisms. On-site treatment, therefore, appears to be a viable approach for the larger corporations with adequate land resources.

Cooperation between waste generators and waste management firms to build on-site incinerators is being suggested as a beneficial approach for both parties.

Some advantages would be: the generator's off-site transporation costs would be eliminated, liability would be reduced, and the dependable supply of wastes generated on-site would enable the waste management firm to attract as regular clients those generators producing a waste stream that the "host" generator's wastes could neutralize. Furthermore, the public would tend to perceive the facility as another stage in an already established production process, thus lessening opposition.

Economic Factors

There is a lack of consistent information on unit costs of incineration, as well as that of other current hazardous waste management practices. Unit costs vary widely depending on such factors as: (1) the physical state, the British thermal unit (Btu) content, ash content, and the toxicity of the particular waste, (2) the form of the waste (i.e., drum versus bulk), and (3) the competition at the regional level among alternative waste management facilities. For ocean incineration, reliable estimates of costs are further complicated because the availability of ocean incineration capacity has not yet been firmly established in the commercial market. For all hazardous waste management practices, costs are also strongly affected by changes in the federal regulatory program.

Recently, commercial incineration has become increasingly cost competitive with land disposal. The price of landfilling hazardous waste has increased substantially because of increased operating costs and the requirements for pretreating and special handling of many waste streams prior to landfilling. For example, the prices charged by waste management firms generally increased about 25 to 40% per yr from 1978 to 1984, depending on the type and form of waste and the geographical location; at some facilities, the price increase exceeded 50%. This cost increase for landfilling of hazardous waste was the highest reported for any treatment or disposal option, a trend tending to make incineration a more attractive disposal alternative. Recent changes in the RCRA program and the potential long-term liability costs associated with landfills provide additional incentives for incineration.

Nevertheless, the cost of incineration, which may be reduced as technology advances, is generally higher than many other disposal alternatives. Although this cost may be considerable, it is the NIMBY syndrome, and not economic factors, that present the main barrier to incineration as a more widely used hazardous waste management option.

1.5   WASTE CHARACTERIZATION

Waste characterization is a major factor in assessing the feasibility of converting a hazardous waste material by process incineration. It effects the design of the process incinerator and its emission control system, and helps determine the compatibility of a waste with a proposed or available facility. It also plays a part in determining process incineration operating conditions for complete conversion of a specific waste.

Hazardous wastes can be classified under five categories: (1) waste oils and chlorinated oils, (2) flammable wastes and synthetic organics, (3) toxic metals, etchants, pickling, and plating wastes, (4) explosive, reactive metals, and compounds, and (5) salts, acids, and bases. A particular waste may overlap into any number of these five categories. Flammable wastes are comprised mainly of contaminated solvents; this category also includes many oils, pesticides, plasticizers, complex organic sludges, and off-specification chemicals. Synthetic organic compounds include halogenated hydrocarbon pesticides, polychlorinated biphenyls, and phenols. Only *combustible* wastes are candidates for incineration; if a waste contains a significant amount of heavy metals, the metals should be concentrated and removed prior to incineration.

A number of wastes present special problems when incinerated. Metals, for example, become extremely fine metal oxide particles that may not be collectable by conventional air pollution control equipment. Resins may polymerize, coating incinerator surfaces and plugging nozzles designed to atomize liquid wastes. Polyolefins and nitrocellulose may detonate rather than burn. When the wastes being incinerated contain a significant concentration of halogenated compounds, the formation of undesirable combustion products such as hydrogen chloride, hydrogen fluoride, and hydrogen bromide results. For design purposes, it is necessary to know (a) whether the waste is a gas, liquid, or solid; (b) the fraction of the waste that is organic and whether the inorganic material includes water; and (c) why the waste is hazardous and to what degree.

Characterization of waste is usually accomplished using available analytical techniques, although problems do arise because almost all of the traditional analytical methods are applicable for the analysis of either pure chemicals or nearly homogeneous materials.[9] Chemical wastes are typically highly heterogeneous and the sample matrix is almost always complex. Hence, many of the usual methods or instruments do not work well on waste samples. This problem is further complicated because almost all of the state-of-the-art instruments are designed to analyze microsized specimens. It is very unlikely that the few milligrams of a heterogeneous waste specimen are representative of the whole sample.

Proper sample collecting and handling are obviously critical steps in determing waste characteristics. Because sampling situations vary widely, no universal sampling procedure is recommended. The EPA published a checklist of suggested steps to be followed to help maximize the safety of sampling personnel, minimize sampling time and cost, reduce errors in sampling, and protect the integrity of samples after sampling.[10]

To properly characterize the waste for incineration the following data must be determined during the analysis.[10]

The Percentages of Carbon, Hydrogen, Oxygen, Nitrogen, Sulfur, Halogens, Phosphorus, Ash, Metals, and Salts in the Waste, as Well as Its Moisture Content. The amounts of these components need to be known in order to calculate stoichiometric combustion air requirements and predict combustion gas flow and composition. (Part II covers these stoichiometric calculations in detail.) The presence of halogenated and sulfur-bearing wastes can result in the formation of HCl, HF, H_2, and SO_2 in the incinerator gases. These must be removed with suitable scrubbing equipment before discharge into the atmosphere. (Chapter 10 covers the topic of scrubbing equipment in detail.) Also, in the incineration of organic waste containing chlorine, sufficient hydrogen should be provided by either the waste or the auxiliary fuel so that hydrogen chloride, rather than chlorine gas, is formed. The nitrogen content of waste material is generally low, but the presence of nitrogen-containing materials (nitrates, ammonium compounds, etc.) can greatly increase the NO_x emissions.

Trace Metals. Trace metals are a potential cause for concern in incineration emissions and analysis for them should generally be performed unless it is known that they are not present in the waste. As mentioned earlier, wastes containing significant amounts of metals will generally be poor candidates for incineration. Such wastes will require post-combustion emission control of a special type. The effluent or solid waste from the emission control device must in turn be treated as a hazardous waste, even though it will be considerably reduced in volume and weight from the original hazardous waste.

Ash Content. The ash content of the waste should be determined in order to evaluate the potential for excessive slag formation as well as potential particulate emissions from the incinerator. Kinematic viscosity and the size and concentration of solids in a liquid waste are the most important physical properties to consider in evaluating a liquid waste incineration design. The physical handling system and burner atomization techniques are dependent on viscosity and solid content of the waste. Chemically complex sludges may contain such elements as sodium, potassium, magnesium, phosphorus, sulfur, iron, aluminum, calcium, silicon, oxygen, nitrogen, carbon, and hydrogen. Several chemical reactions can be expected to take place in the high temperature oxidizing atmosphere of an incinerator operation treating such sludges. Resulting ash may contain Na_2SO_4, Na_2CO_3, $NaCl$, and soon. Sludges containing substantial amounts of sodium can cause defluidization of a fluidized bed by forming low melting eutectic mixtures. (The fluidized-bed incinerator is described in Chapter 7.) Furthermore, if the particles of the fluidized bed are silicasand, Na_2SO_4 will react with the silica to form a viscous sodium silicate glass that will cause rapid defluidization.

Heating Value. The heating value of a waste corresponds to the quantity of heat released when the waste is burned and is commonly expressed in units of British thermal units per pound (Btu/lb). This quantity must be considered in establishing an energy balance for the combustion chamber and in assessing the need for auxiliary fuel firing. As a rule of thumb, a minimum heating value of about 5000 Btu/lb is required to sustain combustion.

Special Characteristics. Special characteristics of the waste such as extreme toxicity, mutagenicity or carcinogenicity, corrosiveness, fuming, odor, pyrophoric properties, thermal instability, shock sensitivity, and chemical instability should be considered in incinerator facility design. Thermal or shock instability are of particular concern from a combustion standpoint, since wastes with these properties pose an explosion hazard. Other special properties relate more directly to the selection of waste-handling procedures and air pollution control requirements, topics to be considered in later chapters.

In Table 1.5.1 a scheme for grouping combustible hazardous wastes according to elemental content is shown.[11] The four groups represent varying degrees of ease of incineration. For example, Classes 3 and 4 are more difficult to burn by virtue of their content of chlorine, bromine, fluorine, sulfur, phosphorus, silicon, and sodium. Combustion of these elements generates acids and oxides and hence, special emission control techniques are required.

It may not be necessary to follow the complete, elaborate analysis protocol for each shipment of waste from the same source, unless the material is entirely different from earlier shipments. How often the shipments should be sampled and for what

TABLE 1.5.1. Chemical Waste Classification

Waste Class	Elemental Composition	Example
1	C, H and/or C, H, O	Tars from production of styrene Off-specification phenol
2	C, H, N and/or C, H, N, O	Solid residue from manufacture of aromatic amines TDI manufacture reactor tar bottoms[a]
3	C, H, Cl and/or C, H, Cl, O	Vinyl chloride monomer manufacturing wastes Phenolic tar from 2,4-D manufacture[b]
4	C, H, N, Cl and/or C, H, Cl, N, O	Nitrochlorobenzene manufacturing wastes
	C, H, S and/or C, H, S, O	Petroleum refining sour waste
	C, H, F and/or C, H, F, O	Fluorinated herbicide wastes
	C, H, Br and/or C, H, Br, O	Ethylene bromide manufacturing wastes
	C, H, P and/or C, H, P, O	Malathion
	C, H, Si and/or C, H, Si, O	Tetraethyl orthosilicate wastes
	C, H, Na and/or C, H, Na, O	Refinery spent caustic

[a] TDI is toluene diisocyanate.
[b] 2,4-D is 2,4-dichlorophenoxyacetic acid.

parameters samples should be analyzed is best determined on a case-by-case basis using good engineering judgment.

1.6 THE NEED TO INCINERATE

From a regulatory point of view, incineration now appears to be the most viable means of waste treatment for organics. The next several decades should increase the demand for incineration because of limitations placed on other hazardous waste management options as a result of the 1984 RCRA amendments.

Incineration of wastes offers the following advantages or potential advantages.

- *Volume reduction*, especially for bulky solids with a high combustible content.
- *Detoxification*, especially for combustible carcinogens, pathologically contaminated material, toxic organic compounds, or biologically active materials that would affect sewage treatment plants.

- *Regulatory compliance*, especially for fumes containing odorous compounds, photoreactive organics, carbon monoxide, or other combustible materials subject to regulatory emission limitations.
- *Environmental impact mitigation*, especially for organic materials that would leach from landfills or create odor nuisances.
- *Energy recovery*, especially when large quantities of waste are available and reliable markets for by-product fuel or steam are nearby.

These advantages have justified development of a variety of incineration systems, of widely different complexities and functions to meet the needs of municipalities, commercial and industrial firms, and institutions. The most developed and commonly used incinerators are the liquid injection and rotary kiln. Although liquid injection incinerators have been the most popular in Europe and the United States, rotary kiln incinerators are better suited to handle all physical forms of hazardous wastes. New high temperature incinerators are capable of destroying highly toxic organic wastes (e.g., organochlorine, herbicide orange, and PCBs) with high efficiency. Multiple hearth, fluidized bed, and other technologies such as cement kilns and molten salt combustion appear promising but need further demonstration to prove that they are economically, technically, and environmentally acceptable. The various types of incinerators are further discussed in Chapter 7.

Incineration is not a new technology and has been used for treating organic hazardous waste for many years. The major benefits of incineration are that the process actually destroys most of the waste rather than just disposing of or storing it. Incineration can be used for a variety of specific wastes and is becoming competitive in cost compared to other disposal methods.

Although incineration has become more attractive in hazardous waste management, it is not without its problems. Among these are[12]

- *Cost*. In most instances, incineration is a costly processing step, both in initial investment and in operation.
- *Operational Problems*. Variability in waste composition and the severity of the incinerator environment result in many practical waste handling problems, high maintenance requirements, and—at some installations—equipment unreliability.
- *Staffing Problems*. The low status often accorded to waste disposal sometimes makes it difficult to obtain and retain qualified supervisory and operating staff.
- *Secondary Environmental Impacts*. Many waste combustion systems result in the emission of odors, sulfur dioxide, hydrogen chloride, carbon monoxide, carcinogenic polynuclear hydrocarbons, nitrogen oxides, fly ash and particulate fumes, and other toxic or noxious materials into the atmosphere. Waste waters from residue quenching or scrubber-type air pollution control are often highly acidic and contain high levels of dissolved solids, abrasive suspended particles, biological and chemical oxygen demand, heavy metals, and pathogenic organisms. Finally, residue disposal can present a variety of aesthetic, water pollution, and health-related problems.
- *Public Sector Reaction*. Few incinerators are installed without arousing concern, close scrutiny, and, at times, hostility from the public at large and/or regulatory agencies.
- *Technical Risks*. Since changes in waste character are common and process

analysis is difficult, there is a risk that a new incinerator may not work well or, in extreme cases, not at all.

With all these disadvantages, incineration has persisted as an important concept in waste management.

REFERENCES

1. The Comptroller General, *How to Dispose of Hazardous Waste—a Serious Question that Needs to Be Resolved*, U.S. General Accounting Office, Washington, D.C., CED 79–13, December 1978.
2. P.W. Powers *How to Dispose of Toxic Substances and Industrial Wastes*, Noyes Data Corp., Park Ridge, NJ, 1976.
3. L.M. Thomas, "EPA Fights Hazardous Waste," *EPA Journal*, **10**, 4–7, October (1984).
4. H.L. Hickman, "Why We Have a Hazardous Waste Problem," *EPA Journal*, **10**, 10–11, October (1984).
5. D. White and B. Burke, "Choices in Disposal of Hazardous Waste," *EPA Journal*, **10**, 20–21, October (1984).
6. W.E. Sweet, R.D. Ross, and G.V. Velde, "Hazardous Waste Incineration: A Progress Report," *J. Air Pollution Control Assoc.*, **35**, 138–143, February (1985).
7. N. Basta, R.V. Hughson, and C.F. Mascone, "What Are Your Views on Hazardous Wastes?," *Chem. Eng.*, 58, March 4 (1985).
8. The Comptroller General, *Hazardous Waste Sites Pose Investigation, Evaluation, Scientific, and Legal Problems*, U.S. Government Accounting Office, Washington, D.C., CED 81–57, April 1981.
9. K-C. Lee, J.L. Hasen and G.M. Whipple, "Characterizing Petrochemical Wastes for Combustion," Union Carbide Corporation, So. Charleston, WV; ASTMs Hazardous and Industrial Solid Waste Testing: Fourth Symposium, Arlington, VA, May 2–4, 1984.
10. E.R. DeVera, B.P. Simmons, R.D. Stephens, and D.L. Storm, *Samplers and Sampling Procedures for Hazardous Waste Streams*, Municipal Environmental Research Lab, Cincinnati, OH, EPA-600/2-80/018, January, 1980.
11. T. Shen, U. Choi, and L. Theodore, *Hazardous Waste Incineration Manual*, U.S. EPA Air Pollution Training Institute, Research Triangle Park NC, to be published.
12. W.R. Niessen, *Combustion and Incineration Processes*, Dekker, New York, 1978.

2

Other Options

Contributing authors: Elizabeth Girardi Schoen and Catherine Frega

2.1 INTRODUCTION

Although incineration is presently considered the method of choice to manage organic hazardous waste, there are other options. Incineration provides essentially complete destruction and produces relatively small volumes of waste, most of which is nonhazardous. However, capital and operating costs for incineration are high, and the permitting process is lengthy, expensive, and complicated. Pretreatment methods such as concentration, volume reduction, and removal of solids or inorganics may be required for incineration. Modification of the process generating the waste may also be necessary in order to reduce the quantity of waste.

This chapter discusses other options to incineration as well as pretreatment and process modification techniques associated with incineration. These options are

Chemical treatment
Biological treatment
Physical treatment
Ultimate disposal method
Process modification(s)

The chapter concludes with a discussion of the selection process involved in choosing a form of hazardous waste treatment.

Typically, when options other than incineration are chosen, more than one unit process (or step) is required. For example, a typical process sequence that could serve as an alternative to incineration might be sedimentation with neutralization, biodegradation, followed by sludge separation, and finally landfilling the waste. These alternative options are often less costly and are possibly easier to permit and start up than incineration.

2.2 CHEMICAL TREATMENT

There are various chemical processes used to treat hazardous waste. Relatively few are used as the primary or the only treatment step. Most of the chemical processes prepare the waste for further processing and are but one step in a scheme of waste processing. The following chemical processes will be discussed: calcination or sintering, catalysis, chlorinolysis, electrolysis, hydrolysis, microwave discharge, neutralization, oxidation, photolysis, precipitation, and reduction. Some of the processes such as neutralization and precipitation are commonly used; others, microwave discharge, for example, are still in the development stage. For more detail on these processes the reader is referred to the literature.[1,2]

Calcination

Calcination is a thermal decomposition process without any interaction with a gaseous phase. Operating temperatures of about 1800°F and atmospheric pressure can produce dry powder from slurries, sludges, and tars as well as aqueous solutions, by driving off volatiles. Typical calciners are open hearth, rotary kiln, and fluidized bed. Calcination is a well-established process and recommended as a one-step process for the treatment of complex wastes containing organic and inorganic components. It is also one of the few processes that can satisfactorily handle sludges. A real advantage of calcination is that it can concentrate, destroy, and detoxify in a single step. Organic components are usually destroyed while inorganics are reduced in volume and leachability, thus making them suitable for landfill.

Treatment applications include recalcination of lime sludges from water treatment plants, coking of heavy residues and tars from petroleum refining operations, concentration and volume reduction of liquid radioactive wastes, and treatment of refinery sludges containing hydrocarbons, phosphates and compounds of calcium, magnesium, potassium, sodium, iron, and aluminum. The use of calcination is likely to expand in the future since it handles tars, sludges, and residues that are difficult to handle by other methods.

The energy requirements for calcination are high; depending upon the amount of water and organics in the waste stream, the energy needed can be as much as 20 times greater than that used in other chemical methods. If the waste stream contains a combustible organic, the energy requirement would be reduced. Operating costs are variable and dependent upon the content of the waste stream. Investment costs for calcination must also include air pollution control equipment such as particulate-removal devices and gas absorbers.

Catalysis

Catalysis is a modification of the mechanism and rate of a chemical reaction. If a waste stream can be modified by a chemical reaction, the successful use of a catalyst will reduce operating cost and energy requirements. Catalysts can be *selective* such as in the detoxification of chlorinated pesticides by dechlorination, or *versatile* as in the complete destruction of cyanides by air oxidation. While catalytic oxidation is commonly used as an alternative to incineration in the decomposition of organic wastes, it is not widely used in the treatment of hazardous waste. The technology is not well developed, and because of catalyst specificity, progress in the development of one application does not necessarily mean progress in the development of another. In addition, the many components in a waste stream, in varying concentrations, can inhibit the effectiveness of a catalyst on the target reaction. Some laboratory research has been done on the following catalytic processes: oxidation of cyanides, sulfides and phenols, and decomposition of sodium hydrochlorite solutions. A catalyst will normally reduce operating temperature and therefore conserve energy. Catalytic processes may be higher in capital costs, but lower in operating costs, and thus result in lower overall expenditure.

Chlorinolysis

Chlorinolysis is a process that converts most liquid chlorinated hydrocarbons to carbon tetrachloride. Organic feedstock and preheated excess chlorine are fed into a preheated reactor; at operating conditions of 900°F and 20 atm, this mixture reacts to form, among other products, carbon tetrachloride, which is removed by distillation.

This process is primarily a production process, not a waste treatment process. Handling and transporting of feedstocks is a hindrance to the common use of chlorinolysis as a waste recovery method. The future growth of this technique depends on the market for carbon tetrachloride and a feed process stream of the proper wastes. For this type of production process, the energy requirements are rather small. Chlorinolysis produces hydrochloric acid and phosgene gas as effluents, which must be treated. The chlorine is highly reactive necessitating the use of reactors made of high-purity nickel surrounded by stainless steel jackets; this results in a high capital investment.

Electrolysis

Electrolysis is the reaction of either oxidation (loss of electrons) or reduction (gain of electrons) taking place at the surface of conductive electrodes immersed in an electrolyte under the influence of an applied potential. Electrolytic processes can be used for reclaiming heavy metals, including toxic metals from concentrated aqueous solutions. Electrolysis is not useful for organic waste streams or viscous tarry liquids. Application to waste treatment has been limited because of cost factors. A frequent application is the recovery for recycle or reuse of metals, like copper, from waste streams. Pilot applications include oxidation of cyanide waste and separation of oil–water mixtures. Electrolysis has energy costs that can be 10 to 35% of operating costs. Gaseous emissions may occur and, if these are hazardous and cannot be vented to the atmosphere, further treatment such as scrubbing is required. Wastewater from the process may also require further treatment. Costs are dependent upon concentration of the feed and desired output.

The most common waste treatment application of electrolysis is the partial removal of heavy metals from spent copper pickling solutions. When the typical concentration of the spent pickling solution is 2 to 7% copper, the system design is similar to that of a conventional electroplating bath. When recovering from more dilute streams, mixing and stirring are necessary to increase the rate of diffusion; it may also be necessary to use a large electrode surface area and a short distance between electrodes.

Another consideration with this method is the removal of collected ions from the electrodes. This removal may or may not be difficult, but must be addressed. The material collected must be ultimately disposed of if it is not suitable for reuse.

Hydrolysis

Hydrolysis is the reaction of a salt with water to produce both an acid and a base:

$$XY + H_2O \longrightarrow HY + XOH$$

Reactions generally require high temperature and pressure, acid or alkali, and sometimes catalysts. Feed streams can be aqueous, slurries, sludges, or tars. The process could be useful for a variety of wastes, but few applications have been used. The petroleum industry uses hydrolysis to recover sulfuric acid from the sludge of acid treatment of light oils. Hydrolysis has also been used to detoxify waste streams of carbamates, organophosphorus compounds, and other pesticides.

Energy requirements vary with application, but are generally high. Some products of hydrolysis are toxic and so require further treatment. An additional drawback is that, depending on the waste stream, the products may not be predictable and the mass of *toxic* substance discharged may be greater than that of the waste originally inputted for treatment. Operating costs are completely dependent upon application and can be extremely high.

Microwave Discharge

In this technique, high frequency microwave power is used to establish a plasma or gaseous discharge in which neutral molecules are partially decomposed into metastable, free radical, and ionic species. This technique is not yet used for waste treatment. One problem is that, when the decomposition products recombine to form stable molecules, the products may be of higher molecular weight or greater toxicity than the original reactants. Some research has been done on organic vapors.

Power requirements and capital cost appear to be high. A possible future use could be for small quantities of highly toxic materials.

Neutralization

Neutralization is a reaction that adjusts either acid (low pH) or base (high pH) by addition of the other to achieve a pH of 7 (neutrality). (For a definition and discussion of the pH, see Chapter 4.) The treated stream undergoes essentially no change in physical form, except possible precipitation or gas evolution. The process has extremely wide applications to aqueous and nonaqueous liquids, slurries, and sludges. It is very widely used in waste treatment. Some applications include pickle liquors, plating wastes, mine drainage, and oil emulsion breakage. It is commonly used in industries such as batteries, aluminum, coal, inorganic and organic chemicals, photography, explosives, metals, pharmaceuticals, power plants, and textiles.

The reaction can take place in batch or continuous reactors (see Chapter 6 for details on reactors). The flow can be cocurrent or countercurrent depending upon desired results and waste stream characteristics. The reaction can be in the liquid or gaseous phase, or both. Limitations include temperature dependence and residual formation. The reaction may be exothermic and cause a temperature increase that may be undesirable; for example, higher temperature may cause the formation of other toxins or hazardous wastes, or the high temperature stream may require special handling. Residuals such as precipitates or gases may also form and then must be treated by other types of processing such as clarifying or absorbing. Neutralization may not make the waste less hazardous, it may only be a treatment for the next step of waste handling. For example, changing the pH may help precipitate heavy metals, prevent metal corrosion or damage of process equipment, affect operation of biological treatment, or provide a neutral recycle material.

Energy requirements are fairly low and are related to pumping and agitation. Capital and operating costs are highly variable depending upon specific process requirements. Neutralization is a very common process step but rarely comprises the sole hazardous waste treatment.

Some common methods of neutralizing wastes are mixing acidic and alkaline streams together, passing acid wastes through packed beds of limestone, mixing acid waste with lime slurries, adding solutions of concentrated bases such as caustic (NaOH) or soda ash (Na_2CO_3) to acid streams, blowing waste flue gas from a boiler through alkaline waste liquids, adding compressed CO_2 to basic wastes, adding acid such as H_2SO_4 or HCl to basic (alkaline) waste streams. The choice of the acid or alkali used is often based on process requirements but cost is the primary consideration. Lime and sulfuric acid are relatively inexpensive. Lime will, in the process of neutralizing sulfate-bearing wastes, form calcium sulfate, which will precipitate and may be undesirable. Caustic and soda ash are more expensive, but may be the better choice of base for wastes containing sulfates. If sulfides or cyanides are present in the waste stream, toxic gas may be evolved during neutralization; these gases require special handling by gas treatment devices such as scrubbers. A schematic of a neutralization system is given in Fig. 2.2.1.[2]

Oxidation

Oxidation is a process in which one or more electrons are transferred from the chemical being oxidized to the chemical initiating the transfer (*oxidizing agent*). The main use in treating hazardous wastes is detoxification. For example, oxidation is used to change cyanide to less toxic cyanate or completely oxidize cyanide to carbon dioxide and nitrogen. Oxidation can also aid in the precipitation of certain ions in cases where the more oxidized ion has a lower solubility.

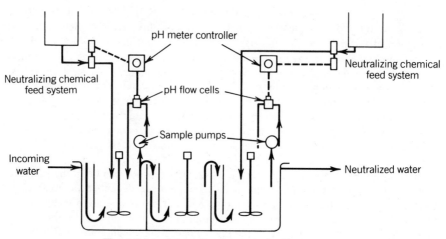

Figure 2.2.1. Schematic of a neutralization system.

Some common oxidizing agents and their uses are:

- *Chlorine* as a gas or in hypochlorite salt is often used to oxidize cyanide. It can also be used on phenol-based chemicals, but this application is somewhat limited because of the formation of toxic chlorophenols when the process is not adequately controlled.
- *Ozone* is used in the oxidation of cyanide to cyanate and to oxidize phenols to nontoxic compounds. Because ozone is an unstable molecule, it is highly reactive as an oxidizing agent. There are no inherent limitations to oxidation with ozone, but components of the wastewater should be examined to determine that enough ozone is being added to also oxidize other components that could compete with the toxic components for ozone.
- *Ozone with ultroviolet (UV) light* can oxidize some compounds. Halogenated organic compounds are resistant to ozone, but in the presence of UV light can be oxidized.
- *Hydrogen peroxide* (H_2O_2) is a powerful oxidizing agent and is successful with phenols, cyanides, sulfur compounds, and metal ions. The process is sensitive to pH, which must be controlled. The optimum pH is $\sim$ 3 or 4, with process efficiency decreasing at higher or lower pH.
- *Potassium permanganate* ($KMnO_4$) is a excellent oxidizing agent and reacts with aldehydes, mercaptans, phenols, and unsaturated acids. It has been used to destroy organics in wastewater and potable water. The reduced form is managenese dioxide, which is mostly insoluble and can be easily removed by filtration. The process is pH-sensitive and is faster at high pH (up to 9.5).

Oxidation is used in the treatment of waste, the most common applications being dilute waste streams. The energy consumption is relatively low since only pumps and mixing equipment are needed. One disadvantage is that some oxidizing agents add other metal ions that may have to be removed. The costs are lower when *continuous* rather than *batch* processes are used. The oxidation processes are not too successful with sludges; because of the difficulty in achieving good contact mixing, incomplete reaction occurs.

An additional limitation to the use of ozone oxidation is a result of its high instability—it must be generated on-site, which can add considerably to its expense. On the positive side, ozone possesses antibacterial and antiviral properties; this disinfecting characteristic has been widely used to scrub gases to remove odor. Oxidation with ozone is very effective for a wide variety of components and as a result is used extensively, despite the expense.

Photolysis

Photolysis is the breakage of chemical bonds under the influence of UV or visible light. The extent of the degradation varies according to the compound and the light used. Complete conversion of an organic to CO_2 and H_2O by this method is highly improbable and partly degraded components are likely to be hazardous. Large-scale applications of photolysis have not been developed, and are not likely to be developed in the future. There are some UV-assisted ozonation and chlorination applications that have been successful. The process has not been studied in great detail, but would surely be highly energy intensive. Specialized application to detoxification of pesticide-contaminated solvents is a possible future use.

Precipitation

Precipitation involves the alteration of the ionic equilibrium to produce insoluble precipitates. Chemical precipitation is allied with solids separation processes such as filtration, in order to remove the sediment. Undesirable metal ions and anions are commonly removed from waste streams by converting them to an insoluble form. The process is sometimes preceded by chemical reduction of the metal ions to a form that can more easily be precipitated.

Chemical equilibrium can be affected by a variety of means in order to change the solubility of certain compounds. For example, precipitation can be induced by:

- *Alkaline Agents*. An alkaline agent such as lime or caustic soda, when added to waste streams, raises the pH. The higher pH decreases the solubility of the metal ions that precipitate out of solution as hydroxides. The optimum pH for precipitation varies depending on the metal ion to be removed. The hydroxide precipitation can also be affected by organic radicals that form chelates. Chelates mask the typical precipation.
- *Sulfides*. Soluble sulfides, such as hydrogen or sodium sulfide, and insoluble sulfides, such as ferrous sulfide, are used to precipitate heavy metal ions as insoluble metal sulfides. Sodium sulfide and sodium bisulfide are most commonly used for this purpose. The sulfides have lower solubilities than hydroxides in the high pH range and also have low solubility around the pH 7 range or below. Both the sulfide ion and hydrogen sulfide are toxic so that sulfide precipitation usually requires pretreatment and close control. Pretreatment is normally an adjustment to a pH of 7 or 8. Posttreatment usually involves aeration or chemical oxidation.
- *Sulfates*. Zinc sulfate or ferrous sulfate are used to precipitate cyanide complexes. Precipitation of cyanide complexes does not destroy the cyanide molecules and the collected sludge must therefore be appropriately handled. Reports have shown that this type of sludge in the presence of sunlight can break down to free cyanide.
- *Carbonates*. Carbonate precipitation is successful in removing metals. Direct precipitation of metal ions with calcium carbonate, or the conversion of hydroxides into carbonates with carbon dioxide, is common since carbonates are easily filtered out. The solubilities of carbonates are usually between those of hydroxides and sulfides.

Precipitation with chemicals is a common waste stream treatment process and is effective and reliable. Energy requirements are fairly low as compared to other processes. The disposal of the resulting sludge can propose a serious problem, however. Some examples of industries using precipitation for waste removal include: inorganic chemicals, metal finishing, copper forming, foundries, explosive manufacturing, pharmaceutical manufacturing, electronic component manufacturing, and textile mills.

Reduction

In chemical *reduction*, one or more electrons are transferred from the reducing agent to the chemical being reduced. The lowering of the valence state may reduce toxicity or encourage a particular chemical reaction. The first step in a reduction process is normally pH adjustment. The reducing agent can be a gas, solution, or finely divided powder. Mixing is critical in order to assure contact between the reducing agent and the waste for complete reaction. The reduction reaction is normally followed by a separation step, such as precipitation, to remove the reduced compound.

Some common reducing agents are sulfur dioxide, sodium metabisulfite, sodium bisulfite, and ferrous salts for reducing chromium; sodium borohydride to reduce mercury; and alkali metal hydride to reduce lead. Sulfur dioxide is normally used as a gas. The reduction of chromium with sodium metabisulfite and sodium bisulfite is highly dependent on the pH and temperature. Sodium hydroxide must be added to control the pH during the reaction. Since the reduction of chromium with ferrous sulfate works best when the pH is below 3, acid must be added during the reaction.

The major application of chemical reduction is for the treatment of chromium waste. Chromium is very toxic in its hexavalent state (Cr^{6+}); trivalent chromium (Cr^{3+}) is much less toxic and will precipitate in an alkaline solution by forming a hydroxide. Chromium reduction is used for waste treatment in industries such as: metal finishing, inorganic chemical manufacture, coil coating, battery manufacture, iron and steel manufacture, aluminium forming, electronic manufacture, porcelain enameling, and pharmaceutical manufacturing.

Since energy is required only for pumping and mixing, energy requirements for reduction are low. Continuous processes are more economical than batch processes. Treatment by reduction is effective and reliable. The main disadvantage is the potential formation of compounds as toxic or hazardous as the compounds originally in the waste. In the treatment of mixed wastes, interference of oxidizing agents can decrease the efficiency of the reducing agent. The use of sulfur dioxide as the reducing agent has the added disadvantage of special handling and storage requirements, since sulfur dioxide is somewhat hazardous. An illustration of hexavalent chromium reduction with sulfur dioxide is shown in Fig. 2.2.2.

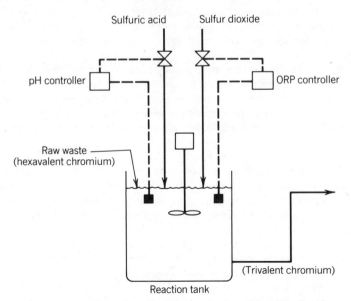

Figure 2.2.2. Hexavalent chromium reduction with sulfur dioxide.

2.3 BIOLOGICAL TREATMENT

Biological processes also involve chemical reactions, but are differentiated from the chemical category in that these reactions take place in or around microorganisms. The most common use of biological processes in waste treatment is for the decomposition of organic compounds.

The different biological processes described here are activated sludge, aerated lagoons, anaerobic digestion, composting, enzyme treatment, trickling filter, and waste stabilization ponds. All of these, except enzyme treatment, use microorganisms to decompose the waste. Enzyme treatment generally involves extracting the enzyme from the microorganism and using it to catalyze a particular reaction. With proper control, these processess are reliable and environmentally sound; additive chemicals are usually not needed and operational expenses are relatively low.

Activated Sludge

The *activated sludge* process uses microoganisms to decompose organics in aqueous waste streams. The microorganisms take the organics into the cell, through the cell wall, and into the cytoplasm where the organics are broken down by enzyme oxidation and hydrolysis to produce energy and other cellular material. Besides taking in the organics as food, the biomass acts as a filter to collect colloidal matter and suspended solids. Volatile organics can be driven off somewhat by the aeration process. Some metals are collected in the organisms and in the sludge.

The controlling factors in the process are the extent of mixing to insure contact of the microorganisms with the waste, the amount of dissolved oxygen for the respiration of the organisms, the concentration of metals that are toxic to the microorganisms, and the types and concentration of the organics to be decomposed. Organics that have been successfully decomposed by microorganisms include: polysaccharides, proteins, fats, alcohols, aldehydes, fatty acids, alkanes, alkenes, cycloalkanes, and some aromatics. Requiring more careful process control, isoalkanes, halogenated hydrocarbons, and lignin have also been decomposed by microorganisms.

The activated sludge process must usually be preceded by neutralization, some metal removal, and possible solids removal. Most activated sludge processes work with < 1% suspended solids. The process is normally followed up by a separation step, usually sedimentation, to remove the biological sludge from the waste liquid stream. Much of the biological sludge is recycled, although some portions are disposed of in landfills.

Activated sludge is the most widely used of the biological waste treatment processes. The energy requirements are a large part of operating cost. The pumping, mixing, and aeration require power. The system typically requires no chemical additions and degradation is natural. The main process disadvantage is that it does not handle slurries, tars, or streams with a high concentration of suspended solids. The reason it is best suited for liquid streams is the critical need for good mixing to assure contact of the waste with the microorganisms. The activated sludge process has been used to treat the waste streams from iron and steel manufacture, canneries, pulp and paper mills, petroleum refining, organic chemical manufacture, and pharmaceutical manufacture.

Aerated Lagoon

The *aerated lagoon* is an earthen basin that is artificially aerated; the basin is generally lined to make it impermeable. The microbial reactions are the same as those that take place in the activated sludge process except the biological sludge is not recycled. In this process microorganisms are used to decompose organics in waste streams containing < 1% solids; solid contents > 1% usually result in the settling of some solid matter because of limited mixing and aeration. These solids often undergo anaerobic microbial decomposition.

The aerated lagoon process has been used for petrochemical, textile, pulp and paper mill, cannery, leather tanning, gum and wood processing, and some other industrial waste streams. The aerated lagoons usually require less energy and are therefore less expensive to operate since recycle systems are not needed; more land area is a requirement, however. The removal efficiencies are not as high as those for the activated sludge process since mixing and aeration are less effective. The retention time in an aerated lagoon is typically longer than that for the activated sludge process. One other disadvantage is that aerated lagoons do not handle streams with variable organic and metal constituents as well as activated sludge.

Anaerobic Digestion

In *anaerobic digestion*, microorganisms that do not require oxygen for respiration are employed. This process is useful for the degradation of simple organics. The microbiology is not well understood but essentially the cells use part of the organic compounds for cell growth and part is converted to methane and carbon dioxide. A delicate equilibrium is required and this makes the process less suitable for industrial waste streams.

Anaerobic digestion has been used for sewage treatment, but few applications for hazardous waste have been implemented. There has been some study done for the treatment of brewery, alcohol distillery, and cotton kiering waste streams.

The anaerobic digestion takes place in a closed vessel with no agitation mechanisms. The gases rise to the top and are collected for possible use as a heating fuel. The digested sludge settles to the bottom and once it is stable and inert, can be disposed of by landfilling.

The power requirements are low and consequently operating costs are fairly low. There are disadvantages because of the need for control of pH, temperature, and waste concentration and constituents. There is potential for application to the degradation of specialized wastes from pharmaceutical or other processes.

Composting

Composting is aerobic digestion by microorganisms in the soil. The organisms decompose organics and multiply. This process, unlike other biological processes, can tolerate some toxicants and metals. This is accomplished essentially by piling waste in the ground and aerating it occasionally by turning and moving the soil. The collection of leachate and runoff is normally required to protect the ground water.

Composting is commonly used for organic wastes and complete digestion requires about 3 or 4 months. Since there is a market for the soil and the decomposed organics (humus) in Europe, it is frequently used there. Most of the attempts to use this technique in the United States have failed because of a lack of this market. Composting has been successful with municipal refuse, high concentration organic sludges, and some petroleum refineries.

Energy requirements are somewhat higher than for other biological processes. Equipment for landmoving and possibly pumping is required. Composting occasionally requires the addition of some nutrients as well as alkali for pH control. The few possible emissions are carbon dioxide, steam, liquid effluent of partly oxidized organics and some unpleasant odors, all of which are fairly simple to control. The main disadvantages are the large land requirements, and the necessity of a market for the humus.

Enzyme Treatment

Enzyme treatment involves the application of specific proteins (simple or combined), which act as catalysts in degrading the waste. Enzymes work on specific types of compounds, specific molecules, or a specific bond in a particular compound. Enzymes are inhibited by the presence of insoluble inorganics, are sensitive to pH and temperature fluctuations, and do not adapt to variable concentrations.

Enzyme treatment may prove useful for specific industrial applications, but at present is not commonly used in waste treatment. The specificity of enzymes and their fragile nature make them fairly useless in most waste applications. Addition of enzyme in an attempt to catalyze activated sludge processes have failed. Since it is not used on a large scale for waste treatment, little data are available on the economics of this process.

Trickling Filters

Trickling filters are also used for the decomposition of organic waste streams. The microbiological reactions are similar to those that occur with activated sludge and aerated lagoons. The microorganisms are held in a support media and waste water is trickled over them. Usually waste streams are sprayed to absorb oxygen before passing through the support media. Microorganisms and filter design vary depending upon the waste and the efficiency required.

Some industries that use trickling filters are refineries (with oil, phenol, and sulfide wastes), canneries, pharmaceutical manufacture, and petrochemical manufacture. The energy requirements for this process are low, although recycled pumps are required for recirculation of the filter effluent. Efficiency of removal is not high, about 50 to 85%; because of this, trickling filters are normally used in combination with other methods, such as activated sludge.

Waste Stabilization Ponds

Waste stabilization ponds are shallow basins into which wastes are fed for biological decomposition. The chemical reactions involved are the same as those that occur in the other biological processes. Aeration is provided by the wind, and anaerobic digestion may also occur near the bottom of deeper ponds. This method can handle only low concentration of organics with $< 0.1\%$ suspended solids.

The ponds are very commonly used for sewage treatment and dilute industrial wastes. Some industries using them include: steel, textiles, oil refineries, paper and pulp, and canneries. Waste stabilization ponds are normally used as a final treatment step for effluent because they are not efficient enough to be used on their own. Some advantages are low-energy needs and, assuming appropriate feeds, few effluent handling problems. Disadvantages include a high sensitivity to inorganics and suspended solids and large land acreage requirements. Occasionally the basin must be lined for leaching control.

2.4 PHYSICAL TREATMENT

There are more than 20 known types of physical treatment processes used in the handling of hazardous wastes; however, very few of these are fully developed or commonly used in industry. Some treatments have been found to have little potential use so that further research in these areas is unlikely. Zone refining, freeze drying, electrophoresis, and dialysis all fall into this category. The difficulty of the operation and/or the high cost of these processes overshadows any future use they may have. The most common processes used today are sedimentation, filtration, flocculation, and solar evaporation. Most other processes fall in between these two extremes, that is, they show some potential for future use but are not presently used to any great extent.

Physical treatments may be separated into two categories: *phase separation* and *component separation* processes; in the latter, a particular species is separated from a single-phase, multicomponent system. The various physical treatments may fall into one or both of these categories. Sedimentation and centrifugation are used in phase separation, liquid ion exchange and freeze crystallization are used for component separation, and distillation and ultrafiltration are used in both.

Phase separation processes are employed to reduce waste volume and to concentrate the hazardous waste into one phase before further treatment and material recovery are performed. Such waste streams as slurries, sludges, and emulsions, which contain more than one phase, are the usual candidates for this category.

Filtration, centrifugation, and flotation may be used on slurries that contain larger particles. If the slurry is colloidal, flocculation and ultrafiltration are generally used. If the slurry or sludge is known to contain any volatile components, evaporation or distillation is used to remove them from the waste stream. Because emulsions are so difficult to separate, the type of physical treatment required is usually decided on a case-to-case basis.

Component separation processes remove particular ionic or molecular species without the use of chemicals. Most of these are used in wastewater treatment and include such techniques as: liquid ion exchange, reverse osmosis, ultrafiltration, air stripping, and carbon adsorption. The first three are used to remove ionic and inorganic components; the last two techniques are used to remove volatile components and gases.

Many factors need to be considered when selecting a particular type of physical treatment. These include: the characteristics of the waste stream and the desired characteristics of the output stream, the technical feasibility of the different physical treatments when applied to a particular case, regulatory constraints, and economic, environmental, and energy considerations. The selection of waste treatment methods is discussed in Section 2.7.

In this section, the more established types of physical treatment used today are examined; these include; carbon adsorption, resin adsorption, centrifugation, distillation, electrodialysis, evaporation, liquid–liquid extraction, filtration, flocculation, flotation, freeze crystallization, high-gradient magnetic separation, ion exchange, liquid solidification, air stripping, steam stripping, and ultrafiltration.

Carbon Adsorption

Carbon adsorption is a recommended and established process used in separating organic and certain inorganic species from aqueous waste streams. It is suggested that the concentration of the waste species be 1% or less so that carbon regeneration is less frequent. A good number of carbon adsorption systems are being used today in municipal and industrial wastewater treatment facilities.

In carbon adsorption, a filter bed of activated carbon is placed in a vessel and used to adsorb certain components. A large adsorptive surface area, about 2.5 to 7.5 million ft^2/lb, is needed to accomplish the process. Best results are achieved when the component is slightly soluble in water, has a high molecular weight, high polarity, and low ionization capability, and when the concentration of any suspended solids is < 50 parts per million (ppm).[2] Once the carbon has been fully used, regeneration of the carbon may be done simultaneously with either the destruction (thermal regeneration) or recovery (nonthermal regeneration) of the *adsorbate* (i.e., the material adsorbed by the carbon).

Energy costs for this process range from 5 to 25% of the total operating cost depending upon whether thermal regeneration is used. If the carbon is thermally regenerated, equipment costs increase since a furnace, afterburner, and scrubber are needed. If the carbon is not regenerated a means of carbon disposal is needed.

Resin Adsorption

Resin adsorption is used in the removal of organic solutes from aqueous waste streams. Solute concentration in the wastewater may be as high as 8%. It is the preferred method when recovery of the adsorbate is desired since thermal regeneration of carbon destroys the organic material. It is also useful when there is a high concentration of dissolved inorganic salts in the waste stream.

Synthetic resins may be used to remove hydrophobic or hydrophylic solutes, which may be recovered by chemical means. Their applications include phenol recovery, fat removal, and color removal from wastewaters. Possible future uses include the removal of pesticides, carcinogens, and chlorinated hydrocarbons.

A schematic of a typical resin adsorption system is shown in Fig. 2.4.1.[2] Resin adsorption is similar to carbon adsorption except that two filter beds are used—one bed is used for adsorption while the other is being regenerated. The waste stream flows downward in the system at a rate of 1 to 10 gal/min · ft^2 of cross section and adsorption stops when the bed becomes saturated or the effluent concentration reaches a certain level.

Energy costs for this system depend on whether the resin is regenerated or not. If the resin is not regenerated it must be disposed. Because the high cost of resins results in a high capital cost, regeneration is usually desirable to keep overall costs low.

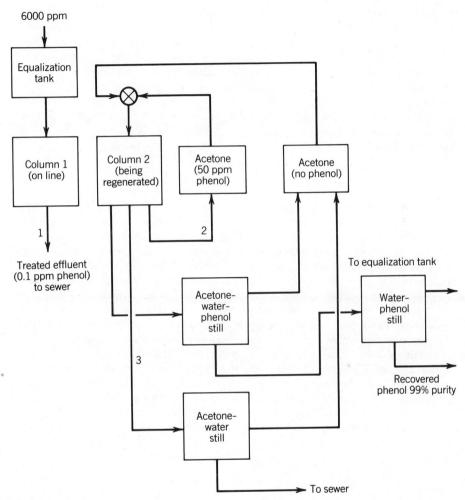

Figure 2.4.1. Schematic of a resin adsorption system. *Note:* The three wash streams resulting from regeneration are sequential cuts, taken in the order listed. The last cut (No. 3) is a post rinse with water.

Centrifugation

Centrifugation is a well-established, popular liquid–solid separation process used in commercial and municipal waste treatment facilities. It is usually used to reduce slurry and sludge volumes (*sludge dewatering*) and to increase the solids concentration in these waste streams. It is a technically and economically competitive process and is commonly used on waste sludges produced from water pollution control systems and on biological sludges produced in industry and municipal treatment facilities. Possible future uses include the removal of soluble metals from wastewaters and the dewatering of sludges produced in sulfur dioxide air pollution control systems.

Centrifugation is performed in a closed system and is therefore an excellent choice when treating volatile or toxic fluids. The liquid and solid are mechanically separated by centrifugal force. The removal of most of the liquid increases the solid concentration in, and reduces the volume of, the waste stream. The collected solid waste may then be treated and disposed or recovered. Three types of units are avilable for centrifugation; these are solid bowl, disk type, and basket type. The first two are used in large plants, the third, in smaller plants. A diagram of the solid-bowl centrifuge in shown in Fig. 2.4.2.[2]

Distillation

Distillation is a liquid phase separation process used to recover organic components from hazardous waste product streams. Because it is a nondestructive process with no effluent problems, rising chemical prices and stricter emissions regulations are making distillation a more competitive and popular recovery process. Some of its uses include recovering resins from resin adsorption systems after regeneration, and recovering solvents from carbon adsorption systems and paint wastes. Distillation cannot be used to separate thick wastes such as sludges or slurries.

There are five types of distillation processes: batch, continuous fractional, azeotropic, extractive, and molecular. In general, the basis for separation is the difference in vapor pressures of the components in the mixture. The waste mixture fed to the vessel is heated to produce a liquid phase and a vapor phase. Components with relatively high volatilities rise to the top and those with lower volatilities flow to the bottom. After the components are separated, the vapor is usually condensed. The condensed vapor (*overhead product*) and the *bottoms product* stream may be recovered or treated and disposed. This process is illustrated in Fig. 2.4.3.[2]

Capital, operating, and utility costs for distillation systems depend upon the nature of the feed and product streams.

Electrodialysis

Electrodialysis is a fully developed process used in separating ionic components and is commonly used in the desalination of brackish waters. Its use for hazardous waste treatment, however, is still in the pilot plant stage. The potential uses of electrodialysis are in acid mine drainage treatment, the desalting of sewage plant effluents, and in sulfite-liquor recovery. One disadvantage of this process is that it may produce low level toxic or flammable gases during operation. A schematic of an electrodialysis unit is shown in Fig. 2.4.4.[2]

In electrodialysis, separation of an aqueous stream is achieved through the use of synthetic membranes and an electrical field. The membrane allows only one type of ion to pass through and may be chosen to remove either anions or cations. The electric field causes the positive and negative ions to move in opposite directions and hence produces one stream rich in a particular ion and one depleted of that ion. Both streams may subsequently be recycled or disposed. It is important that pH and concentration be controlled for efficient operation.

Capital cost for an electrodialysis process is $\sim$ 20 to 25% of the total cost of a

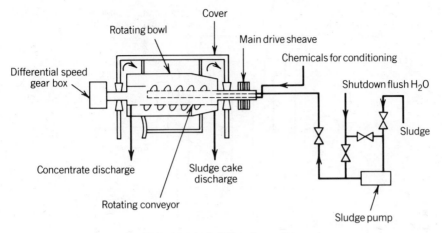

Figure 2.4.2. Solid-bowl centrifuge.

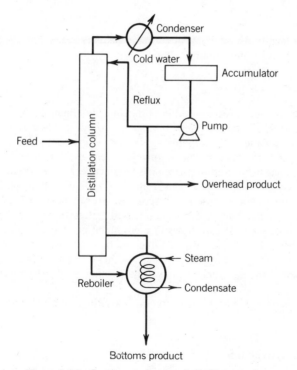

Figure 2.4.3. Continuous fractional distillation column.

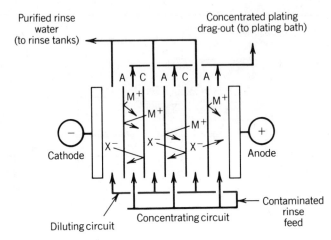

Purified rinse
water
(to rinse tanks)

Concentrated plating
drag-out (to plating bath)

Cathode

Anode

Diluting circuit

Concentrating circuit

Contaminated
rinse
feed

Figure 2.4.4. Schematic of an electrodialysis unit. C—cation selective membrane; A—anion selective membrane; M^+—cations; X^-—anions.

waste treatment facility. Capital and operating costs vary depending upon the initial ion concentration and water volume.

Evaporation

Evaporation is a common process used in both water desalination and the treatment of hazardous wastes. It may be used to treat a variety of waste streams: liquids, slurries, sludges, organic and inorganic streams, streams containing suspended or dissolved solids, and streams containing nonvolatile dissolved liquids. In waste treatment facilities, it has been used in processing radioactive wastes, wastes from paper mills and molasses distilleries, and wastes from (TNT) trinitrotoluene manufacturing.

The process and equipment used in evaporation are similar to that of distillation, except the vapor is not collected and condensed unless organic components are present. Usually the waste stream flows through metal pipes that are heated by low pressure steam outside the pipe walls. Other modes of operation that have been used are the solar evaporation from ponds or the heating of open vessels. The process concentrates the hazardous wastes of the original feed stream and reduces its volume.

Evaporation is an energy-intensive process and utility and equipment costs may be high.

Liquid–Liquid Extraction

Liquid–liquid extraction is used for the removal and recovery of organic solutes from aqueous and nonaqueous waste streams. Concentrations of solute in these streams range from a few hundred parts per million to a few percent. Most types of organic solutes may be removed by this process; though the technology is well developed, however, it is not commonly used in treating hazardous wastes. As environmental regulations on industrial wastewaters are made more stringent, its usage is likely to increase. Extraction has been used in removing and recovering phenols, oils, and acetic acid from aqueous streams and in removing and recovering freons and chlorinated hydrocarbons from organic solvent streams.

Liquid–liquid extraction may be performed in a single stage or a multistage unit. In this process, the waste stream comes into contact with another liquid stream in which the waste stream solvent is immiscible. The organic solutes are soluble in the extracting solvent and are distributed between the two streams. Increased contact of the waste and solvent streams increases removal of the organic solutes. Recovery of the solute and solvent from the product stream is carried out by stripping or distillation. The recovered solute may be either treated and disposed, reused, or resold. Capital investment in this type of process depends on the particular waste stream to be processed.

Filtration

Filtration is a popular liquid–solid separation process commonly used in treating wastewater and sludges. In wastewater treatment, it is used to purify the liquid by removing suspended solids. This is usually followed by flocculation or sedimentation for further solids removal. In sludge treatment, it is used to remove the liquid (*sludge dewatering*) and concentrate the solid waste, thereby lowering the sludge volume. This method is highly competitive with other sludge dewatering processes. Filtration may also be used in treating nonaqueous liquid waste streams. It is likely to come into greater use as industrial wastewater regulations become more stringent.

In the filtration process, a liquid containing suspended solids is passed through a porous medium. The solids are trapped against the medium and the separation of solids from the liquid results. For large solid particles a thick barrier such as sand may be used. This is known as *granular media* filtration; Fig. 2.4.5 illustrates this process.[2] For smaller particles, a fine filter such as a filter cloth may be used. Fluid passage may be induced by gravity, positive pressure, or a vacuum. The filter is cleaned and the solids collected by passing a stream of water in the opposite direction of the waste stream flow; this is known as *backwashing*.

Flocculation, Precipitation, and Sedimentation

Flocculation, precipitation, and sedimentation are three processes used to separate hazardous waste streams that contain both a liquid and a solid phase. All three are well-developed, highly competitive processes, which are often used in the complete treatment of hazardous waste streams. They may also be used instead of, or in addition to, filtration. Some applications include the removal of suspended solid particles and soluble heavy metals from aqueous streams. Many industries use all three processes in the removal of pollutants from their wastewaters.

These processes work best when the waste stream contains a low concentration of the contaminating solids. Although they are applicable to a wide variety of aqueous waste streams, these processes are not generally used to treat nonaqueous waste streams or semisolid waste streams such as sludges and slurries.

Flocculation and precipitation are physicochemical processes. In *flocculation*, fine suspended particles, which are difficult to settle out of the liquid, are brought together by flocculating chemicals to form larger, more easily collected particles. In *precipitation*, metals soluble in the liquid are made insoluble by precipitating

Filtration cycle

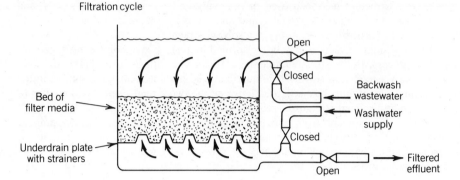

Bed of
filter media

Underdrain plate
with strainers

Open

Closed

Backwash
wastewater

Washwater
supply

Closed

Open

Filtered
effluent

Backwash cycle

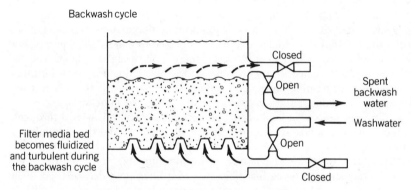

Filter media bed
becomes fluidized
and turbulent during
the backwash cycle

Closed

Open

Spent
backwash
water

Washwater

Open

Closed

Figure 2.4.5. Granular media filtration

chemicals. *Sedimentation* is a physical process in which suspended solids are settled out through the use of gravitational and inertial forces acting on the solids. The solids removed by these processes form a sludge that may be disposed of or treated to recover any valuable material.

The operating costs for a hazardous waste treatment system using all three processes are dependent on the amount and type of flocculating and precipitating chemicals used. Systems using only sedimentation have much lower operating costs.

Flotation

Flotation, or dissolved air flotation, is basically a physical process used to remove organic or inorganic solids suspended in waste streams, such as wastewaters or slurries. The process is used to concentrate waste streams by forming a sludge to remove valuable or toxic solids. Since chemical reagents are sometimes used to remove specific inorganic solids, the process can be chemical as well as physical.

As a physical process, flotation has been used in removing oils and greases from waste streams. Flotation with chemical addition may be a future application for the removal of heavy metal ions, cyanides, fluorides, and carbonyls from hazardous

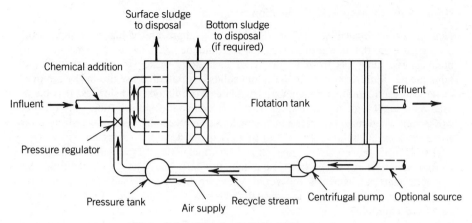

Figure 2.4.6. Dissolved air flotation system.

wastes. Flotation used to remove specific inorganics operates best when the desired product makes up < 10% of the total solids in the stream.

The flotation process allows suspended solids to be removed from a hazardous waste stream through the use of agitation and, in some cases, reagents. The waste stream is agitated by air bubbles that reduce the density of the solid and carry the particles to the top of the liquid or slurry. A froth forms at the surface of the liquid or slurry, which is then removed by skimmers or scrapers. Individual materials or combinations of similar materials may be removed in this way if the proper reagents are used. Air emissions by the air bubbles cause no environmental problems. The sludge removed from the surface may be recovered or disposed. A typical air flotation system is shown in Fig. 2.4.6.

Operating costs for flotation are dependent on: the amount of waste treated, whether or not a reagent is used, and whether the solids are recovered and sold or disposed.

Freeze Crystallization

Freeze crystallization is a phase separation process not yet fully developed for commercial use in treating hazardous waste. Most of the development has been for use in water desalination. On a laboratory scale, this process has been used in the treatment of wastewater containing ammonium nitrate, paper mill bleach solutions, plating liquors, and arsenal redwater. The freeze crystallization process also has potential use in the removal of 1 to 10% total dissolved solids (TDS) in aqueous waste streams.

The contaminated wastewater is subcooled during freeze crystallization in order to form purified ice crystals. The remaining liquid is more concentrated in soluble organic or inorganic materials. The crystals are subsequently removed, washed, and melted to recover the water. The contaminated liquid could be freeze crystallized again to further concentrate the wastes and make waste disposal and treatment much easier.

High-Gradient Magnetic Separation

High-gradient magnetic separation (HGMS) is a phase and component separation process still in the development stage. It could be used to remove magnetic or nonmagnetic materials from a variety of hazardous waste streams: aqueous and nonaqueous liquids, slurries, sludges, and solids. Its potential use lies in the removal of ferromagnetic and paramagnetic particulates from liquids and slurries. It is being used on a small scale in wastewater treatment and coal desulfurization. The process works well when magnetic components make up only a small percentage of the total concentration and solids make up 10 to 15% of the total waste stream.

The HGMS process uses a magnetized ferromagnetic filter to separate magnetic or paramagnetic particles from nonmagnetic particles. The filter captures the magnetic material, which is recovered when the filter is cleaned. By treating the waste stream with a magnetic seed, the filter can also be used to remove non-magnetic material.

Capital investment may be low when only ferromagnetic material is removed, but higher for the removal of nonmagnetic material. High volume applications make the process more economically attractive. Full commercial development is not expected for at least 5 yr.[1]

Ion Exchange

Ion exchange is generally used for the removal of dilute concentrations of heavy metals and anions from aqueous hazardous waste streams. Though the process is fully developed, it is not commonly used in industry. It has been used in recovering effluents from fertilizer manufacturing, the de-ionization of water, treating electroplating wastewaters, and looks promising for the removal of cyanides and selected heavy metals from waste streams.

Ion exchange is a two-step process. First, a solid material, the ion exchanger, collects specific ions after coming into contact with the aqueous waste stream. The exchanger is then exposed to another aqueous solution of a different composition that picks up the ions originally removed by the exchanger. The process is usually accomplished by sending the two aqueous streams through one or more fixed beds of exchangers. An illustration of this process is shown in Fig. 2.4.7. The ion-rich product stream may be recovered or disposed and the ion-poor stream is usually dilute enough for discharge to sewers.

Liquid Ion Exchange

Liquid ion exchange (LIE) serves the same purpose as ion exchange, the removal of heavy metals and anions from aqueous hazardous waste streams. However, LIE may be used for treating a higher concentration of contaminants. This process is also well-developed though not widely used in hazardous waste treatment. It has promising uses in the removal of cyanide from wastewaters and in treating hydroxide slimes produced in electroplating.

LIE uses an organic stream to carry out the transfer of ions from the aqueous waste stream to a second aqueous stream of different composition. The organic stream is immiscible in both aqueous streams. It removes the ionic or inorganic material from the waste feed stream and then passes it on to the second aqueous

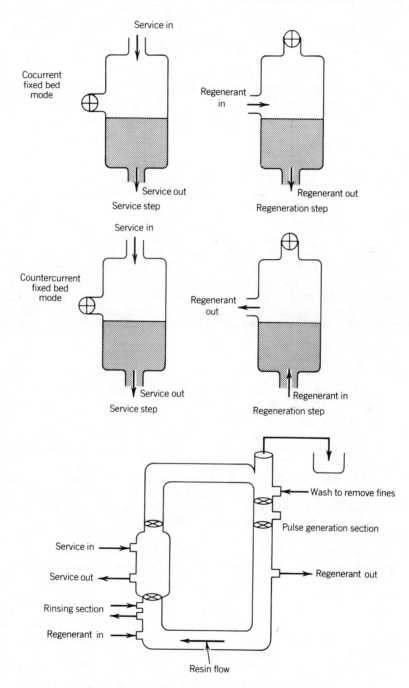

Countercurrent continuous mode (Higgins downflow type)

Figure 2.4.7. Operational modes for ion exchange.

stream. The processing equipment is similar to that of liquid–liquid extraction. The ion-rich stream is either treated further for recovery or disposal. The purified aqueous stream usually contains a dilute concentration of organic solvent and may therefore require further treatment.

Capital investment is dependent upon the type of waste stream being treated. High volume applications improve the economic aspects of this process.

Reverse Osmosis

Reverse osmosis is used in removing dissolved organic and inorganic materials from aqueous waste streams. Concentrations of these dissolved components are usually about 34,000 ppm or less. This process is more frequently used in water desalination but is becoming a more popular form of treating hazardous wastes. It is a good method for removing and concentrating waste streams, recovering valuable organic and inorganic material, and reducing wastewater volume.

Reverse osmosis employs a semipermeable membrane that allows passage of the solvent molecules, but not those of the dissolved organic and inorganic material. A pressure gradient is applied to cause separation of the solvent and solute. Any components that may damage or restrict the function of the membrane must be removed before the process is performed. Capital investment and operating costs depend on the waste stream composition.

Solidification

Solidification is a waste disposal process that transforms a hazardous waste into a nonhazardous solid product by *fixation* or *encapsulation*. In *fixation* a chemical or physical process and a solidifying agent are used to solidify the waste. *Encapsulation* is a process in which the waste is surrounded by a binder after it has been solidified by a chemical agent. Both produce a durable, impermeable, and supposedly environmentally safe product.

Solidification has only recently become a popular form of waste disposal. Previously, it had been used solely for treating radioactive wastes because of its high cost compared to other disposal means. Stricter regulations have made solidification a more popular choice of waste disposal. Solidification processes may be separated into five general categories depending on the solidifying agent used; these are silicate- and cement-based, lime-based, thermoplastic-based, organic polymer-based, and encapsulation techniques. All five of these processes are used to treat hazardous inorganic wastes. The cement- and lime-based methods are also used for sludges from stack-gas scrubbers after the sludges are chemically treated to precipitate out the metals. Solidification is unsuitable for treating organics, oxides, and toxic anions because these materials are difficult to solidify. Solidification has been used in treating wastes produced from steel mills, plating and lead-smelting plants, food production sludges, and sulfur residues.

The *silicate-* and *cement-based* methods may be used as batch or continuous processes. Portland cement and other additivies are added to either a wet or sludge-like waste stream forming an impermeable, rocklike solid. The solidified product may then be used for land reclamation. The degree of solidification and strength of the product depends on the concentration of metals and organics present in the waste stream. This particular solidification process is the one most used in the United

States. Some of its advantages are the additives are not expensive, the process is well developed, equipment is available, and end-product strength and permeability are easily controlled by the amount of cement used. The disadvantages are waste weight and bulk are increased by the cement and pretreatment or special additives may be needed to insure proper solidification.

The *lime-based* process is both chemical and physical. Lime, water, and siliceous material are added to the waste. A product known as *pozzolanic cement* is formed. Additives such as fly ash are also added to increase the strength of the end product. The degree of solidification depends on the reaction of the lime with the other components forming the cement. Once the product is formed, it is often used for landfill, mine reclamation fill, or capping material. The advantages and disadvantages of this process are similar to those of the cement-based process.

Thermoplastic solidification is a two-step process. Initially, the waste, which is first dried, is combined with paraffins, bitumen, and polyethylene at a temperature above 212°F; upon cooling, the mixture begins to solidify. In the second step, the solid waste is thermoplastically coated and then disposed. The thermoplasticized product is fairly permeable to most aqueous solutions. In addition to toxic inorganics, the process is also used in treating nuclear wastes. The equipment used in thermoplastic solidification is more specialized and therefore more costly than the equipment used in the cement-based or lime-based processes. Besides the end-product's resistance to aqueous solutions, contaminant migrations are lower for this process than for the other four solidification methods. On the negative side, thermoplastics are flammable, and the waste stream must be dried before processing takes place.

In the *organic polymer* process, a monomer and a catalyst are combined with the waste stream, and the polymer allowed to form. The product is then containerized and disposed. One popular technique is the urea–formaldehyde process that is used in treating nuclear wastes. Two advantages of the organic polymer process are that it may be used on wet or dry sludges and that the end product weighs less than those produced from the cement-based, lime-based, or thermoplastic-based techniques. The disadvantages are that some organic polymers are biodegradable, the end product must be containerized, and the end product may not be as durable as the end products produced from the other methods.

The *encapsulation* process is somewhat similar to the thermoplastic process. The dried waste is first chemically treated and then coated with a binder, usually polyethylene. The advantages of this process are that the encapsulated waste is very durable and resistant to water and deterioration, and the final product does not need to be stored in containers. The disadvantages are that the process is expensive, only small volumes of waste can be treated, and the sludge must be dried before processing.

The various solidification treatments may be performed on-site, which requires a large investment for the waste generator, or at regional facilities, which requires transportation costs. As with any other option, the choice and final cost of the solidification process will depend upon the type of waste involved, the desired end product, and the investment capabilities of the waste generator.[3]

Air Stripping

Air stripping is generally used in removing ammonia from biologically treated wastewater in order to reduce nitrogen concentrations. It may also be used to remove

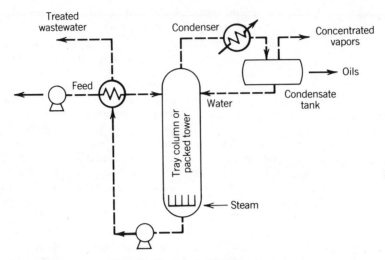

Figure 2.4.8. Steam stripping process.

volatile organics and other gases. As a result of these restrictions, only a small number of strippers are currently being used.

The packed tower (see Chapter 10) is the most compact and efficient device for air stripping. The wastewater in a lime slurry (for phosphate removal and pH control) is sent first to a mixing tank and then to a settling tank to settle out calcium phosphate and calcium carbonate. The clarified wastewater is introduced near the tops of two packed towers, while air is sent countercurrently through the towers to remove the ammonia. The concentration of ammonia in the air product should be about 3.7×10^{-7} lb/ft^3, which is below emission standards. The wastewater product is then sent to a recarbonation basin so that calcium carbonate may be precipitated, removed, and reused.

Steam Stripping

Steam stripping is used to remove dilute concentrations of ammonia, hydrogen sulfide, and other volatile components from aqueous waste streams. It has been used to recover sulfur from refinery waste and organics, such as phenol, from industrial wastes. Work is currently being done on its possible use in removing chlorinated light hydrocarbons from industrial wastewater.[1]

The stream stripping process is carried out in a distillation column, which may be either a packed or tray tower (see Chapter 10). Steam is sent up from the bottom of the column and is used to remove volatile components from the waste stream, which is flowing downward (see Fig. 2.4.8). The product stream, rich in volatile components, may be further treated to recover these components. If the volatiles contain sulfur, the stream may need treatment for sulfur dioxide emissions. The purified aqueous stream may be discharged if a sufficient amount of the volatiles has been removed.

Ultrafiltration

Ultrafiltration is a process used on aqueous wastewaters to remove both dissolved and suspended high molecular weight solids or colloids that are difficult to remove by

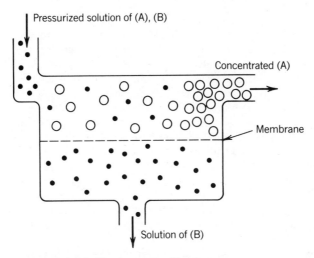

Figure 2.4.9. Membrane ultrafiltration process.

other methods. This treatment can remove particles as small as 10^{-3} to 10^{-2} microns (μm) and may be used to purify wastewaters by concentrating wastes or to fractionate waste streams. It is a fully developed process quite competitive with adsorption and evaporation in treating single-phase waste streams. It is currently being used in treating wastewaters produced by pharmaceutical, chemical, and electronics manufacturing companies. Its future uses may be in treating dye, pulpmill, and industrial laundry wastes.

Ultrafiltration uses a selective membrane to remove solutes or colloids from a pressurized waste stream passed through it (see Fig. 2.4.9).[2] The larger particles are retained by the membrane and the smaller particles pass through. The retained solids may be recovered or undergo further treatment before being disposed.

Capital and operating costs depend on the capacity of the system, the type of wastewater being treated, and the type of membrane used.

2.5   ULTIMATE DISPOSAL

Ultimate disposal is described by many as the final process in the treatment and management of hazardous wastes.* Before disposal most wastes undergo the various treatments previously described (biological, chemical, and/or physical) in order to concentrate, detoxify, and reduce the volume of the wastes. This has not always been the case. Before the early 1970s, most wastes were haphazardly landfilled or ocean dumped with little concern to the environmental effects of these practices. The passage of certain environmental protection laws over the last two decades (see Chapter 3) has helped to eliminate the irresponsible handling of hazardous wastes by requiring that proper measures be taken before they are disposed.

---

* The term "ultimate disposal methods" was coined by the USEPA and originally assigned to the four processes discussed in this section. The authors disagree with this nomenclature, preferring that "ultimate disposal" be reserved for a process in which the wastes are chemically or biologically rendered innocuous or a process that has been modified so that hazardous wastes are no longer generated.

Landfarming, deep-well injection, landfilling, and ocean dumping are four methods used in the so-called "ultimate" disposal of hazardous wastes. All four methods are discussed in this section. The latter three may be used on a variety of wastes; landfarming is used only on organic wastes. Landfarming is also a biological treatment method of treating wastes and could have been listed in Section 2.4. In this technique, nutrients in the soil convert the hazardous waste into nonhazardous materials that may enrich the soil. Wastes that are ocean dumped or landfilled also undergo some degree of biological conversion and degradation but not to the same extent as landfarmed wastes. The main purpose of ocean dumping, landfilling, and deep-well injection is the containment of the hazardous wastes.

Landfarming

Landfarming is one of several terms used to describe the process of disposing hazardous and nonhazardous wastes in the upper layer of the soil. The process is not a new one; it has been used for almost 30 yr in the disposal of oily petroleum wastes. This application has increased in usage over the years and is now employed to treat up to 50% of these petroleum wastes as well as a number of other biodegradable wastes.

Landfarming has a number of advantages: it is an effective and low cost disposal method; it is an environmentally safe and simple process not dependent upon processing equipment; and it is a natural form of waste disposal which can, in some cases, improve the fertility and nature of the soil.[4] There are also some limitations to landfarming. The bulk of the disposed waste must contain organic components. It is not recommended for use on inorganics, particularly when the pH of the waste is below 7. Wastes containing materials that could pollute the air, groundwater, or the soil itself are not candidates for this method. Some wastes may need pretreatment in order to make them suitable for landfarming; this additional treatment adds to the total cost of the waste disposal.

The mechanism behind landfarming is the breakdown of organic and inorganic material present in the waste by microorganisms residing in the soil. Once biodegraded, the material may then be used by other life forms. The upper layer is used in landfarming because it is more densely populated by microorganisms, resulting in a higher rate of biodegradation.

After its application to the soil, the waste may undergo decomposition, leaching of water-soluble components, volatilization, and finally, incorporation into the soil matrix.[4] Microorganisms, such as bacteria and yeasts, decompose the waste. Some of the carbon in the waste is released as carbon dioxide and the remainder is converted to natural organic matter that benefits the soil. In order that maximum conversion take place, it is important that enough nitrogen and/or phosphates be present. It is also important that the degree of volatilization and leaching of the waste is low so that air and water are not polluted.

There are a number of factors that affect the degree of biodegradation of the wastes. These include: waste composition, microorganism and waste contact within the soil, temperature, pH, oxygen, inorganic nutrients (specifically nitrogen and phosphates), and moisture content.

- *Waste Composition.* Waste composition is the most important factor affecting biodegradation. Earlier in this section it was mentioned that wastes consisting mostly of organic material (such as petroleum wastes) are the best candidates for landfarming and therefore biodegradation. Certain components are more readily biodegraded than others and their rate of biodegradation is therefore faster. By removing most nonorganic components before landfarming, the biodegradation rate may be increased.

- *Contact between Microorganism and Waste.* The second factor that increases the degree of biodegradation is a larger contact area between the soil microorganisms and the waste. A mechanical device may be needed to achieve this in the case of sludges and semisolid wastes.

- *Soil Temperature.* Soil temperature is one factor that is not easily controllable. Since biodegradation increases with increasing temperature, this process tends to work better in warmer climates and during the spring and summer months.

- *Soil pH.* It was mentioned earlier that if the soil is too acidic (pH < 7) excessive leaching of metals can occur, resulting in possible contamination of the groundwater. A pH between 7 and 9 is optimum for this process and to achieve this, lime is often added to the soil.

- *Oxygen.* Because the rate of aerobic biodegradation is greater than that of anaerobic biodegradation, an aerobic environment is preferred in the soil. If the soil is saturated with water and overloaded with biodegradable organic waste, it can become anaerobic. Control of the amount of waste and proper drainage are therefore required for this process to be successful.

- *Inorganic nutrients.* Sufficient amounts of nitrogen and phosphates are needed for a high degree of biodegradability. If not enough is present in the soil, these chemicals should be added in the form of fertilizer. Too much fertilizer, however, can cause groundwater contamination by the nitrates.

- *Moisture content.* The soil's moisture content is the final factor affecting the rate of biodegradation. A moisture content between 6 to 22 wt % is enough to produce good biodegradation.

The major steps involved in the landfarming process are as follows: site selection, site preparation, waste analysis, waste application, soil-waste blending, and post waste-addition care. A landfarming site should be flat enough to prevent erosion problems but also slightly sloped to avoid flooding problems; the soil on the site should be uniform for better waste application; the waste should be well tested before application and pretreated if necessary; the waste should be applied to and mixed into the soil uniformly; and, after applying the waste, the land should be inspected periodically to make sure the previous factors are complied with.

Deep-Well Injection

Deep-well injection is an ultimate disposal method that transfers liquid wastes far underground and away from freshwater sources. Like landfarming, this disposal process has been used for many years by the petroleum industry and is used to dispose of salt waters in oil fields. When the method first came into use, very often the injected brine would eventually contaminate groundwater and freshwater sands because the site was poorly chosen. The process has been improved and laws such as the *Safe Drinking Water Act of 1974* insure that sites for potential wells are better surveyed. Today, injection wells are placed as far away as possible from drinking water sources, usually at least $\frac{1}{4}$ mile.

Many factors go into the selection of a deep-well injection site. For example, the rock formation surrounding the disposal zone must be strong but permeable enough to absorb the liquid wastes and the site must be far enough from drinking water sources to prevent contamination. Once a site is selected, it must be tested by drilling a pilot well. The performance data from the pilot well, besides testing permeability and water quality, also aid in the design of the final well and in determining the proper injection rate.[5]

The type of waste injected into the well plays a role in how deep the injection will be made. The more toxic the waste the farther down the disposal zone must be. Disposal zones have been classified into five different types:

1. *Zone of Rapid Circulation.* This zone describes the area that runs from the soil surface down to only a few hundred feet. Injection is not done into this zone.

2. *Zone of Delayed Circulation.* This zone contains circulating fresh water and may be used for certain wastewaters if properly monitored. The water circulation is slow enough so that residence times of a few decades to a few centuries can be achieved for the waste.

3. *Subzone of Lethargic Flow.* The liquid flowing in this zone is very slow moving and saline. More concentrated hazardous wastes are injected here.

4. *Stagnant Subzones.* The liquid contained in this region is hydrodynamically trapped and the zone is generally several thousand feet below the soil surface. Highly toxic wastes may be injected here if the zone can properly accept and keep the waste.

5. *Dry Subzones.* Salt beds fall under this classification. This zone does not contain water and is nearly impermeable. Because there is a possibility that liquid movement could occur through hydrofractures, this zone must be monitored. Otherwise, any waste injected here would be isolated from any water sources.

Salt beds have also been considered for storing nuclear wastes. One possibility is to inject water to remove some of the salt and thereby form an underground cavern. The cavern could then be filled with the waste, plus possibly a solidifying agent, and then cement-sealed.

In order for wastewaters to be deep-well injected, the following criteria must apply: the wastewaters must have a low volume and high concentration of waste, must be difficult to treat by other methods, cannot cause an unfavorable reaction with material in the disposal zone, must be biologically inactive, and must be noncorrosive. Only wastes untreatable by other means are considered for deep-well injection because of the limited amount of available space for injection and its possible pollution of drinking waters.

Figure 2.5.1 shows an illustration of a typical injection well. The surface casing runs from the surface to about 200 ft below the freshwater table and is well sealed to prevent the degradation of freshwater sands. The protection casing is inside the surface casing and extends from ground level to the disposal zone; it is also sealed. The injection tubing is used to pass the waste stream through the well. It is sealed at the well head and just above the disposal zone. The annular space between the injection tubing and protection casing is filled with a noncorrosive fluid and kept at a pressure 0.5 psi (pounds per square inch) above the injection pressure.[5]

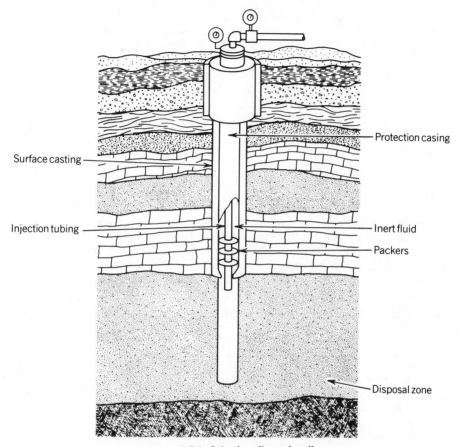

Figure 2.5.1. Injection disposal well.

Landfilling

Landfilling is the third ultimate disposal method and is generally used on wastes in the form of sludges. There are two types of landfilling, area fill and trenching. *Area fill* is essentially accomplished above ground while *trenching* involves burying the waste. Trenching is the more established and popular form of the two. Since trenching requires excavation, however, area fill has the advantage that it requires less manpower and machinery. Area fill is also less likely to contaminate ground-water since the filling is above ground. Trenching, however, may be used for both stabilized and unstabilized sludges and makes more efficient use of the land. Both techniques require the use of lime and other chemicals to control odors and with both methods, cold and wet weather can cause problems. Both methods also produce gas, which can cause explosions or harm vegetation, and leachate, which can contaminate ground and surface water.

Most hazardous wastes must be subjected to one or more pretreatments such as solidification, degradation, volume reduction, and detoxification before being land-filled. This practice stabilizes the waste and helps decrease the amount of gas and leachate produced from the landfill. Landfilling is similar to landfarming in that both ultimate disposal methods combine wastes and soil. Landfarming however involves the biochemical reaction between soil nutrients and wastes to degrade and stabilize the waste; as a result, only specific types of wastes can be landfarmed. A larger variety of wastes may be handled by landfilling.

Area fill may be done in one of three ways: by mixing the waste with soil and forming a mound with the mixture and covering with soil; by spreading alternate layers of soil and soil–waste mixture over the area; or by filling a containment area surrounded by dikes with the waste and then covering with a soil layer. These forms of area fill are known as area fill mound, area fill layer, and dike containment, respectively. In *area fill mounds*, the sludge–soil mixture may be stacked in mounds as high as 6 ft. A soil covering, usually 3 to 5 ft, is placed on top of the mound. The amount of covering and soil-to-sludge ratio are dependent on the concentration of solids in the waste and on the degree of mound stability required. An earthen containment may be constructured for better mound stability. In an *area fill layer*, the sludge–soil mixture is spread out over an area in consecutive layers each having a thickness form 0.5 to 3 ft. Layers of soil 0.5 to 1 ft thick are often placed between the mixture layers. Final heights for these landfills will vary depending upon the amount of waste being disposed and the size of the site. Sites for these landfills are usually flat or slightly sloped. *Dike containment* landfills employ dikes to surround the four sides of the containment area. The site is often on a hill, which can serve as one of the containment sides. In this method, access roads are needed in order to bring the waste to the top of the landfill. Layers of soil, 1 to 3 ft thick, are usually placed at certain levels in the landfill, as well as a final layer on top, 3 to 5 ft thick. The final height of a containment area can range from 10 to 30 ft with widths of 50 to 100 ft and lengths of 100 to 200 ft.

Trenching involves placing the waste in an evacuated trench and covering it with one or more layers of soil. There are two types of trenching: *narrow* trenching operations, with trench widths of 10 ft or less, and *wide* trenching operations with trench widths > 10 ft. The type of operation used depends upon the solids content of the waste. In either case, it is necessary that the width between trenches is large enough to achieve sidewall stability and space for soil stockpiles and equipment. There must be a soil thickness of 2 to 5 ft between the trench bottom and nearest groundwater level in order to provide adequate leachate control.

Wastes may be placed in *narrow* trenches in one application, with one soil layer placed on top of the waste. The soil removed from the trench may be used for the covering. Narrow trenches of 2 to 3 ft are best for sludges with low solids content (15 to 20%) because the support from the side walls prevents the soil layer from sinking to the bottom. Sludges with higher solids content should be placed in wider trenches.

In *wide trench* operations, excavating and filling equipment are placed within and at the top of the trench. The excavated soil is used to cover the sludge and to provide an intermediate soil layer in order to support the equipment. The sludge must have a fairly high solids content in order to support the equipment and soil covering at the top of the trench. If the tench is very wide, dikes may be used to confine the sludge to certain areas.

When a landfill site is to be selected, a number of factors must be evaluated; these factors fall into three categories: technical, economic, and public acceptance considerations. The following factors are technical considerations: haul distance, site size and life, topography, surface water, soils and geology, groundwater, soil quantity and suitability, vegetation, environmentally sensitive areas, archaeological or historical significance, site access, and land use. Economic considerations involve: site capital cost, site operating cost, and hauling cost. Public acceptance considerations include: local laws and public opinion; a public hearing is usually held in order to obtain local government and public input.

In comparing the costs of the different types of landfills, wide trench landfilling has the lowest capital and operating costs; area fill layer has the highest capital cost, and area fill mound has the highest operating costs.[6] The other types of landfills fall somewhere in between these extremes. Capital cost in this case involves land, site preparation, equipment purchase, and engineering. Overall costs for landfills are dependent upon the efficiency of the operation with respect to land use, equipment, and manpower. Hauling costs are independent of the landfill method and hence do not play a role in an economic comparison of the different techniques.

Once a landfill has been completed and closed, it must still be maintained in order to prevent soil erosion and other damaging effects of weather. It must also be monitored for gas and leachate as well as for any other potential environmental hazards.

Ocean Dumping

Ocean dumping is the last "ultimate" disposal method to be discussed. Although, it is probably the simplest of the four techniques, its aftereffects and long-term consequences are more complex and less understood then those of other ultimate disposal methods. Regulations passed over the last 20 yr have limited the types and amounts of wastes that may be dumped. Therefore, in some cases, ocean dumping may be no more attractive than any other ultimate disposal method, despite its simplicity.

Currently, only certain types of wastes may legally be ocean dumped. In some cases, such as industrial wastes, the disposal is regulated and may even require some treatment before disposal. *Uncontrolled* ocean dumping of untreated wastes occurs in the form of oil spills, rain carrying air pollutants, and runoff carrying land pollutants.

Until the early 1970s, ocean dumping in the United States was not strictly controlled; between 1945 and 1970, ocean dumping of wastes was used in ever-increasing amounts. This was done in order to avoid contaminating inland bodies of water. As a result of the passage of such regulations as the *Ocean Dumping Act* and the *Clean Water Act*, the practice of ocean dumping was severely restricted. The Ocean Dumping Act prohibited the dumping of any radioactive wastes and any biological or chemical warfare agents. It also gave the EPA and the Army Corps of Engineers the power to regulate ocean dumping of any wastes that were not dredge spoils.

Ocean dumping actually describes two forms of waste disposal: one into shallow offshore waters and the other into deep ocean waters. Each of these methods has advantages and disadvantages. The advantages of dumping offshore are more information and experience are available, transportation costs to the site are lower, and any resulting pollution is localized. The disadvantage is that dumping close to land could lead to problems with fishing industries and ruin offshore mineral deposits.[7] Ocean dumping into deep waters offers the advantage of spreading the waste over large areas; the resulting lower waste concentration should have a less harmful effect on the marine environment. The disadvantage is that there is less information available and greater difficulty in monitoring the effects.

Ocean dumping is used not only as a final means of containment for hazardous wastes but also as a means of converting the hazardous waste to nonhazardous material. This is accomplished by dilution and biological conversion of the waste in the ocean. The mechanism for the biological degradation is not as well defined as that for landfarming, which is limited to specific types of hazardous wastes. Because such a variety of wastes are still ocean dumped and the long-term effects to the marine environment have not been well monitored, there is the possibility of very damaging, long-term consequences from this practice. The severity of the effects of ocean dumping are dependent on the type of waste, the input rate, the concentration of the waste in the ocean, the rate of dispersion, and the rate at which it undergoes biological conversion.

In recent years, criteria and monitoring procedures have been developed by the EPA that are used to measure the effects of ocean dumping, particularly for the offshore method. While this resulted in a better understanding of ocean dumping, more research is still needed to access the impact of present and possible future wastes (incineration residue, radioactive wastes) on the ocean environment.[7] There is still much to be learned about the various physical, chemical, and environmental processes that take place in the marine environment in order to understand and utilize ocean dumping fully.

2.6   PROCESS MODIFICATION

As an alternative to treatments that change or dispose of a hazardous waste, the process that generates the waste can often be modified to reduce or eliminate the waste effluent. Although this method is being presented after the waste treatment and disposal methods (Sections 2.2–2.5), in practice it is usually the first line of action. The types of process modifications discussed here are waste minimization–abatement, waste exchange and waste recycle–reuse. For more extensive detail on this topic the reader is directed to the literature.[5,8–12]

Waste Minimization–Abatement

In general, engineers are quite knowledgeable about the processes that generate wastes; few engineers are well informed on the systems and processes for the disposal of wastes. An option when considering alternatives for hazardous waste treatment is the redesign of the process that generates the waste. This can be accomplished either by substituting a new process for an old one, or by revamping an old process. Some examples of such techniques are changing the solvent used from a hazardous to a nonhazardous solvent; changing the catalyst used from a hazardous to a nonhazardous catalyst; controlling the temperature and pressure of the reaction to inhibit hazardous side-product formation; using higher quality raw materials that do not contain hazardous impurities; plus many others. Some specific examples where such methods have been used are

> In the manufacture of nitric acid, the air supply and temperature are carefully controlled to minimize the amount of NO_x emissions. (NO_x represents various oxides of nitrogen.)

In the pulp and paper industry, direct-contact evaporators are used in preference to the noncontact type to reduce particulate emissions.

In the purification of explosives where there are numerous washing steps, the effluent washes from one stage are used in later stages of the process.

The major drawback to this option is that most generators cannot afford to redesign the process; however, as disposal of hazardous waste becomes more difficult and more expensive, this option may gain in popularity.

Waste Exchange

The waste from one industry is often an acceptable raw material for another. Waste exchanges were first implemented and are now fairly common in Europe; there are few in the United States. A major stumbling block to widespread adoption of the method in this country has been the lack of an intermediary or an information bank to help match supply and demand. Even if a suitable industry (i.e., one that can use the waste) is found, it must be located fairly close to the waste generator. Transportation of the waste generally requires permitting and special handling considerations; this increases the cost.

Waste Recycle

In order to eliminate or significantly reduce waste discharge, recycle or reuse of waste streams can also be employed. In this method, however, treatment of the waste stream is usually required to regain the appropriate quality before the stream is used again. Recycle of waste streams is fairly common in many industries, for example: iron and steel manufacturing, explosive manufacturing, nonferrous metal manufacturing, steam and electric power plants, coal mining, inorganic chemical manufacturing, aluminum forming, pharmaceutical manufacturing, battery manufacture, textiles, and paint-ink formulations. Usually, waste recycle is only used by large generators because of the high cost.

Methods of recycle are related to the particular reuse of the waste stream. The waste stream normally undergoes some processing to recover the liquid or solvent, and to remove metal ions for disposal. Some examples of common recovery techniques are evaporation, reverse osmosis, and activated carbon adsorption. Whenever a recycle stream is used, the quality of the product produced should be carefully monitored.

2.7   THE SELECTION OF HAZARDOUS WASTE TREATMENT PROCESSES

If the hazardous waste cannot be eliminated through process modification, a viable hazardous waste management method must be chosen. The procedure used in this selection can be a long and complicated one. Before a specific treatment or means of disposal is chosen, the waste generator must determine how toxic and/or hazardous the waste is and if the treatment or disposal is to be done on- or off-site. Once these questions are answered, an appropriate method for treating the waste must be chosen; this choice is based not only on technical considerations, but on economic and regulatory factors as well.

As described in Chapter 1, wastes may be classified in one of three ways: *highly toxic* or *highly hazardous* waste, which requires extra care in managing; *hazardous* waste, which may be managed by standard methods; and *nonhazardous* waste, which does not present much of a health or environmental threat. To determine the classification for a particular waste, federal (EPA) and state criteria should be used.

The next decision a waste generator must make is whether to manage the waste directly or have it done by a commercial firm. Companies that dispose of their wastes on-site do so because they have the technical and physical resources, prefer complete control over the waste management and its costs, and wish to avoid the specter of future liability associated with off-site disposal. According to the EPA, over 80% of all waste generators dispose of their wastes on-site for these reasons, or simply because there are no commercial firms within their area.

If a waste generator decides to manage the waste directly, a method for treating and/or disposing of the waste must be chosen with regard to technical, economic, regulatory, and public opinion considerations.

Technical Considerations

Technical considerations include: waste characteristics, technical suitability of the available treatments, and treatment objectives. The waste must be analyzed so that the proper form of treatment may be selected. Should the waste contain more than one type (e.g., a mixture of heavy metals and organics), more than one method of treatment may be needed. Waste treatment may be divided into five categories: physical, chemical, biological, thermal (incineration), and solidification or encapsulation (a subcategory of physical treatment). *Physical treatments* (Section 2.4), may be used alone or with the other types of treatment; their function is to concentrate wastes, reduce waste volume, and separate different waste components for continued treatment or disposal. *Chemical treatments* (Section 2.2) are used to convert hazardous wastes into other less hazardous forms; they are generally used on one substance or similar substances because the reactions involved are specific.[8] *Biological treatments* (Section 2.3) are more specific than the other types and are used mainly on organic wastes. *Solidification* and *encapsulation* (Section 2.4) are also treatments for specific types of waste; these methods work well on inorganic wastes but physical and/or chemical pretreatment is usually necessary.

Incineration is one of the most popular forms of waste treatment, especially when organic wastes are involved. It destroys most organic components in the waste and the resulting sludge and ash are almost always less hazardous and easier to dispose of than the original waste. It is difficult to obtain a permit for incineration, however, and many stringent regulations govern its use. Table 2.7.1 shows the applicability of various treatment processes, excluding incineration, to a wide variety of different waste types.[1]

The *objective* of the treatment is another important factor that influences treatment selection. It may be desired, for example, to destroy the hazardous components in the waste, reduce the hazardous components and then isolate them, or merely to separate the hazardous components. The end product of the waste treatment must be compatible with the type of ultimate disposal to be used for the waste: sewage discharge, landfilling, deep-well injection, and so on. For example, if the waste is to be disposed of by landfarming, the end product of the waste treatment should consist of biodegradable organic components. Table 2.7.2 shows the output stream charac-

TABLE 2.7.1. Applicability of Treatment Processes to Various Types of Wastes[a]

| Processes | Single Liquid Phase | | | | Two Phases | |
	Solid	Inorganic	Organic	Mixed	Slurry[b]	Sludge
Phase Separation						
Filtration	n	n	n	n	y	n
Sedimentation	n	n	n	n	y	n
Flocculation	n	n	n	n	n	n
Centrifugation	n	n	n	n	y	n
Distillation	n	n	n	n	y	y
Evaporation	n	n	n	n	y	y
Flotation	n	n	n	n	y	n
Ultrafiltration	n	n	n	n	n	n
HGMS	p	n	n	n	n	n
Precipitation	n	n	n	n	n	n
Component Separation						
Ion exchange	n	y	y	y	n	n
Liquid ion exchange	y	y	y	n	y	y
Freeze crystallization	n	y	y	n	y	y
Reverse osmosis	n	y	n	y	n	n
Carbon adsorption	n	y	n	y	n	n
Resin adsorption	n	y	n	y	n	n
Electrodialysis	n	y	n	y	n	n
Air stripping	n	n	y	y	y	n
Steam stripping	y	n	y	y	y	n
Ammonia stripping	y	y	n	n	n	n
Ultrafiltration	n	y	y	y	n	n
Solvent extraction	p	y	y	y	n	n
Distillation	n	n	y	y	y	y
Evaporation	n	y	y	y	y	y
Chemical Transformation						
Neutralization	n	y	y	y	y	y
Precipitation	n	y	y	y	n	n
Hydrolysis	p	n	y	y	y	y
Oxidation	n	y	y	y	n	n
Reduction	n	y	y	y	n	n
Ozonolysis	n	y	y	y	n	n
Calcination	y	y	y	y	y	y
Chlorinolysis	n	n	y	y	n	n
Electrolysis	n	y	n	y	n	n
Microwave	n	n	y	y	n	n
Biological	n	n	y	y	y	y
Catalysis	n	y	y	y	n	n
Photolysis	n	n	y	y	n	n

[a] y = yes, workable; n = no; p = possible.
[b] Slurry is defined here as a pumpable mixture of solids and liquids.

TABLE 2.7.2. Characteristics of the End Products of Various Waste Treatments

Treatment		Code Number	Form	Output Streams	
Type	Process			Characteristics	Possible Follow-On Steps
Phase separation	Filtration	F 1	Sludge	15–20% solids	Landfill, calcination
		F 2	Liquid	500–5000 ppm total dissolved solids (TDS)	Component separation
	Sedimentation	S 1	Sludge	2–15% solids	Decantation
	Centrifugation	S 2	Liquid	10–200 ppm suspended solids (SS)	
	Flotation	Fl 1	Stabilized	Particle-bearing froth	Skimming
		Fl 2	Liquid	Solution	Component separation
	HGMS	H 1	Slurry	Magnetic and paramagnetic particulates	Recovery
		H 2	Liquid	Solution	Component separation
	Flocculation	Fc	Sludge or slurry	Flocculated particulates	Sedimentation, filtration, centrifugation
	Distillation	Dis 1	Sludge	Still bottoms	Calcination
		Dis 2	Liquid	Pure solvent	Sale
	Evaporation	E 1	Solid		Resource recovery
			Liquid	Condensate	Recovery or disposal
Component separation					
Inorganics	Ion exchange	IE 1	Liquid	Concentrated solution of hazardous components	Precipitation, recycle, electrolysis
		IE 2	Liquid	Purified water with hazardous components at ppm levels	Discharge
	Liquid ion exchange	Similar to ion exchange			
	Carbon adsorption	CA 1	Solid	Adsorbate on carbon	Chemical regeneration
		CA 2	Liquid	Purified water	Discharge

54

	Reverse osmosis	RO 1	Liquid	Concentrated solution of hazardous components	Precipitation, electrolysis, recycle
		RO 2	Liquid	Purified solution, TDS >5 ppm	To water treatment
	Electrodialysis	ED 1	Liquid	Concentrated stream, 100–500 ppm salts	Precipitation, metal recovey
		ED 2	Liquid	Dilute stream	To water treatment
	Freeze crystallization	FC 1	Sludge	Concentrated brine	Recovery
		FC 2	Liquid	Purified stream, ~100 TDS	To water treatment
Organics	Carbon adsorption	CA 1	Solid	Adsorbate on carbon	Thermal or chemical regeneration
		CA 2	Liquid	Purified water	Discharge
	Resin adsorption	RA 1	Solid	Adsorbate on resin	Solvent regeneration
			Liquid	Purified water, <10 ppm organics	To water treatment
	Steam stripping	SS1	Liquid	Aqueous steam concentrated in volatile organics	Recovery; incineration
			Liquid	Dilute aqueous stream with 50–100 ppm organics	To water treatment
	Solvent extraction	SE 1	Liquid	Concentrated solution of hazardous components in extraction solvent	Recovery of extraction solvent
		SE 2	Liquid	Purified liquid; hazardous component concentration <10 ppm	Recycle or discharge
	Distillation	Dis 1	Sludge	Still bottoms	Incineration
		Dis 2	Liquid	Pure liquid	Sale
Chemical transformation	Neutralization	N	Unchanged	Stream of altered pH	Component separation
	Precipitation	Ppt 1	Sludge	Supernatant with concentrations governed by solubility of precipitate	Landfill, calcination
		Ppt 2	Liquid		Depends on product stream composition
	Oxidation Ozonation	O	Liquid	CO_2, H_2O and other oxidation products	Depends on product stream composition
	Reduction	R	Slurry	Heavy metals and residual reducing agent	Filtration

55

TABLE 2.7.2. (*Continued*)

Treatment		Output Streams			
Type	Process	Code Number	Form	Characteristics	Possible Follow-On Steps
	Calcination	Cal 1	Solid	Oxide and/or other residue	Landfill or recovery
		Cal 2	Gas	Volatiles (CO_2, NO_x, SO_x, hydrocarbons, fine particules)	Wet scrubbing
	Hydrolysis	Hy	Liquid, slurry, or sludge	Mixture of products that may or may not be toxic	Resource recovery
	Electrolysis	El 1		Cathode reaction products	Often a recovered metal
		El 2		Anode reaction products	Depends on nature of products
	Photolysis	P	Liquid	Solution of photodecomposition products	Depends on nature of products
	Microwave discharge	M	Gas	Similar to incinerator emissions	Wet scrubbing
	Catalysis	Depends on reaction catalyzed			
Biological treatment	Activated sludge	AC 1	Liquid	Clean water	Discharge
		AC 2	Sludge	Heavy metals and refractory organics	Calcination
	Aerated lagoon				
	Trickling filter				
	Waste stabilization pond	W	Liquid	Clear water	Discharge
	Anaerobic digestion	An 1	Sludge	CO_2, methane	Incineration
		An 2	Gas		Fuel recovery
	Composting	Com 1	Sludge	Concentrated in metals	Calcination or recovery
		Com 2	Liquid	Leachate solution of partially degraded organics	Component separation

56

TABLE 2.7.3. Comparison of Waste Treatment Processes

Process	State of the Art[a]	Required Feed Stream Properties[b]	Characteristics of Output Stream(s)
(a) Processes that Separate Heavy Metals from Liquid Waste Streams			
Physical removal			
Ion exchange	Used but not common (Catalyst IV)	Concentration <4000 ppm; aqueous solutions, low SS	One concentrated in heavy metals; one purified
Reverse osmosis	Catalyst IV	Concentration >400 ppm; aqueous solution; controlled pH; low SS; no strong oxidants	One concentrated in heavy metals; one with heavy metal concentrations >5 ppm
Electrodialysis	Catalyst IV	Aqueous solutions; neutral or slightly acidic; Fe and Mn <0.3 ppm; Cu <400 ppm	One with 1000–5000 ppm heavy metals; one with 100–500 ppm heavy metals
Liquid ion exchange	Catalyst IV	Aqueous solutions; no concentration limits; no surfactants; SS <0.1%	Extraction solvent concentrated in heavy metals; purified water or slurry
Freeze crystallization	Needs development (Catalyst III)	Aqueous solutions; TDS ~10%	Concentrated brine or sludge; purified water, TDS ~100 ppm
Chemical removal			
Precipitation	Common (Catalyst V)	Aqueous or low viscosity nonaqueous solutions; no concentration limits	Precipitated heavy metal sulfides, hydroxides, oxides, etc.; solvent with TDS governed by solubility products of precipitates
Reduction	Catalyst IV	Aqueous solutions; concentrations of heavy metals <1%; controlled pH	Acidic solutions with reagent (oxidized $NaBH_4$ or Zn); metallic precipitates
Electrolysis	Catalyst III	Aqueous solutions; heavy metal concentrations <10%	Recovered metals; solution with 2–10 ppm heavy metals
(b) Processes that Destroy Organics			
Biodegradation	Catalyst V	Dilute aqueous streams with soluble organics	Pure water

TABLE 2.7.3. (*Continued*)

Process	State of the Art[a]	Required Feed Stream Properties[b]	Characteristics of Output Stream(s)
Oxidation	Catalyst IV	Dilute aqueous solutions of phenols, organic sulfur compounds, chlorinated hydrocarbons, etc.	Oxidation products in aqueous solutions
Ozonation	Catalyst IV	Aqueous solutions; concentrations <1%	Oxidation products in aqueous solutions
Calcination	Catalyst IV	Organics and inorganics that decompose thermally	Solid oxides; volatile emissions
Hydrolysis	Needs development (Catalyst III)	Aqueous or nonaqueous streams; no concentration limits	Hydrolysis products
Photolysis	Needs research (Catalyst II)	Aqueous streams; transparent to light; components that absorb radiation	Photolysis products
Chlorinolysis	Catalyst III	Chlorinated hydrocarbon waste streams; low sulfur; low oxygen; can contain benzene and other aromatics; no solids; no tars	CCl_4; HCl, and phosgene
Microwave discharge	Catalyst II	Organic liquids or vapors	Discharge products; not accurately predictable

(*c*) *Processes that Separate Toxic Anions from Liquid Waste Streams*

Physical removal Ion exchange	Catalyst IV	Inorganic or organic anions in aqueous solution	Concentrated aqueous solutions
Liquid ion exchange	Catalyst IV	Inorganic or organic anions in aqueous solution	Concentrated solutions in extraction solvent

Process	Catalyst/Status	Feed	Products
Electrodialysis	Catalyst IV	Aqueous stream with 1000–5000 ppm inorganic salts; and pH; Fe and Mn <0.3 ppm	Concentrated aqueous stream (10,000 ppm salts); dilute stream (100–500 ppm salts)
Reverse osmosis	Catalyst IV	Aqueous solutions with up to 34,000 ppm total dissolved solids	Dilute solution (~5 ppm TDS); concentrated solution of hazardous components
Freeze crystallization	Needs development (Catalyst III)	Aqueous salt solutions	Purified water; concentrated brine
Chemical removal Oxidation	Catalyst III–IV	Aqueous solutions of cyanides, sulfides, sulfites etc.; concentrations <1%	Oxidation products
Ozonation	Catalyst IV	Aqueous solutions of cyanides; concentrations <1%	Cyanate solutions
Electrolysis	Catalyst III	Alkaline aqueous solutions of cyanides or concentrated HCl solutions (>20%)	Cyanides to ammonium and carbonate salt solutions; HCl to Cl_2 gas

(d) Processes that Can Accept Slurries or Sludges

Process	Catalyst/Status	Feed	Products
Calcination	Used but not common (Catalyst IV)	Waste stream with components that decompose by volatilization (hydroxides, carbonates, nitrates, sulfites, sulfates)	Solid greatly reduced in volume; volatiles
Freeze crystallization	Needs development (Catalyst III)	Low viscosity aqueous slurry or sludge	Brine sludge; purified water
HGMS	Needs research (Catalyst II)	Magnetic or paramagnetic particles in slurry	Particles adsorbed on magnetic filter
Liquid ion exchange	Catalyst IV	Solvent extractable inorganic component	Solution in extraction solvent
Flotation	Catalyst III	Flotable particles in slurry	Froth
Hydrolysis	Catalyst II	Hydrolyzable component	Hydrolysis products
Anaerobic digestion	Catalyst IV	Aqueous slurry; <7% solids; no oils or greases; no aromatics or long chain hydrocarbons	Sludges; methane and CO_2

TABLE 2.7.3. (*Continued*)

Process	State of the Art[a]	Required Feed Stream Properties[b]	Characteristics of Output Stream(s)
Composting	Catalyst IV	Aqueous sludge; <50% solids	Sludge; leachate
Steam distillation	Catalyst IV	Sludge or slurry with volatile organics	Volatile, solid residue
Solvent extraction	Catalyst IV	Solvent extractable organic	Solution of extracted components; residual sludge
(e) Processes that Can Accept Tars or Solids			
Calcination	Used but not common (Catalyst IV)	Tars or solids that can be volatilized	Volatiles; char and/or metal oxides
Hydrolysis	Needs development (Catalyst III)	Tars or solid powders	Hydrolysis products
Steam distillation	Catalyst IV	Solids contaminated with volatile organics	Purified solids; condensed organics
Dissolution	Catalyst IV	Tars, solids, or solid powders that will dissolve in some reagent	Liquid solution for further treatment; solid residue
Crushing and grinding	Catalyst V	Bulk solid	Powdered solid
Cryogenics	Needs research (Catalyst II)	Tars, bulk solid	Reduced particle size

[a] Catalyst I   = Process is not applicable in a useful way to waste treatment.
Catalyst II   = Process might work in 10–15 yr, but needs more research.
Catalyst III = Process is useful to hazardous wastes, but needs more development work.
Catalyst IV = Process is developed but not commonly used in industry.
Catalyst V   = Process commonly used in industry.
[b] SS = suspended solids; TDS = total dissolved solids

teristics for many treatment processes, excluding incineration, as well as possible follow-up measures.[1] In Table 2.7.3, processes to perform various waste treatment tasks are compared. These tasks are (a) separation of heavy metals from liquid waste streams, (b) destruction of organics, (c) separation of toxic anions from liquid waste streams, (d) processes that can handle slurries or sludges, and (e) processes that can handle tars or solids. Not all of these processes are currently feasible for hazardous wastes; the second column describes the present "state of the art" for each process.

Economic Considerations

Economic feasibility is an obviously important factor to be considered in treatment selection. Cost is a strong function of the amount of waste to be treated and the simplicity and degree of commercial usage of the treatment process. The various costs include capital investment, operating costs, utility costs, and expense of the final disposal of the waste product.

Regulation Considerations

Environmental factors, in the form of state and local environmental regulations, also influence waste treatment selection. The EPA RCRA regulations especially control the type of waste treatment that may be used and the characteristics of the waste that is to finally be disposed. These regulations are the topic of Chapter 3.

Public Opinion

The attitude of local residents and politicians also has an influence on hazardous waste management. As described earlier, the NIMBY syndrome often arises when a hazardous waste facility is about to be constructed, in spite of adherence to EPA regulations that have been established to protect the community. If the waste generator works with the community early in the planning stages for a new treatment facility, there is likely to be less public opposition later on.

REFERENCES

1. National Technical Information Service, U.S. Department of Commerce, *Physical, Chemical, and Biological Treatment Techniques for Industrial Wastes, Vol. 1*, PB-275-054, 1977.
2. Office of Research and Development, U.S. Environmental Protection Agency, *Treatability Manual: Volume III. Technologies for Control/Removal of Pollutants*, EPA-600/8-80-042c, July 1980.
3. R.B. Pojasek, "Solid-Waste Disposal: Solidification," *Chem. Eng.*, 140–145 August 13 (1979).
4. R.L. Huddleston, "Solid-Waste Disposal: Landfarming," *Chem. Eng.* 119–124 February 26 (1979).
5. G.W. Dawson and B.W. Mercer, *Hazardous Waste Management*, Wiley, New York, 1986.
6. M. Sittig, *Landfill Disposal of Hazardous Wastes and Sludges*, Noyes Data Corporation, Park Ridge, NJ, 1979.
7. B.H. Ketchum, D.R. Kester, and P.K. Park, *Ocean Dumping of Industrial Wastes*, Plenum, New York, 1981.

8. J.A. Mackie and K. Niesen, "Hazardous-Waste Management: The Alternatives," *Chem. Eng.*, 50–64, August 6 (1984).

9. R.E. Kenson, "Recovery and Reuse of Solvents for VOC Air Emissions," *Environmental Progress*, **4** (3), 161–164, August (1985).

10. H. Bhatt, R. Sykes, and T. Sweeney, *Management of Toxic and Hazardous Wastes*, Lewis Chelsea, MI, 1985.

11. K. Noll, C. Haas, C. Schmidt, and P. Kodukula, *Recovery. Recycle, and Reuse of Industrial Wastes*, Lewis Chelsea, MI, 1985.

12. L.L. Tavlarides, *Process Modifications for Industrial Source Reduction*, Lewis, Chelsea, MI, 1985.

3

Standards and Regulations

Contributing author: Madho Ramnarine Singh

3.1 INTRODUCTION

The emergence of hazardous waste as a major environmental issue has forced society to reexamine the ultimate ends of the whole pollution control enterprise. The hazardous waste problem is a by-product of this nation's rapid economic growth over the last 30 or 40 yr. As industry has expanded, we have developed more and more reliance on a broader range of chemicals and chemical by-products in plastics, automobiles, and other major manufacturing sectors. When chemicals are used to develop new products, however, wastes are generated and unfortunately the phenomenal growth in the production of waste was not mirrored by growth in the field of waste management. Much of the waste produced made its way into the environment where it poses a serious threat to ecological systems and public health. It was not until the mid-1970s that Congress enacted the *Resource Conservation and Recovery Act* (RCRA), the first major hazardous waste regulatory document to address waste generation and disposal effectively.

This chapter is organized in historical fashion. In Section 3.2, hazardous waste legislation prior to the RCRA of 1976—what little there was—is discussed. Section 3.3 examines the 1976 RCRA with particular emphasis on Subtitle C of the act, which lays the groundwork for the current standards and regulations pertaining to the management of hazardous waste. In Section 3.4, the *1984 Hazardous and Solid Waste Amendments* (HSWA) to the RCRA, also known as the *New RCRA*, are discussed. Along with key elements of the amendments, a discussion of Subtitle I, which expands the regulation domain of Subtitle C, is included. Section 3.5 covers the *Permitting Process* for incinerators, treatment/storage tanks, and land treatment, with special emphasis on incinerators. Finally in Section 3.6, the future of incineration, in the light of economic factors and future restrictions on land disposal treatment, is projected and compared with that for the other available alternatives.

3.2   EARLY LEGISLATION

Prior to the passage of RCRA, little solid waste legislation existed. Practices that permitted technologies to remove pollutants efficiently from air and water streams, only to dispose of them indiscriminantly on land, were commonplace. Legislative authority to control solid waste disposal dates as far back as the 1899 *River and Harbor Act*, the main concern of which was keeping the waterways *navigable*, rather than keeping them *clean*. The following is a quotation from the *Refuse Act*, which comprises Section 13 of the River and Harbor Act[1]:

> ⋯ it shall not be lawful to throw, discharge or deposit, or cause, suffer or procure to be thrown, discharged or deposited either from or out of any ship, barge, or other floating craft of any kind, or from the shore, wharf, manufacturing establishment, or mill of any kind, any refuse matter of any kind or description whatever other than that flowing from streets and sewers and passing therefrom in a liquid state, into any tributary of any navigable water from which the same shall float or be washed into such navigable water; and it shall not be lawful to deposit, or cause, suffer, or procure to be deposited material of any kind in any place on the bank of any navigable water, or on the bank of any tributary of any navigable water, where the same shall be liable to be washed into such navigable water; either by ordinary or high tides, or by storms or floods, or otherwise, whereby navigation shall or may be impeded or obstructed: provided that nothing herein contained shall extend to, apply to, or prohibit the operations in connection with the improvement of navigable waters or construction of public works, considered necessary and proper by the United States officers supervising such improvement or public work: and provided further, that the Secretary of War, whenever in the judgment of the Chief of Engineers, anchorage and navigation shall not be injured thereby, may permit the deposit of any material above mentioned in navigable waters within limits to be defined and under conditions to be prescribed by him, provided application is made to him prior to depositing such material; and whenever any permit is so granted the conditions thereof shall be strictly complied with, and any violation thereof shall be unlawful.

The first specific solid waste legislation aimed at protecting the environment did not pass through Congress until the mid-1960s. The *Solid Waste Disposal Act* of 1965 fostered a research program concerned with utility-generated solid waste and the recovery of valuable fractions of municipal solid waste. This research and development (R&D) program concentrated on the improvement of collection, transport, recycling, reuse, processing, and disposal methods of solid waste. The program was jointly administered by the Department of Interior (Bureau of Mines) and the Department of Health, Education, and Welfare (Bureau of Solid Waste Management).

The *Resource Recovery Act* of 1970 provided a new directive to the 1965 *Solid Waste Disposal Act*. While the 1965 act was limited mainly to R&D, the 1970 act expanded the federal role to the planning of solid waste management programs and training; and while the 1965 act made no mention of *hazardous wastes*, the 1970 act registered concern for the adverse health effects of such wastes and directed the Department of Health, Education, and Welfare to study the problem of storage and disposal of hazardous wastes.

The 1970 act intiated a change in the nation's attitude toward waste disposal. For one thing, the emphasis shifted from simply waste disposal to include waste management and resource recovery; for another, it became evident that the existing regulations were inadequate to assure the proper management of hazardous wastes. The result of this latter concern was the passage by Congress in 1976 of the *RCRA*, an amendment to the Resource Recovery Act of 1970.

3.3   THE RESOURCE CONSERVATION AND RECOVERY ACT OF 1976

The Resource Conservation and Recovery Act (RCRA) of 1976 completely replaced the previous language of the *Solid Waste Disposal Act* of 1965 to address the enormous growth in the production of waste. The objectives of this act were to promote the protection of health and the environment and to conserve valuable materials and energy resources by[1,2]:

- Providing technical and financial assistance to state and local governments and interstate agencies for the development of solid waste management plans (including resource recovery and resource conservation systems) that promote improved solid waste management techniques (including more effective organizational arrangements), new and improved methods of collection, separation and recovery of solid waste and the environmentally safe disposal of nonrecoverable residues.
- Providing training grants in occupations involving the design, operation, and maintenance of solid waste disposal systems.
- Prohibiting future open dumping on the land and requiring the conversion of existing open dumps to facilities that do not pose danger to the environment or to health.
- Regulating the treatment, storage, transportation, and disposal of hazardous wastes that have adverse effects on health and environment.
- Providing for the promulgation of guidelines for solid waste collection, transport, separation, recovery, and disposal practices and systems.
- Promoting a national research and development program for improved solid waste management and resource conservation techniques, more effective organizational arrangements, new and improved methods of collection, separation, recovery, and recycling of solid wastes and environmentally safe disposal of nonrecoverable residues.
- Promoting the demonstration, construction, and application of solid waste management, resource recovery and resource conservation systems that preserve and enhance the quality of air, water and land resources.
- Establishing a cooperative effort among federal, state, and local governments and private enterprises in order to recover valuable materials and energy from solid waste.

Structurewise, the RCRA is divided into eight subtitles. These subtitles are (A) General Provisions; (B) Office of Solid Waste; Authorities of the Administrator; (C) Hazardous Waste Management; (D) State or Regional Solid Waste Plans; (E) Duties of the Secretary of Commerce in Resource and Recovery; (F) Federal Responsibilities; (G) Miscellaneous Provisions; and (H) Research, Development, Demonstration, and Information. Portions of Subtitles A and C are discussed here; in Subtitle A, key words and terms used in the act are defined; Subtitle C lays the framework for the RCRA hazardous waste management program.

Subtitle A: General Provisions

In Section 1004, legal definitions for many key terms used in waste management are presented. Some of the more pertinent of these are[3]

Hazardous waste: A solid waste or combination of solid wastes, which because of its quantity, concentration, or physical, chemical or infectious characteristics may (a) cause, or significantly contribute to an increase in mortality or an increase in serious irreversible, or incapacitated reversible illness, or (b) pose a substantial present or potential hazard to human health or the environment when improperly treated, stored, transported or disposed of, or otherwise managed.

Hazardous waste generation: The act or process of producing hazardous waste.

Hazardous waste management: The systematic control of collection, source separation, storage, transportation, processing, treatment and recovery and disposal of hazardous waste.

Solid waste: Any garbage, refuse, sludge from a waste treatment plant, water supply treatment plant, or air pollution control facility and other discarded material, including solid, liquid, semisolid or contained gaseous material resulting from industrial, commercial, mining and agricultural operations and from community activities, but does not include dissolved material in domestic sewage or solid or dissolved material in irrigation return flows or industrial discharges which are point sources subject to permits under the federal *Water Pollution Control Act* or source, special nuclear, or byproduct material as defined by the *Atomic Energy Act.*

Storage: When used in connection with hazardous waste, means the containment of hazardous waste, either on a temporary basis, or for a period of years, in such a manner as not to constitute disposal of such hazardous waste.

Disposal: The storage, deposit, injection, dumping, spilling, leaking or placing of any solid waste or hazardous waste into or in any land or water so that such solid waste or hazardous waste or any constituent thereof may not enter the environment or be emitted into the air or discharged into any waters including ground waters.

Treatment: When used in connection with hazardous waste, means any method, technique or process, including neutralization designed to change the physical, chemical or biological character or composition of any hazardous waste so as to neutralize such waste or as to render such waste non-hazardous, safer for transport, amenable for recovery, or reduced in volume. Such terms include any activity of processing designed to change the physical form or chemical composition of hazardous waste so as to render it non-hazardous.

Subtitle C: Hazardous Waste Management

Subtitles C and D generate the framework for regulatory control programs for the management of hazardous and solid nonhazardous wastes, respectively. The hazardous waste program outlined under Subtitle C is the one most people associate with the term *RCRA*. It is divided into 11 subsections, each of which is summarized here, using much of the language contained in the original document.[2,4]

1. *Section 3001. Identification and Listing of Hazardous Waste.* Section 3001 required EPA to promulgate the criteria to be used in determining how hazardous a particular waste is. Factors such as flammability (ignitability), corrosivity, toxicity, persistance, and degradability in the environment as well as potential for accumu-

lation in tissue are to be considered in defining the criteria. In addition, other hazardous characteristics may be added to the list in the future if the need arises.

2. *Section 3002. Standards Applicable to Generators of Hazardous Waste.* The administrator (EPA) is required to promulgate regulations, after consultation with appropriate federal and state agencies, establishing standards for generators of hazardous waste. These standards are to establish record-keeping practices that accurately identify the waste stream's constitutents and quantities; the disposition of the waste; labeling the waste for storage, transport, or disposal identification; use of appropriate containers; furnishing information to treatment facilities, storage, or disposal facilities on the chemical composition of the waste; use of a manifest system to reconcile the disposition of the waste and to assure waste only proceeds to a permitted facility; submitting reports to EPA regarding the quantities of waste generated during this reporting period; and the disposition of all hazardous wastes during that same period.

3. *Section 3003. Standards Applicable to Transporters of Hazardous Waste.* Under Section 3003, standards for the transporters of hazardous wastes are established. The standards involve record keeping of the waste transported, including source and delivery points; proper labeling of the wastes; compliance with the prescribed manifest system; and limiting transportation of all such hazardous waste only to permitted facilities designated on the manifest form. This section also stipulates that EPA must coordinate its efforts with the Department of Transportation in those cases where the hazardous wastes fall under the Hazardous Materials Transportation Act.

4. *Section 3004. Standards Applicable to Owners and Operators of Hazardous Waste Treatment, Storage and Disposal Facilities.* This section provides for the establishment of standards and regulations for facilities that treat, store, and dispose of hazardous wastes. These standards and regulations pertain to maintaining records of all hazardous waste handled with respect to treatment, storage, and disposal; reporting, monitoring, and inspection of manifests for wastes received to assure compliance; maintaining and operating facilities in accordance with procedures satisfactory to EPA; designating the location, design, and construction of the facility; maintaining contingency plans to minimize unanticipated damage; maintaining the operation of facilities to provide training, continuity of ownership, and to provide standards for financial responsibility.

5. *Section 3005. Permits for Treatment, Storage, and Disposal of Hazardous Waste.* This section covers permits for the treatment, storage, or disposal of hazardous waste. A permit application must include estimates of the quantities, compositions, and concentrations of the waste anticipated, as well as the time, frequency, or rate of receiving and handling the waste, and description of the site. The permit process is discussed in detail in Section 3.5.

6. *Section 3006. Authorized State Hazardous Waste Program.* State hazardous waste programs under Section 3006 will receive approval from EPA unless the state program (a) is not equivalent to the federal programs, (b) is not consistent with federal or state programs of other states, or (c) does not provide adequate enforcement of compliance. Any action taken under a hazardous waste program authorized under this section has the same force and effect as actions taken by EPA under this subtitle.

7. *Section 3007. Inspections.* This section provides one of the means of

monitoring compliance. Through it, EPA is given the right to gain entry to facilities and records where hazardous wastes are generated, treated, stored, or disposed of for the purpose of inspection and sampling. Records, reports, and information resulting from these inspections are available to the public, unless the operator can demonstrate that release of these data would divulge information privileged to protection.

8. *Section 3008. Federal Enforcement.* When EPA determines that a facility is in violation of the regulations, a failure-to-comply notification is issued. If the violation extends beyond 30 days after notification, EPA may: (a) issue a second notification requiring compliance within a specified time limit, (b) assess civil penalties, or (c) revoke the RCRA permit. Criminal penalties may be assessed to transporters delivering to nonpermitted facilities, persons disposing of waste without a permit, or persons found to have falsified information on permit applications.

9. *Section 3009. Retention of State Authority.* In order to make the federal program the minimum standard for hazardous waste regulation, states are prohibited through Section 3009 from enacting regulations less stringent than those promulgated under RCRA.

10. *Section 3010. Effective Date.* Section 3010 requires that, 90 days after the promugation of Section 3001, all generators, transporters, and disposers must have completed the hazardous waste notification procedure. This section also sets a period of 6 months between effective date and promulgation date of Subtitle C regulations.

11. *Section 3011. Authorization of Assistance to States.* This section authorizes funding to assist states in the development and implementation of hazardous waste programs at the state level.

The RCRA regulations are continuously being developed and published according to the following procedure. When a regulation is first proposed, it is published in the *Federal Register* as a *proposed* regulation, along with a discussion of EPAs rationale for the regulation. For a period of time, normally 60 days, public reaction is invited. After this period, EPA revises the proposed regulation and finalizes it by again publishing it in the *Federal Register*. Annually, all such regulations are compiled and placed in the *Codes of Federal Regulations* (CFR); this process is called *codification*. Most of RCRA has been codified in this manner and can be found in Volume 40 of the CFR.[4]

3.4 THE HAZARDOUS AND SOLID WASTE AMENDMENTS OF 1984

Since 1976 the RCRA has been twice amended by Congress, once in 1980 and again in 1984. Thus it has, since its inception, undergone a process of continuous evaluation in order to keep pace with society's changing needs. The 1984 amendments, known by the title *Hazardous and Solid Waste Amendments* (HSWA or the New RCRA) significantly expanded the scope of RCRA and has resulted in numerous changes in EPAs policy and practice. Key elements of these amendments are given here:[4]

1. *Landfilling of Liquids.* The placement of bulk or noncontainerized liquid hazardous waste or hazardous waste containing free liquids in any permitted or

interim status landfills is banned after May 8, 1985. As of November 8, 1985, the placement of nonhazardous liquids in permitted or interim status landfills is also banned; however, an owner or operator may obtain an exemption from the ban on nonhazardous liquids by demonstrating to EPAs satisfaction that the only reasonably available alternative for these nonhazardous liquids is a landfill or outlined surface impoundment that already contains hazardous waste, and that the disposal of the nonhazardous liquids in the owner's or operator's landfill will not present a risk of contamination to any underground source of drinking water.

2. *Minimum Technological Requirements.* This provision establishes minimum standards for hazardous waste management facilities such as double liners or their performance equivalent for new land disposal facilities or interim status facilities.

3. *Corrective Action.* Any permit issued after November 8, 1984 must require corrective action for all releases of hazardous waste or constituents from any solid waste management unit regardless of when the waste was placed in such a unit; financial assurance for the completion of such corrective action is required as well. This provision applies to any solid waste management unit, including inactive units, at any treatment, storage, or disposal facility seeking a permit under RCRA.

4. *Cleanup Beyond a Facility's Property Boundary.* The owner or operator of a permitted facility must institute corrective action beyond the facility boundary whenever necessary to protect human health and the environment, unless the owner or operator shows that permission to undertake such action on adjacent property cannot be obtained.

5. *Groundwater Monitoring Variances.* The new regulations allow surface impoundments, waste piles, and landfills to waiver groundwater monitoring requirements if such units are fitted with a double liner, leak detection system, leachate collection system (for landfills and waste piles), and are located entirely above the seasonal high water table. This section also exempts any lined waste pile from groundwater monitoring requirements even if the pile does not have leak detection and leachate collection systems as long as the pile is located entirely above the seasonal high water table and the waste is removed periodically so that the liner can be examined.

6. *Groundwater Monitoring Variance for Certain Engineered Structures.* This variance is applicable on a case-by-case basis, only to an engineered structure that (a) does not receive or contain liquid waste (or waste containing free liquid), (b) is designed and operated to exclude liquid from precipitation, or other runoff, (c) has multiple leak detection systems within the outer layer of containment that are operated and maintained throughout the life of the unit, including the closure and postclosure care periods, and (d) prevents the migration of hazardous constitutents beyond the outer layer of containment prior to the end of the post-closure period.

7. *Salt Dome Formations, Salt Bed Formations, Underground Mines, and Caves.* For all noncontainerized (or bulk) liquid hazardous waste, the placement of waste in these four settings is prohibited until (a) EPA has determined and promulgated performance and permitting standards and (b) a permit has been issued for the facility. For containerized liquid and all other nonliquid hazardous waste, the placement of such waste in the four settings is prohibited until a permit has been issued to the facility.

8. *Dust Suppression.* This provision prohibits the use of waste oil, used oil, or other material that is contaminated with dioxin or any other hazardous waste (other

than a waste that is identified as hazardous solely on the basis of ignitability) for dust suppression or road treatment.

9. *Underground Injection.* The injection of hazardous waste into or above any underground formation that contains an underground source of drinking water within $\frac{1}{4}$ mile of the injection well is prohibited. This statutory ban on so-called *Class IV* wells is effective automatically, 6 months from the date of enactment of the amendments, in any state that does not have identical or more stringent prohibitions in effect under an applicable underground injection control (UIC) program. The ban does not apply to the injection of contaminated groundwater into the aquifer from which it was withdrawn if (a) the injection is part of a federally supervised cleanup action under RCRA, (b) contaminated groundwater is treated to substantially reduce hazardous constituents prior to injection, and (c) such cleanup, when completed, will be sufficient to protect human health and the environment.

10. *Small Quantity Generators.* Small quantity generators (between 100 to 1000 kg/month) of hazardous waste must either: (a) treat, store, or dispose of the waste at a permitted facility, (b) dispose of the waste at a facility authorized by a state to manage municipal or industrial solid waste, (c) store no more than 1000 kg on-site, or (d) recycle the waste. These requirements remain in effect until EPA promulgates new standards applicable to small quantity generators.

11. *Preconstruction Ban/TSCA Exception.* Construction of a new hazardous waste management facility without a RCRA permit is prohibited. However, this statute makes exception for any facility with EPA approval under the *Toxic Substances Control Act* (TSCA) for the incineration of polychlorinated biphenyls (PCBs). Under TSCA regulation, any PCB article or PCB container may be stored for disposal for up to 1 yr. The facility need not obtain a permit for construction of the storage unit but must ensure that the unit meets certain criteria specified in the regulation.

12. *Permit Life.* Any permit for a treatment, storage, or disposal facility is for a fixed term, not to exceed 10 yr. Each permit for a land disposal facility is to be reviewed 5 yr after the date of issuance or reissuance and modified as necessary to assure that the facility complies with currently applicable requirements necessary to protect human health and the environment.

13. *Authority to Add Conditions.* The intent of this amendment is to authorize the agency (EPA) to impose permit conditions beyond those mandated by the regulations, where necessary to protect human health and the environment.

14. *Expansion of Interim Status for Newly Regulated Units.* (The term *interim* is defined in detail in Section 3.5. A facility that has been granted such status is temporarily regarded by EPA as being permitted and allowed to operate even though Part B of the two-part permit application has not yet received EPA approval.) A facility may qualify for interim status if it is in existence on the effective date of a statutory or regulatory change under RCRA that requires the facility to have a permit (e.g., a facility which treats, stores, or disposes of a newly listed hazardous waste).

15. *Loss of Interim Status for Failure to Submit Part B.* Interim status for owners and operators of land disposal facilities terminates within a 12-month period unless the owner or operator submits Part B (see previous paragraph) prior to that date and demonstrates compliance with the applicable ground water monitoring and financial responsibility requirements. Interim status for owners and operators of incinerators

terminates by November 8, 1989 unless Part B of the application was submitted by November 8, 1986. For all other facilities, interim status terminates by November 8, 1992, unles. Part B is submitted by November 8, 1988.

16. *Ban on Hazardous Waste in Certain Cement Kilns.* Cement kilns located in incorporated cities with populations > 500,000 are prohibited from burning hazardous waste, or any fuel containing a hazardous waste, unless these kilns comply with the regulations applicable to hazardous waste incinerators. This prohibition remains in effect until EPA develops standards for cement kilns burning hazardous waste.

17. *Labeling of Hazardous Waste Fuels.* Any fuel containing a hazardous waste, or a hazardous waste burned directly as a fuel, must include a warning label in the invoice or bill of sale for the fuel. The warning label must state that the fuel contains hazardous wastes and must list the hazardous wastes contained therein. The statute contains three exceptions to the labeling requirements; these are for hazardous waste-derived petroleum coke and for two types of hazardous waste-derived fuels from petroleum refining operations.

18. *Household Wastes.* A resource recovery facility generating energy by burning municipal waste is not considered to be managing hazardous waste, as long as the facility receives and burns only household waste or solid waste from other sources that contains no hazardous waste. Such a facility may not accept hazardous wastes from any nonhousehold sources.

19. *Minimum Technological Requirements for Incinerators.* Incinerators that receive permits after enactment of the amendments must meet a minimum destruction and removal requirement; namely, that the principal organic hazardous constitutents (POHCs) of the waste feed be incinerated with a minimum destruction and removal efficiency (DRE) of 99.99%.

20. *Exposure Information and Health Assessments.* Part B permit applications for landfills and surface impoundments (described in detail in Section 3.5) must be accompanied by information on the potential for exposure of the public to hazardous wastes through releases related to the unit.

21. *Additional Criteria and Delisting Procedures.* In evaluating a delisting petition, EPA must consider factors other than those for which the waste was listed, if there is reason to believe that such additional factors could cause the waste to be a hazardous waste.

22. *Research, Development, and Demonstration Permits.* The HSWA provide EPA with authority to issue permits for research, development, and demonstration treatment activities for which permit standards have not been established. As authorized, EPA reserves the right to waive or modify procedures to expedite permitting as long as human health and the environment are protected. The permit is to be issued for a maximum of 360 days of operation and may be renewed up to three times.

23. *Hazardous Waste Exports.* In general, the export of hazardous waste out of the country is prohibited unless: (a) the person exporting such waste provides notification to EPA, (b) the government of the receiving country has consented to accept the waste, (c) a copy of the receiving country's written consent is attached to the manifest that accompanies the shipment, and (d) the shipment conforms to the terms of the consent. In lieu of meeting these requirements, a person may export hazardous waste if the United States and the government of the receiving country

have entered into an agreement and the shipment conforms to the terms of the agreement.

24. *Waste Minimization.* Except for small quantity generators, waste generators must certify that waste volume and/or quantity and toxicity have been minimized and that the generator's proposed treatment, storage, or disposal method minimizes the threat to human health and the environment.

The HSWA amendments differ from previous hazardous waste management legislation in that Congress not only placed *explicit* requirements in the statute, but also instructed EPA in general language to develop *specific* regulations. Another factor that distinguishes the HSWA is the ambitious timetable they established; many of the HSWA provisions are already in place; others go into effect within very short time frames. The HSWA also established what are called *hammer* provisions—statutory requirements ultimately banning the landfilling of hazardous wastes that go into effect automatically if EPA fails to issue regulations by certain dates.[5]

3.5   THE PERMIT PROCESS

The permitting of any hazardous or toxic waste management site or process is a highly complex, lengthy, and costly procedure. The RCRA imposes legally enforceable requirements on owners or operators of *TSD* facilities (TSD is an acronym for the treatment, storage, or disposal of solid wastes considered hazardous under RCRA). A permit is granted only after compliance with federal standards has been demonstrated. Under this permit, the facility has legal authorization to treat hazardous waste within the limits specified in the permit.

A RCRA permit application is composed of two parts: A and B. To satisfy application requirements for *Part A*, Consolidated Permit Application Forms 1 (EPA Form 3510–1) and 3 (EPA Form 3510–3) must be completed and submitted to the appropriate federal or state agency. To satisfy application requirements for *Part B*, a wide variety of procedures to safeguard human health and the environment, for example, groundwater monitoring, contingency plans, and emergency response plans, must be described in detail. The EPA has provided listings and general descriptions of what must be included in this part. However, because there is no specific form to complete, the success of the Part B application depends on how well the applicant can follow and interpret EPA directives to the satisfaction of the agency (federal, state, and/or local) that has RCRA permitting authority over the facility.[6]

Definitions of Terms

Definitions of selected terms associated with the RCRA permit application process are given here.[6] Again, most of the language of the original RCRA document is used.

- *Closure.* The process in which the owner/operator of a hazardous waste management facility discontinues active operation by treating, removing from the site, or disposing of on-site, all hazardous wastes in accordance with an approved closure plan. *Closure* entails specific financial guarantees and technical tasks which are included in a permittee's closure plan and which the permittee must implement as part of his/her permit obligations. Under RCRA, the facility must be closed in such a way that (1) minimizes the need for further facility maintenance and (2) controls, minimizes, or eliminates post-closure escapes of hazardous waste, its constitutents, or by-products to ground or surface waters, or to the atmosphere. Generally, closure must occur within 90 days after the facility receives its final shipment of hazardous wastes.

- *Closure Plan.* A plan, subject to approval by authorized regulatory agencies, which must be submitted by the owner/operator of a hazardous waste management facility as part of the RCRA permit application. The approved plan becomes part of the permit conditions subsequently imposed on the applicant. The plan identifies steps required to (1) *completely* or *partially close* the facility at any time during its intended operating life, and (2) *completely close* the facility at the end of its intended operating life.

- *Acutely Hazardous Wastes.* Commercial chemical products and manufacturing intermediates which are listed (under their generic names) by RCRA as *acutely hazardous substances.* Also included are off-specification commercial chemical products and manufacturing chemical intermediates which, if they met specifciations, would have the generic names listed; and any residue or contaminated soil, water, or other debris resulting from the cleanup of a spill of any of these substances.

- *Post-Closure Care Period.* The period following closure of a hazardous waste management facility during which the owner/operator must continue to secure the facility sufficiently to protect human health and the environment. The post-closure care period entails specified and guaranteed financial obligations plus the performance of certain technical tasks such as groundwater monitoring. Typically, this period continues for 30 yr after the date of completing closure of a TSD facility for which post-closure care is required.

- *Post-Closure Plan.* Similar to the closure plan, except that it identifies activities (monitoring, maintenance, etc.) to be performed after closure of a TSD facility.

- *TSD Facility.* Any facility that treats, stores, and/or disposes of hazardous wastes.

- *Delisting.* The process of exclusion (or petitioning for exclusion) of a solid waste from the definition of *hazardous waste* even though that waste is listed as *hazardous* under RCRA. To have a waste delisted, the owner/operator of a facility must petition for a regulatory amendment, the granting of which is contingent on demonstrating to EPA and/or state authorities that the waste does not meet any of the criteria under which the waste was originally listed as *hazardous.* The amendment, if granted, excludes only the waste generated at the specific facility and for which the applicant has provided sufficient demonstration by way of sampling, testing, and other procedures required by RCRA.

- *Existing TSD Facility.* A treatment, storage, or disposal facility that began operation or construction on or before November 19, 1980; such a facility may qualify for *interim status.* (See definition of *interim status* below.)

- *New TSD Facility.* One that began operation or construction after Novembe 19, 1980; such a facility cannot qualify for *interim* status.

- *Final Authorization.* A final permission granted by federal RCRA authorities for a state program to supersede the federal RCRA program. In such a case, all hazardous waste activities in the state are subject only to the state and not to the federal program.

- *Notification or Notification Form.* A form which notifies the regulatory agencies of any hazardous waste management activities. Owners/operators of existing TSD facilities must complete and submit this form to the authorized federal or state RCRA regulatory authorities. New facilities are not required to file notification, but both Parts A and B of the RCRA permit application must be filed before these facilities can legally begin operation. Existing facilities that handle hazardous wastes identified in RCRA regulations must file notification no later than 90 days after the waste is listed in the *Federal Register*.

- *Interim Status.* The period during which the owner/operator of an existing TSD facility is treated as having been issued a RCRA permit even though Part B of the permit application has not yet been submitted. Owners/operators of a new facility cannot, by definition, qualify for interim status. An existing facility may automatically qualify for interim status if the owner/operator files the notification form with Part A. Interim status continues until Part B is requested by regulation officials, submitted by the applicant, and processed. The applicant has at least 6 months to submit Part B after it is requested by the authorities. *Interim status* should not be confused with *interim authorization*, (see below) a term which relates specifically to state programs and has nothing to do with permit applications.

- *Interim Authorization.* Temporary authorization granted by federal RCRA authorities to a state that allows some components of the state requirements to apply in place of federal RCRA requirements. However, federal requirements still apply for those components not covered by the *interim authorization*.

- *Finally Effective RCRA Permit.* A set of legally enforceable requirements imposed on the owner/operator of a TSD facility whose completed RCRA permit application (Parts A and B) has been approved and duly processed by authorized regulatory agencies. In return for complying with permit conditions, the permittee receives legal authorization, within the limits specified in the permit, to handle hazardous waste. The *finally effective permit* is generally issued for a fixed period not to exceed 10 yr, after which the permittee must reapply for a new one at least 180 days before the current permit expires. A *finally effective permit* may also be subject to updating (via an application to be filed by the permittee) if changes are made at the permitted facility.

Application for a Permit

The specific steps that must be taken to secure a RCRA permit depends on (a) whether the facility is existing or new, (b) its location, and (c) its operational characteristics. The facility is considered by RCRA to be an *existing facility* if it was operating or under construction on or before November 19, 1980.

Submission of both notification and Part A of the application automatically qualifies such a facility for interim status; this facility is treated by EPA as if it were permitted, until a final determination is made on the permit application. If the facility is *new*, that is, one which began operation or construction after November 19, 1980, both Parts A and B of the RCRA application must be filed with the appropriate agencies at least 180 days before construction of the facility may begin. To operate the new facility, both parts of the application must first be accepted and a finally effective permit issued.

Since the completion of Part A is normally simple and routine, it will not be discussed further. Part B is much more complicated and warrants more complete discussion.

Part B

Two types of information are required for Part B of the application.[6] The first type is *general information* that is required on all RCRA permit applications; the second type covers *facility-specific information* on the particular type of treatment, storage, disposal operations employed at the facility. These may include either one or a combination of the following methods: landfill, surface impoundment, tanks, containers, incinerators or land treatment. The effort involved in completing the facility-specific section of Part B can be relatively easy or difficult depending on the specific methods of disposal employed. For example, if only temporary storage facilities such as tanks or containers are involved, the task is fairly simple; if treatment such as incineration or land disposal is involved, the task is probably quite complex.

General Information The *general information* requirements for Part B are fairly extensive and can involve a variety of tasks and documentation such as engineering studies; chemical, hydrogeological, and geological analysis; and planning and financial analysis. A public advisement program through the mechanism of public hearings is also necessary. Other pertinent general requirements imposed on virtually all RCRA permit applicants are general facility description, chemical and physical waste characterization, waste analysis plan, security procedures, general inspection schedule, preparedness and prevention requirements, contingency plan, traffic patterns and traffic control procedures, facility location information, outline of personnel training programs designed to meet RCRA requirements, closure plans (and where applicable, post-closure plans), closure cost estimate and related financial assurance mechanism (and where applicable, post-closure cost estimate plus related financial assurance mechanism), proof of coverage by a state financial mechanism (where appropriate), a topographic map for the facility including the land area defined by a 1000-ft margin perimeter of the groundwater monitoring and ground-waste installations (where applicable).

Facility-Specific Information Besides the *general information* section, which is required on virtually all RCRA permit applications, a detailed section for *facility-specific information* must also be submitted for Part B. The type of information requested varies from facility to facility and depends on whether or not the facility (1) uses tanks to store or treat hazardous waste, (2) stores containers of hazardous waste, (3) incinerates hazardous waste, (4) uses land treatment to dispose of hazardous waste, (5) disposes of hazardous waste in landfills, (6) stores, treats, or disposes of hazardous waste in surface impoundments, or (7) treates or stores hazardous waste piles.

If it can be demonstrated that the requested information cannot be provided, at least to the extent required, federal EPA or authorized state agencies may modify Part B application requirements. For example, in a case where seasonal treatment monitoring information is required, the alloted 6-month period may not be sufficient. In such a case, agencies may grant a permit that includes special provisions for continued monitoring, analysis, and review of the treatment technology.

Permitting of Hazardous Waste Incinerators

The permitting of hazardous waste incinerators is a complex multifaceted program conducted simultaneously on the federal, state, and local levels. Because of the variety of state and local regulations for the handling, transportation, treatment, and disposal of hazardous wastes, as well as those concerning the operation of incinerators, each start-up has a unique set of permit requirements.

Generally speaking, hazardous waste incinerators require the following permits: federal RCRA, state RCRA, TSCA (for PCBs), state air quality, state waste water discharge, and federal waste water discharge permits. A variety of local permits may also be necessary. Each of these require data substantiating an incinerator's operation at or above performance levels determined by environmental legislation. Each, as well, requires a public hearing and discussion of environmental impacts.

The basic thrust of current incineration regulations requires that all hazardous waste incinerators meet three performance standards:

1. *Principal Organic Hazardous Constituents (POHC)*. The DRE for a given POHC is defined as the mass percentage of the POHC removed from the waste. The POHC performance standard requires that the DRE for each POHC *designated* in the permit be 99.99% or higher. The DRE performance standard implicity requires sampling and analysis to measure the amounts of the *designated* POHC(s) in both the waste stream and the stack effluent gas during a trial burn. (The term *designated POHC* is described in more detail later in this section.)

2. *Hydrochloric Acid*. The limitation for HCl emissions from the stack of an incinerator is specified in two ways: (a) < 4 lb/h (1.8 kg/h) of HCl can be emitted from the incinerator (i.e., *before* the flue gas reaches any scrubbing equipment); and (b) if the 4-lb/h requirement is not met, a minimum scrubbing efficiency of 99% must be achieved for the HCl (i.e., a maximum emission rate to the atmosphere of 1% of the HCl in the stack gas entering any pollution control equipment is allowed). This sampling standard implicitly requires, in some cases, sampling and analysis to measure the HCl in the stack gas.

3. *Particulates*. Stack emissions of particulate matter are limited to 0.08 grains per dry standard cubic foot (gr/dscf) for the stack gas *corrected to 50% excess air*. (The term *excess air* is discussed in detail in Part II. It is a measure of the amount of air supplied to the incinerator over and above that needed for complete combustion of the waste and/or auxiliary fuel.) If > or < 50% excess air is being used, the measured particulate concentration in grains per dry standard cubic foot must first be adjusted before comparison with the 0.08-gr/dscf standard. This adjustment is made by calculating what the concentration would be if 50% excess air was used. In this way, a decrease of particulate concentration due solely to increasing air flow in the stack is not rewarded, and an increase of particulate concentration due solely to reduction of air flow in the stack is not penalized.

Compliance with these performance standards is documented by a *trial burn* of the facility's waste streams. As part of the Part B RCRA permit application, a trial burn plan detailing waste analysis, an engineering description of the incinerator, sampling and monitoring procedures, test schedule and protocol as well as control information must be developed. After submittal of Part B, EPA reviews the documents for completeness; if the documentation is incomplete, a notice of deficiency requesting further information is issued; if the documentation is complete, EPA conducts a public hearing and issues a temporary or draft permit, allowing the owner or operator to build the incinerator and initiate the trial burn procedure.

The temporary permit covers four phases of operation. During the *first* phase, immediately following construction, the unit is operated for *shake-down* purposes to identify possible mechanical deficiencies and to assure its readiness for the trial burn procedures. This phase of the permit is limited to 720 h of operation using hazardous waste feed. The trial burn is conducted during the *second* phase; this is the most critical component of the permitting process since it demonstrates the incinerator's ability to meet the three performance standards. In addition, performance data collected during the trial burn phase are reviewed by the permitting official and become the basis for setting the conditions of the facility permit. These conditions include: (1) allowable waste analysis procedures, (2) allowable waste feed composition (including acceptable variations in the physical or chemical properties of the waste feed), (3) acceptable operating limits for carbon monoxide in the stack, (4) waste feed rate, (5) combustion temperature, (6) combustion gas flow rate, and (7) allowable variations in incinerator design and operating procedures (including a requirement for shutoff of waste feed during startup, shutdown, and at any time when conditions of the permit are violated).

To verify compliance with the POHC performance standard during the trial burn, it is not required that the incinerator DRE for every POHC identified in the waste be measured. There is a procedure for identifying the one (or more) POHCs with the greatest potential for a low DRE, based on the expected difficulty of thermal degradation (incinerability) and the concentration of the POHC in the waste. This becomes the *designated* POHC for the trial burn. The EPA permit review personnel work with the owners/operators of the incinerator facility in determining which POHCs in a given waste should be *designated* for sampling and analysis during the trial burn.

If a wide variety of wastes are to be treated, a difficult-to-incinerate POHC at high concentration may be proposed for the trial burn. The designated trial burn POHC does not have to be actually present in the normal waste; it does, however, have to be considered more difficult to incinerate than any POHC in the normal waste. The substitute POHC in this strategy is referred to as a *surrogate* POHC.

The *third* phase consists of completing and submitting the trial burn results. This phase can last several weeks to several months, during which the incinerator is allowed to operate under specified conditions. The data to be reported to regulatory agencies after the burn include: (1) a quantitative analysis of the POHCs in the waste feed, (2) a determination of the concentration of particulates, POHCs, oxygen, and HCl in the exhaust gas, (3) a quantitative analysis of any scrubber water, ash residues, and other residues to determine the fate of the POHCs, (4) a computation of the DRE of the POHCs, (5) a computation of the HCl removal efficiency if the HCl emission rate exceeds 1.8 kg/h (4 lb/h), (6) a computation of particulate emissions, (7) the identification of sources of fugitive emissions and their means of control, (8) a measurement of average, maximum, and minimum temperatures and combustion gas velocities (gas flows), (9) a continuous measurement of carbon monoxide (CO) in the exhaust gas, and (10) any other information EPA may require to determine compliance.

Provided that performance standards are met in the trial burn, the facility can begin its *fourth* and final phase, which continues through the duration of the permit. In the event that the trial burn results do not demonstrate compliance with standards, the temporary permit must be modified to allow for a second trial burn.[7]

The costs associated with the trial burn normally include expenditures for surveying, equipment, sampling, analysis, cleanup, and report preparation. These details follow.

- *Site Survey.* These costs include professional services and travel to a local site for inspection of the facility to be tested and discussion of plans for trial burns. Specific characteristics of wastes to be incinerated are assumed known or provided by the waste generator. Information on the compatibility of these wastes with the specific facility characteristics is assumed available.
- *Quality Assurance/Quality Control (QA/QC).* Because every test represents a unique situation, appropriate QA/QC procedures must be employed throughout both the sampling and analytical phases of any test. For example, all measuring devices should be properly calibrated, all solvents and resins should be checked for purity before the sampling program is begun, sampling trains should be all glass and throughly cleaned to remove all sealant greases and any other residues, and ground glass plugs and aluminum foil should be used to cap off the adsorbent trap and other sample-exposed portions of the trains to minimize the potential for contamination. Costs associated with QA/QC include professional services and sampling and testing equipment.
- *Equipment Preparation.* Sampling equipment may have to be leased by the facility or provided by a consulting firm. Certain costs are incurred in calibrating, loading of instruments and transport to facility site.
- *Equipment Setup and Takedown.* Installation of equipment includes the erection of scaffolding, securing of ports, and placement of sampling instruments at facility stacks for the proper monitoring and collecting of trial burn samples.
- *Stack Sampling.* A minimum of three tests involving 3 to 6 h per test is normal. Costs for testing, instrumentation, and adjustments associated with a permit application are considered separate costs not attributable to the trial burn. Development of sampling procedures and verification of test methods are presumed available and accomplished prior to the trial burns.
- *Sample Analysis.* Laboratory analysis costs include the preservation and transporting of samples to an off-site laboratory. Compounds to be analyzed generally include any potential air pollutants. Gas chromatography (GC) and mass spectrometry (MS) testing is conducted for the chlorinated hydrocarbons.
- *Equipment Cleanup.* These costs are the routine expenditures incurred for cleaning and storing various sampling and analysis equipment.
- *Report Preparation.* The written report displays and interprets the trial burn results. Costs for writing, typing, artwork, reproduction, and so on, are included here.

In conducting the trial burn, only approved EPA sampling and analysis methods or their equivalents may be employed. The current EPA sampling methods, VOST

(*Volatile Organics Sampling Train*) and MM-5 (*Modified Method-5*) are commonly used; other possible sampling methods are EPA Method-3, EPA Method-25, High Volume Modified Method-5, and SASS train. Analytical methods for hazardous organic compounds include *gas chromatography* (GC), *flame ionization detection* (FID), *photoionization detection* (PID), *Hall electrolytic conductivity detection* (HECD), *electron capture detection* (ECD), *gas chromatography/mass spectrometry* (GC/MS), *high performance liquid chromatography* (HPLC), and *atmospheric pressure chemical ionization mass spectrometry* (APCI-MS).

The POHC compounds normally have to be concentrated from large amounts of the flue gas sample on special polymer resins; a 3-h collection time is typical. The extraction and concentration techniques required to prepare the sample for analysis are complicated and time consuming. As a result, trial burns are very expensive and it takes about 30 to 60 days before the final results are known.

From start to finish, the permitting of a hazardous waste incinerator usually takes from 18 to 36 months. This process can be considerably longer if such obstacles as a *notice of deficiency*, inconclusive trial burn results, or intense public opposition occur.

3.6   THE FUTURE OF INCINERATION

The choice of a hazardous waste disposal method is limited to basically three alternatives: (1) waste reduction through process modification (elimination, substitution, recycle/reuse, or minimization), (2) conversion of the hazardous waste to a less hazardous waste, or (3) placement of the waste or residuals in the environment. These three are not necessarily exclusive of each other and, more often than not, more than one are used. The first of these, waste reduction, is usually preferable in terms of economic return as well as short- and long-term protection of the environment and public health. The third approach, the placement of waste and waste residuals in the environment, includes secure landfills, above-ground landfills, deep-well injection, ocean disposal, geological isolation, and seabed implacement. The 1984 Amendments (HSWA), coupled with expected future prohibition and financial liability, place severe limitations on these options.

The second approach includes all forms of waste treatment. In the 1976 RCRA, treatment is described as "any method, technique, or process, including neutralization, designed to change the physical, chemical, or biological character or composition of any hazardous waste so as to neutralize such waste or as to render such waste less hazardous, safer for transport, amenable for recovery, amenable for storage, or reduced in volume." With the exception of incineration, these physical, chemical, and biological methods were discussed in some detail in Chapter 2. Of all of these options, incineration is becoming a more and more attractive alternative. This method has the capability of safely converting large volumes of primarily organic hazardous waste to much smaller volumes of ash and other residues plus nontoxic gaseous emissions. Moreover, process incineration can often provide an optimum permanent solution to management of hazardous waste with minimal long-term ecological burden. Based on the 1984 RCRA amendments, incineration now appears—from a regulatory point of view—to be the most viable means of waste treatment for organics. The demand for incineration is expected to increase over the present capacity due, in large extent, to restrictions in the 1984 RCRA amendments regarding land disposal options. For example, EPA predicted that, by the end of this decade, demand for incineration of liquid chlorinated organics alone will exceed current capacity by seven to eight times.

Another reason for an optimistic forecast for the use of incineration is *ocean incineration*. Incineration at sea represents a new horizon in the incineration option. There are two types of ocean incineration: shipboard and offshore platform. Both of these have advantages over land-based facilities: (1) remoteness from populated areas and (2) lower cost, because the environmental controls can be based upon proximity to land and on meteorological and oceanographic conditions. There are problems, however, the most serious of which may be public opposition; these problems are discussed in Chapter 7.

3.7   ILLUSTRATIVE EXAMPLES

The illustrative examples provided here pertain to standards and regulations. Those readers not familiar with technical calculations should bypass these examples and the problems at the conclusion of this chapter until Part II has been reviewed. The reader may then choose to return to this problem set.

Example 3.7.1.   A packed column is operating at an efficiency of 85.5% for the removal of HCl from a waste flue gas. Since current regulations require that no more than 4.0 lb/h be emitted into the atmosphere, what is the maximum inlet HCl rate in the flue gas to operate within the regulations?

SOLUTION. By definition,

$$E = \frac{(w_{in} - w_{out})}{w_{in}}$$

where E = efficiency (expressed as a fraction)
w_{in} = inlet mass flow rate
w_{out} = outlet mass flow rate

This equation may be rearranged and solved for w_{in}:

$$w_{in} = \frac{w_{out}}{(1 - E)}$$

For this problem, w_{out} is the maximum allowable emission rate (4 lb/h) and the fractional efficiency (E) is 0.855:

$$w_{in} = \frac{4.0}{(1.0 - 0.855)}$$

$$= 27.58 \text{ lb/h}$$

This represents the maximum allowable mass flow rate of HCl in the inlet flue gas.

Example 3.7.2. A quench tower operates at a HCl removal efficiency of 65%. This is then followed by a packed tower absorber. What is the minimum collection efficiency of the absorber if an overall HCl removal efficiency of 99.0% is required?

SOLUTION. Select as a basic 100 lb/h of HCl entering the unit. The mass of HCl leaving spray tower is calculated from the following equation.

$$E = \frac{(w_{in} - w_{out})}{w_{in}}$$

$$w_{out} = w_{in}(1 - E) = 100(1.0 - 0.65)$$

$$= 35 \text{ lb/h HCl leaving the spray tower}$$

Use the overall efficiency to calculate the mass flow rate of HCl leaving the packed tower absorber.

$$w_{out} = w_{in}(1 - E) = 100(1.0 - 0.990)$$

$$= 1.0 \text{ lb/h leaving the packed tower}$$

The efficiency of the packed tower can now be calculated

$$E = \frac{(w_{in} - w_{out})}{w_{in}} = \frac{(35.0 - 1.0)}{35.0}$$

$$= 0.971 = 97.1\%$$

Example 3.7.3. A small quantity of a waste mixture is incinerated at 2000°F using 5% ash coal to provide energy necessary for combustion. If 300 ft³ of the flue gas (measured at 2000°F) are produced for every pound of coal burned, what is the maximum effluent particulate loading in gr/ft³ (at 2000°F)? Assume no contribution to the particulates from the waste.

SOLUTION. Based on the problem statement, assume the volume of gas produced on incinerating the waste is negligible. Select as a basis 1.0 lb of coal burned.

$$\text{mass of particulates} = (1.0 \text{ lb})(0.05)$$

$$= 0.05 \text{ lb}$$

The maximum particulate loading is given by

$$W = \frac{0.05}{300}$$

$$= 1.667 \times 10^{-4} \text{ lb/ft}^3$$

where W = particulate loading (mass of particulates/total volume of flue gas).

To convert to gr/ft^3, note that there are 7000 gr/lb. Therefore,

$$W = (1.667 \times 10^{-4})(7000)$$
$$= 1.167\, gr/ft^3$$

Example 3.7.4.   With reference to Example 3.7.3, what dilution factor or particulate collection efficiency is required to achieve the ambient air quality standard of 75 $\mu g/m^3$? What efficiency is require to achieve an outlet loading of 0.08 gr/ft^3?

SOLUTION. First convert W_{in} to $\mu g/m^3$.

$$W_{in} = 1.667 \times 10^{-4}\, lb/ft^3$$
$$= 2.6727 \times 10^6\, \mu g/m^3$$

where W_{in} = inlet loading (mass particulate/volume gas).

The dilution factor (DF) may now be calculated.

$$DF = \frac{W_{in}}{W_{std}} = \frac{2.6727 \times 10^6}{75} = 3.564 \times 10^4$$

where W_{std} = ambient air quality standard.

The particulate collection efficiency (E), is given by

$$E = \frac{(W_{in} - W_{std})}{(W_{in})} = \frac{(2.6727 \times 10^6 - 75)}{(2.6727 \times 10^6)}$$
$$= 0.99997 = 99.997\%$$

Note: This definition of E is equivalent to that given in Example 3.7.1 in terms of mass flow rates (w).

The required efficiency to achieve an outlet loading of 0.08 gr/ft^3 is

$$E = \frac{(1.167 - 0.08)}{1.167}$$
$$= 0.931 = 93.1\%$$

Example 3.7.5.   The waste flow rate into an incinerator is 100 lb/h. Calculate the waste flow rate leaving the unit to achieve a destruction efficiency of (**1**) 95%, (**2**) 99%, (**3**) 99.9%, (**4**) 99.99%, (**5**) 99.9999%.

SOLUTION. Since

$$DRE = \left(\frac{w_{in} - w_{out}}{w_{in}} \right) 100$$

we may rearrange this equation and solve for w_{out}.

$$w_{out} = w_{in}\left[1 - \left(\frac{DRE}{100}\right)\right]$$

1. For a DRE of 95%.

$$w_{out} = 100\left[1 - \left(\frac{95}{100}\right)\right]$$

$$= 5\,lb/h$$

2. For a DRE of 99%,

$$w_{out} = 100\left[1 - \left(\frac{99}{100}\right)\right]$$

$$= 1\,lb/h$$

3. For a DRE of 99.9%,

$$w_{out} = 0.1\,lb/h$$

4. For a DRE of 99.99%,

$$w_{out} = 0.01\,lb/h$$

5. For a DRE of 99.9999%,

$$w_{out} = 0.0001\,lb/h$$

Example 3.7.6.   A waste mixture consisting of xylene, toluene, and chlorobenzene is presently being incinerated with 50% excess air at 2050°F with a 2.2-s residence time. The componential waste inlet feed rate and the stack discharge rate (including particulate catch) are given. The stack gas flow rate has been determined to be 13,250 dscfm (dry standard cubic feet per minute). Is the unit in compliance? Components (2) and (3) are the designated POHCs.

Compound	Inlet (lb/h)	Outlet (lb/h)
1 C_8H_{10} (xylene)	960	0.154
2 C_7H_8 (toluene)	953	0.082
3 C_6H_5Cl (chlorobenzene)	337.5	0.022
HCl	—	2.2
Particulates	—	7.97

SOLUTION. Since

$$DRE_i = \left(\frac{(w_{in})_i - (w_{out})_i}{(w_{in})_i}\right) 100$$

$$DRE_{xylene} = \left(\frac{960 - 0.154}{(960)}\right) 100 = 99.984\%$$

$$DRE_{toluene} = \left(\frac{953 - 0.082}{(953)}\right) 100 = 99.991\%$$

$$DRE_{chlorobenzene} = \left(\frac{337.5 - 0.022}{(337.5)}\right) 100 = 99.993\%$$

Regulations require that the DRE for each designated POHC be at least 99.99%. In this case, the designated POHCs are toluene and chlorobenzene. Both POHCs exhibit a DRE of > 99.99%. Therefore, the unit is in compliance with current regulations. Although the DRE of xylene is below 99.99%, it is not a designated POHC and therefore not subject to the *four nines* (99.99%) requirement.

Note: This result suggests that the choice of toluene and chlorobenzene as the designated POHCs is a poor one since xylene turns out to be more difficult to burn. However, assuming that the regulatory personnel agreed to the designation of toluene and chlorobenzene, the unit would be in compliance. In practice, the xylene DRE would not even have been measured.

Assume all the chlorine in the feed is converted to HCl. The molar feed rate of chlorobenzene ($\dot{n}_3$) is

$$\dot{n}_3 = \frac{w_3}{(MW)_3} = \frac{(337.5)}{(112.5)}$$

$$= 3.0\,lbmol/h$$

where $\dot{n}_i$ = molar flow rate of component i
$(MW)_i$ = molecular weight of component i

Note: Molecular weights, moles, and the units of lbmol and gmol and discussed in Chapter 4.

Each molecule of chlorobenzene contains one atom of chlorine. Therefore,

$$\dot{n}_{HCl} = \dot{n}_3$$

$$= 3.0\,lbmol/h$$

$$w_{HCl} = 109.5\,lb/h$$

The removal efficiency, RE, for HCl

$$RE = \left(\frac{w_{in} - w_{out}}{w_{in}}\right)100 = \left(\frac{109.5 - 2.2}{109.5}\right)100$$

$$= 97.99\%$$

Regulations require that HCl emissions must either achieve a 99% HCl scrubbing efficiency or emit < 4 lb/h of hydrogen chloride. In this case, emissions are < 4 lb/h and therefore in compliance with current regulations.

The outlet loading W_{out} of the particulates is

$$W_{out} = \frac{w_{out}(7000)}{60} = \frac{(7.97)(7000)}{(13,250)(60)}$$

$$= 0.0702 \, gr/dscf$$

Current regulations require that particulate emissions must be < 0.08 gr/dscf corrected to 50% excess air. It is therefore in compliance with regard to particulates.
The unit is therefore in compliance with present regulations.

Example 3.7.7. During a trial burn, the incinerator was operated at a waste feed rate of 5000 lb/h and 50% excess air. The gas flow rate measured in the stack was 19,200 dscfm. Under these conditions, the measured concentrations of the principal organic hazardous components were

Trichloroethylene (1) = $4.9 \, \mu g/dscf$
1,1,1-Trichloroethane (2) = $1.0 \, \mu g/dscf$
Methylene chloride (3) = $49 \, \mu g/dscf$
Perchloroethylene (4) = $490 \, \mu g/dscf$

Each hazardous component listed here constitutes approximately 5% of the total waste feed rate. Calculate the destruction and removal efficiency of the POHCs. Is the unit in compliance with present regulations?

SOLUTION. Each hazardous component constitutes 5% of the total waste feed. Therefore,

$$(w_{in})_i = (0.05)(\text{total feed rate}) = (0.05)(5000 \, lb/h)$$

$$= 250 \, lb/h$$

The mass flow rate of each hazardous component in the stack is given by

$$(w_{out})_i = \frac{Q_s C_i}{7.57 \times 10^6}$$

where Q_s = volumetric gas flow rate (dscfm)

$\quad\quad\quad\quad$ C_i = concentration of component i (μg/dscf)

$\quad$ 7.57×10^6 = conversion factor that converts lb/h to g/min

Therefore,

$$(w_{out})_1 = \frac{(19,200)(4.9)}{(7.57 \times 10^6)} \quad\quad (w_{out})_2 = \frac{(19,200)(1.0)}{(7.57 \times 10^6)}$$

$$= 0.01243\,\text{lb/h} \quad\quad\quad\quad = 0.00254\,\text{lb/h}$$

$$(w_{out})_3 = \frac{(19,200)(49)}{(7.57 \times 10^6)} \quad\quad (w_{out})_4 = \frac{(19,200)(490)}{(7.57 \times 10^6)}$$

$$= 0.1243\,\text{lb/h} \quad\quad\quad\quad = 1.243\,\text{lb/h}$$

The destruction and removal efficiency for component i, DRE_i, is given by:

$$DRE_i = \left(\frac{(w_{in})_i - (w_{out})_i}{(w_{in})_i}\right) 100$$

$$DRE_1 = \left(\frac{250 - 0.01243}{250}\right) 100 \quad DRE_2 = \left(\frac{250 - 0.00254}{250}\right) 100$$

$$= 99.995\% \quad\quad\quad\quad\quad = 99.999\%$$

$$DRE_3 = \left(\frac{250 - 0.1243}{250}\right) 100 \quad DRE_4 = \left(\frac{250 - 1.243}{250}\right) 100$$

$$= 99.95\% \quad\quad\quad\quad\quad = 99.50\%$$

Present regulation dictates that 99.99% destruction and removal efficiency be achieved. In this case, only two of the principal organic hazardous components (1 and 2) achieved this status. Assuming that either component 3 or 4 (or both) was a designated POHC, the unit would be out of compliance.

PROBLEMS

1. A proposed incineration facility design requires that a packed column and a spray tower are to be used in series for the removal of HCl from the flue gas. The spray tower is operating at an efficiency of 65% and the packed column at an efficiency of 98%. Calculate the mass flow rate of HCl leaving the spray tower, the mass flow rate of HCl entering the packed tower, and the overall efficiency of the removal system if 76.0 lb of HCl enters the system every hour.

2. Two packed columns operating in series are used for removing HCl from a flue gas. If the first column is operating at 65% efficiency and the second column is operating at 99.5% efficiency, calculate the overall efficiency and mass flow rate of HCl leaving the system.

3. Two spray columns operating in series are used for removing HCl from a flue gas coming from a hazardous waste incinerator. An overall efficiency of 99.0% is required. What is the required minimum collection efficiency of the second column if the first column is operating at an efficiency of 70%?

4. A hazardous waste incinerator is burning an ash coal to provide the energy necessary for combustion. Calculate the maximum ash content of the coal. A collection efficiency of 98.60% can be achieved for a maximum required outlet load of 0.001 lb/lb coal. Also calculate the maximum particulate loading in gr/ft^3 if 325 ft^3 of total flue gas are produced for every pound of coal burned.

5. A hazardous waste incinerator is burning anthracite coal containing 7.1% ash to provide the necessary energy for combustion. If 300 ft^3 of total flue gas are produced for every pound of coal burned, what is the maximum effluent particulate loading in gr/ft^3? Assume no contribution to the particulates from the waste. The secondary ambient air quality standard for particulates is 75 μg/m^3. What dilution factor or particulate collection efficiency is required to achieve this standard?

6. A proposed incinerator is designed to destroy a hazardous waste at 2100°F and 1 atm. Current regulations dictate that a minimum destruction efficiency of 99.99% must be achieved. The waste flow rate into the unit is 960 lb/h while that flowing out of the unit is measured as 0.08 lb/h. Is the unit in compliance?

7. A hazardous waste incinerator is burning a waste mixture consisting of toluene (1), chlorobenzene (2), and dichlorobenzene (3) with 50% excess air at 2100°F with a 2.5-s residence time. Analyses of the feed rate composition and stack discharge rate are given. The stack gas flow rate has been determined to be 14,280 dscfm. Is the unit in compliance with present regulations? Components (2) and (3) are the designated POHCs.

Compound	Inlet (lb/h)	Outlet (lb/h)
1 C$_7$H$_8$ (toluene)	867	0.112
2 C$_6$H$_5$Cl (chlorobenzene)	442	0.015
3 C$_6$H$_4$Cl$_2$ (dichlorobenzene)	297	0.018
HCl	—	4.1
Particulates		9.65

8. Analysis of a trial burn performance established emission concentrations of the following principal organic components as follows:

1,1,1-Trichloroethane (1) = 2.1 μg/dscf
Perchloroethylene (2) = 10 μg/dscf
Methylene chloride (3) = 6.0 μg/dscf

The incinerator was operated at a waste feed rate of 6200 lb/h and 50% excess air. The gas flow rate measured in the stack was 21,100 dscfm. Each hazardous component constituted approximately 6% of the total waste feed rate. Calculate the destruction and removal efficiency of the POHCs. Is the unit in compliance with present regulations?

REFERENCES

1. P.N. Cheremisinoff, and F. Ellerbusch, *Solid Waste Legislation, Resource Conservation & Recovery Act, A Special Report*, Washington, D.C., 1979.
2. "Resource Conservation and Recovery Act of 1976", *International Environmental Reporter*, Bureau of National Affairs, Washington D.C., Oct. 21, 1976.
3. "Resource Conservation and Recovery Act, 1976", Congressional Records of Oct. 3 and 11, 1984.
4. "Hazardous Waste Management System; Final Codification Rule," *Federal Register*, 40 CFR, Washington, D.C., Monday, July 15 (1985).
5. T. Shen, U. Choi and L. Theodore, *Hazardous Waste Incineration Manual*, U.S. EPA Air Pollution Training Institute, Research Triangle Park, NC, to be published.
6. J. Casler, *On Permitting the Resource Conservation and Recovery Act*, Environmental Research and Technology, Inc., Concord, MA, 1984.
7. E.D. Cooper, "Incineration Permitting," *Hazardous Materials and Waste Management*, **4**, 10, Jan./Feb. (1986).

II

INCINERATION PRINCIPLES

From the authors' point of view, Chapters 4–6 are the "heart" of this text. While Part III presents much valuable information on the description, design, and operation of equipment used in the incineration of hazardous wastes and Part IV investigates overall incineration plant design (including economic considerations), Part II contains the core calculations on which incineration design is based.

The purpose of Chapter 4 is to review certain basic topics that the reader may or may not already be familiar with. Among these are definitions and explanations of concepts and properties that pertain to incineration: thermal conductivity, heat capacity, heating value, viscosity, density, Reynolds number, ideal gas law, psychrometric chart, steam tables, and so on. Chapter 5 discusses a number of scientific principles, primarily from the fields of chemistry, physics, and chemical engineering: topics such as stoichiometry, thermochemistry, chemical equilibrium, and kinetics. Although these topics are pertinent to hazardous waste incineration, they are treated in a general fashion without any attempt to apply them to specific problems. This latter is the subject of Chapter 6, where the same topics studied in Chapter 5 are applied directly to the incineration process. Important calculational procedures are presented in this chapter; the more important among these are methods for predicting (a) incinerator operating temperature based on the properties of the waste or waste–fuel mixture and the amount of excess air employed during combustion, and (b) flue products produced when burning a given waste–fuel mixture in stoichiometric or excess air.

4

Basic Concepts

Partial contribution: Maritza Montesinos

4.1 INTRODUCTION

This chapter is a review of some basic concepts from physics, chemistry, and engineering in preparation for material that is covered in later chapters. These basic concepts include: units and dimensions, some physical and chemical properties of substances, the ideal gas law, the Reynolds number, and phase equilibria. All of these topics are vital to hazardous waste incineration applications. Because many of these topics are unrelated to each other, this chapter admittedly lacks the cohesiveness that chapters covering a single topic might have. This is usually the case when basic material from such widely differing areas of knowledge as physics, chemistry, and engineering is surveyed. Though these topics are widely divergent and covered with varying degrees of thoroughness, all of them will find later use in this text.

If additional information on these review topics is needed, the reader is directed to the literature in the reference section of this chapter.

4.2 FUNDAMENTALS OF MEASUREMENT

Units and Dimensions

The units used in this text are consistent with those adopted by the engineering profession in the United States. For engineering work, SI (*Système International*) and *English* units are most often employed; in the United States, the *English engineering* units are generally used, although efforts are still underway to obtain universal adoption of SI units for all engineering and science applications. The SI units have the advantage of being based on the decimal system, which allows for more convenient conversion of units within the system. There are other systems of units; some of the more common of these are shown in Table 4.2.1. The English engineering units will primarily be used here. Tables 4.2.2 and 4.2.3 present units for

TABLE 4.2.1. Common Systems of Units

System	Length	Time	Mass	Force	Energy	Temperature
SI	meter	second	kilogram	newton	joule	kelvin, degree Celsius
cgs	centimeter	second	gram	dyne	erg, joule or calorie	kelvin, degree Celsius
fps	foot	second	pound	poundal	foot poundal	degree Rankine, degree Fahrenheit
American engineering	foot	second	pound	pound (force)	British thermal unit, horsepower hour	degree Rankine, degree Fahrenheit
British engineering	foot	second	slug	pound (force)	British thermal unit, foot pound (force)	degree Rankine, degree Fahrenheit

TABLE 4.2.2. **English Engineering Units**

Physical Quantity	Name of Unit	Symbol for Unit
Length	foot	ft
Time	second	s
Mass	pound (mass)	lb
Temperature	degree Rankine	°R
Temperature (alternate)	degree Fahrenheit	°F
Moles	pound-mole	lbmol
Energy	British thermal unit	Btu
Energy (alternate)	horsepower-hour	hp-h
Force	pound (force)	lb_f
Acceleration	foot per second squared	ft/s^2
Velocity	foot per second	ft/s
Volume	cubic foot	ft^3
Area	square foot	ft^2
Frequency	cycles per second, Hertz	cycles/s, Hz
Power	horse power, Btu per second	hp, Btu/s
Specific heat capacity	British thermal unit per (pound mass-degree Rankine)	Btu/lb-°R
Density	pound (mass) per cubic foot	lb/ft^3
Pressure	pound (force) per square inch	psi
	pound (force) per square foot	psf
	atmospheres	atm
	bar	bar

TABLE 4.2.3. **SI Units**

Physical Unit	Name of Unit	Symbol for Unit
Length	meter	m
Mass	kilogram, gram	kg, g
Time	second	s
Temperature	kelvin	K
Temperature (alternate)	degree Celsius	°C
Moles	gram-mole	gmol
Energy	joule	$J, kg\text{-}m^2/s^2$
Force	newton	$N, kg\text{-}m/s^2, J/m$
Acceleration	meters per square second	m/s^2
Pressure	pascal, newton per square meter	$Pa, N/m^2$
Pressure (alternate)	bar	bar
Velocity	meters per second	m/s
Volume	cubic meter, liters	m^3, L
Area	square meter	m^2
Frequency	hertz	Hz, cycles/s
Power	watt	$W, kg\text{-}m^2/s^3, J/s$
Specific heat capacity	joule per kilogram-kelvin	J/kg-K
Density	kilogram per cubic meter	kg/m^3
Angular velocity	radians per second	rad/s

TABLE 4.2.4. Prefixes for SI Units

Multiplication Factors	Prefix	Symbol
$1{,}000{,}000{,}000{,}000{,}000{,}000 = 10^{18}$	exa	E
$1{,}000{,}000{,}000{,}000{,}000 = 10^{15}$	peta	P
$1{,}000{,}000{,}000{,}000 = 10^{12}$	tera	T
$1{,}000{,}000{,}000 = 10^{9}$	giga	G
$1{,}000{,}000 = 10^{6}$	mega	M
$1{,}000 = 10^{3}$	kilo	k
$100 = 10^{2}$	hecto	h
$10 = 10^{1}$	deka	da
$0.1 = 10^{-1}$	deci	d
$0.01 = 10^{-2}$	centi	c
$0.001 = 10^{-3}$	milli	m
$0.000\ 001 = 10^{-6}$	micro	μ
$0.000\ 000\ 001 = 10^{-9}$	nano	n
$0.000\ 000\ 000\ 001 = 10^{-12}$	pico	p
$0.000\ 000\ 000\ 000\ 001 = 10^{-15}$	femto	f
$0.000\ 000\ 000\ 000\ 000\ 001 = 10^{-18}$	atto	a

the English and SI systems, respectively. Some of the more common prefixes for SI units are given in Table 4.2.4. Conversion factors between SI and English units and additional details on the SI system are provided in the appendix.

Conversion of Units

Converting a measurement from one unit to another can conveniently be accomplished by using *unit conversion factors*; these factors are obtained from the simple equation that relates the two units numerically. For example, from

$$1 \text{ foot (ft)} = 12 \text{ inches (in.)} \tag{4.2.1}$$

the following conversion factor can be obtained

$$12 \text{ in.}/1 \text{ ft} = 1 \tag{4.2.2}$$

Since this factor is equal to unity, multiplying some quantity (e.g., 18 ft) by the factor cannot alter its value. Hence

$$18 \text{ ft}(12 \text{ in.}/1 \text{ ft}) = 216 \text{ in.} \tag{4.2.3}$$

Note that in Eq. (4.2.3), the old units of *feet* on the left-hand side cancel out leaving only the desired units of *inches*.

Physical equations must be dimensionally consistent. For the equality to hold, each term in the equation must have the same dimensions. This condition can be and should be checked when solving engineering problems. Throughout the text, and in particular in Part II, great care is exercised in maintaining the dimensional formulas of all terms and the dimensional homogeneity of each equation. Equations will generally be developed in terms of specific units rather than general dimensions (e.g. *feet*, rather than length). This aproach should help the reader to more easily attach physical significance to the equations presented in these chapters.

Significant Figures and Scientific Notation

Significant figures provide an indication of the precision with which a quantity is measured or known. The last digit represents, in a qualitative sense, some degree of doubt. For example, a measurement of 8.32 in. implies that the actual quantity is somewhere between 8.315 and 8.325 in. This applies to calculated and measured quantities; quantities that are known exactly (e.g., pure integers) have an infinite number of significant figures.

The significant digits of a number are the digits from the first nonzero digit on the left to either (a) the last digit (whether it is nonzero or zero) on the right if there is a decimal point, or (b) the last nonzero digit of the number if there is no decimal point. For example:

370	has 2 significant figures
370.	has 3 significant figures
370.0	has 4 significant figures
28,070	has 4 significant figures
0.037	has 2 significant figures
0.0370	has 3 significant figures
0.02807	has 4 significant figures

Whenever quantities are combined by multiplication and/or division, the number of significant figures in the result should equal the lowest number of significant figures of any of the quantities. In long calculations, the final result should be rounded off to the correct number of significant figures. When quantities are combined by addition and/or substraction, the final result cannot be more precise than any of the quantities added or subtracted. Therefore, the position (relative to the decimal point) of the last significant digit in the number that has the lowest degree of precision is the position of the last permissible significant digit in the result. For example, the sum of 3702., 370, 0.037, 4, and 37. should be reported as 4110 (without a decimal). The least precise of the five numbers is 370, which has its last significant digit in the *tens* position. The answer should also have its last significant digit in the *tens* position.

Unfortunately, engineers and scientists rarely concern themselves with significant figures in their calculations. However, it is recommended that—at least for this chapter—the reader attempt to follow the calculational procedure set forth in this subsection.

In the process of performing engineering calculations, very large and very small numbers are often encountered. A convenient way to represent these numbers is to use *scientific notation*. Generally, a number represented in scientific notation is the product of a number (< 10 but $>$ or $= 1$) and 10 raised to an integer power. For example,

$$28,070,000,000 = 2.807 \times 10^{10}$$

$$0.000\,002\,807 = 2.807 \times 10^{-6}$$

A nice feature of using scientific notation is that only the significant figures need appear in the number.

4.3   CHEMICAL AND PHYSICAL PROPERTIES

Temperature

Whether in the gaseous, liquid, or solid state, all molecules possess some degree of kinetic energy; that is, they are in constant motion—vibrating, rotating, or translating. The kinetic energies of individual molecules cannot be measured, but the combined effect of these energies in a very large number of molecules can. This measurable quantity is known as *temperature*; it is a macroscopic concept only, and as such does not exist on the molecular level.

Temperature can be measured in many ways; the most common method makes use of the expansion of mercury (usually encased inside a glass capillary tube) with increasing temperature. (On incineration systems, however, thermocouples or thermistors are more commonly employed.) The two most commonly used temperature scales are the Celsius (or Centigrade) and Fahrenheit scales. The Celsius is based on the boiling and freezing points of water at 1-atm pressure; to the former, a value of 100°C is assigned, and to the latter, a value of 0°C. On the older Fahrenheit scale, these temperatures correspond to 212 and 32°F, respectively. Equations (4.3.1) and (4.3.2) show the conversion from one scale to the other.

$$°F = 1.8(°C) + 32 \qquad (4.3.1)$$

$$°C = (°F - 32)/1.8 \qquad (4.3.2)$$

where °F = a temperature on the Fahrenheit scale
       °C = a temperature on the Celsius scale

Experiments with gases at low-to-moderate pressures (up to a few atmospheres) have shown that, if the pressure is kept constant, the volume of a gas and its temperature are linearly related (Charles' Law) and that a decrease of 0.3663% or (1/273) of the initial volume is experienced for every temperature drop of 1°C. These experiments were not extended to very low temperatures, but if the linear relationship were extrapolated, the volume of the gas would *theoretically* be zero at a temperature of approximately −273°C or −460°F. This temperature has become known as *absolute zero* and is the basis for the definition of two *absolute* temperature scales. (An *absolute* scale is one which does not allow negative quantities.) These absolute temperature scales are the kelvin (K) and Rankine (°R) scales; the former is defined by shifting the Celsius scale by 273 Celsius degrees so that 0 K is equal to −273°C; Equation (4.3.3) shows this relationship.

$$K = °C + 273 \qquad (4.3.3)$$

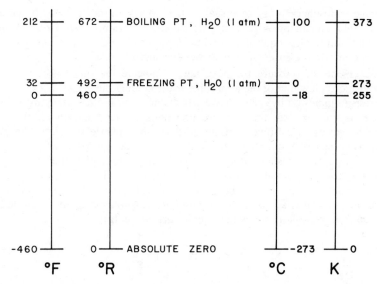

Figure 4.3.1. Temperature scales.

The Rankine scale is defined by shifting the Fahrenheit scale 460 Fahrenheit degrees, so that

$$°R = °F + 460 \qquad (4.3.4)$$

The relationships among the various temperature scales are shown in Fig. 4.3.1.

Pressure

In the gaseous state, the molecules possess a high degree of translational kinetic energy, which means they are able to move quite freely throughout the body of the gas. If the gas is in a container of some type, the molecules are constantly bombarding the walls of the container. The macroscopic effect of this bombardment by a tremendous number of molecules—enough to make the effect measurable—is called *pressure*. The natural units of pressure are force per unit area. In the example of the gas in a container, the *unit area* is a portion of the inside solid surface of the container wall and the *force*, measured perpendicularly to the unit area, is the result of the molecules hitting the unit area and giving up momentum during the sudden change of direction.

There are a number of different methods used to express a pressure measurement. Some of them are natural units, that is, based on a force per unit area, for example, pound (force) per square inch (abbreviated $lb_f/in.^2$ or psi) or dyne per square centimeter (dyn/cm^2). Others are based on a fluid height, such as inches of water (in. H_2O) or millimeters of mercury (mm Hg); units such as these are convenient when the pressure is indicated by a difference between two levels of a liquid as in a *manometer* or *barometer*. *Barometric pressure* and *atmospheric pressure* are synonymous, and measure the ambient air pressure. *Standard barometric pressure* is the average atmospheric pressure at sea level, 45° north latitude at 32°F. It is used to define another unit of pressure called the *atmosphere* (atm). Standard barometric pressure is 1 atm and is equivalent to 14.696 psi and 29.921 in. Hg. As one might expect, barometric pressure varies with weather and altitude.

Measurements of pressure by most gauges indicate the difference in pressure either above or below that of the atmosphere surrounding the gauge. *Gauge pressure* is the pressure indicated by such a device. If the pressure in the system measured by the gauge is greater than the pressure prevailing in the atmosphere, the gauge pressure is expressed positively; if lower than atmospheric pressure, the gauge pressure is a negative quantity; the term *vacuum* designates a negative gauge pressure. Gauge pressures are often identified by the letter g after the pressure unit; for example, psig (pounds per square inch gauge) is a gauge pressure in psi units.

Since gauge pressure is the pressure relative to the prevailing atmospheric pressure, the sum of the two gives the *absolute pressure*, indicated by the letter *a* after the unit [e.g., psia (pounds per square inch absolute)]:

$$P = P_a + P_g \tag{4.3.5}$$

where P = absolute pressure (psia)
$\quad\quad P_a$ = atmospheric pressure (psia)
$\quad\quad P_g$ = gauge pressure (psig)

The absolute pressure scale is *absolute* in the same sense that the absolute temperature scale is *absolute*; that is, a pressure of zero psia is the lowest possible pressure theoretically achievable—a perfect vacuum.

Moles and Molecular Weights

An atom consists of protons and neutrons in a nucleus surrounded by electrons. An electron has such a small mass relative to that of the proton and neutron that the weight of the atom (called the *atomic weight*) is approximately equal to the sum of the weights of the particles in its nucleus. *Atomic weight* may be expressed in *atomic mass units (amu) per atom* or in *grams per gram-atom*. One gram-atom contains 6.02×10^{23} atoms (Avogadro's number). The atomic weights of the elements are listed in Table 4.3.1.

The *molecular weight* (MW) of a compound is the sum of the atomic weights of the atoms that make up the molecule. Units of atomic mass units per molecule (amu/molecule) or grams per gram-mole (g/gmol) are used for molecular weight. One gram-mole (gmol) contains an Avogadro number of molecules. For the English system, a pound-mole (lbmol) contains $454 \times 6.023 \times 10^{23}$ molecules.

Molal units are used extensively in pollutant control calculations as they greatly simplify material balances where chemical (including combustion) reactions are occurring. For mixtures of substances (gases, liquids, or solids), it is also convenient to express compositions in mole fractions or mole percentages instead of mass fractions. The mole fraction is the ratio of the number of moles of one component to the total number of moles in the mixture.

TABLE 4.3.1. Atomic Weights of the Elements[a,b]

Element	Symbol	Atomic Weight
Actinium	Ac	227.0278
Aluminum	Al	26.9815
Americium	Am	(243)
Antimony	Sb	121.75
Argon	Ar	39.948
Arsenic	As	74.9216
Astatine	At	(210)
Barium	Ba	137.34
Berkelium	Bk	(247)
Beryllium	Be	9.0122
Bismuth	Bi	208.980
Boron	B	10.811
Bromine	Br	79.904
Cadmium	Cd	112.40
Calcium	Ca	40.08
Californium	Cf	(251)
Carbon	C	12.01115
Cerium	Ce	140.12
Cesium	Cs	132.905
Chlorine	Cl	35.453
Chromium	Cr	51.996
Cobalt	Co	58.9332
Copper	Cu	63.546
Curium	Cm	(247)
Dysprosium	Dy	162.50
Einsteinium	Es	(252)
Erbium	Er	167.26
Europium	Eu	151.96
Fermium	Fm	(257)
Fluorine	F	18.9984
Francium	Fr	(223)
Gadolinium	Gd	157.25
Gallium	Ga	69.72
Germanium	Ge	72.59
Gold	Au	196.967
Hafnium	Hf	178.49
Helium	He	4.0026
Holmium	Ho	164.930
Hydrogen	H	1.00797
Indium	In	114.82
Iodine	I	126.9044
Iridium	Ir	192.2
Iron	Fe	55.847
Krypton	Kr	83.80
Lanthanum	La	138.91
Lawrencium	Lr	(260)
Lead	Pb	207.19
Lithium	Li	6.939
Lutetium	Lu	174.97

TABLE 4.3.1. (*Continued*)

Element	Symbol	Atomic Weight
Magnesium	Mg	24.312
Manganese	Mn	54.9380
Mendelevium	Md	(258)
Mercury	Hg	200.59
Molybdenum	Mo	95.94
Neodymium	Nd	144.24
Neon	Nc	20.183
Neptunium	Np	237.0482
Nickel	Ni	58.71
Niobium	Nb	92.906
Nitrogen	N	14.0067
Nobelium	No	(259)
Osmium	Os	190.2
Oxygen	O	15.9994
Palladium	Pd	106.4
Phosphorus	P	30.9738
Platinum	Pt	195.09
Plutonium	Pu	(244)
Polonium	Po	(209)
Potassium	K	39.102
Praseodymium	Pr	140.907
Promethium	Pm	(145)
Protactinium	Pa	231.0359
Radium	Ra	226.0254[c]
Radon	Rn	(222)
Rhenium	Re	186.2
Rhodium	Rh	102.905
Rubidium	Rb	84.57
Rutherium	Ru	101.07
Samarium	Sm	150.35
Scandium	Sc	44.956
Selenium	Se	78.96
Silicon	Si	28.086
Silver	Ag	107.868
Sodium	Na	22.9898
Strontium	Sr	87.62
Sulfur	S	32.064
Tantalum	Ta	180.948
Technetium	Tc	(98)
Tellurium	Te	127.60
Terbium	Tb	158.924
Thallium	Tl	204.37
Thorium	Th	232.038
Thulium	Tm	168.934
Tin	Sn	118.69
Titanium	Ti	47.90
Tungsten	W	183.85
Uranium	U	238.03
Vanadium	V	50.942

TABLE 4.3.1. (*Continued*)

Element	Symbol	Atomic Weight
Xenon	Xe	131.30
Ytterbium	Yb	173.04
Yttrium	Y	88.905
Zinc	Zn	65.37
Zirconium	Zr	91.22

[a] Atomic weights apply to naturally occurring isotopic compositions, and are based on an atomic mass of $^{12}C = 12$
[b] For radioactive elements a value given in parenthesis is the atomic mass number of the isotope of longest known half-life.
[c] Geologically exceptional samples are known in which the element has an isotopic composition outside the limits for normal material.

Mass and Volume

The *density* (ρ) of a substance is the ratio of its mass to its volume and may be expressed in units of pounds per cubic foot (lb/ft^3), kilometers per cubic meter (kg/m^3), and so on. For solids, density can be easily determined by placing a known mass of the substance in a liquid and determining the displaced volume. The density of a liquid can be measured by weighing a known volume of the liquid in a volumetric flask. For gases, the ideal gas law (IGL), discussed in Section 4.4, can be used to calculate the density from the pressure, temperature, and molecular weight of the gas.

Densities of pure solids and liquids are relatively independent of temperature and pressure and can be found in standard reference books.[1] The *specific volume* (v) of a substance is its volume per unit mass (ft^3/lb, m^3/kg, etc.) and is, therefore, the inverse of its density.

The *specific gravity* (SG) is the ratio of the density of a substance to the density of a reference substance at a specific condition.

$$SG = \rho/\rho_{ref} \qquad (4.3.6)$$

The reference most commonly used for *solids* and *liquids* is water at its maximum density, which occurs at 4°C; this reference density is 1.000 g/cm^3, 1000 kg/m^3, or 62.43 lb/ft^3. Note that, since the specific gravity is a ratio of two densities, it is dimensionless. Therefore, any set of units may be employed for the two densities as long as they are consistent. The specific gravity of *gases* is used only rarely; when it is, air at the same conditions of temperature and pressure as the gas is usually employed as the reference substance.

Another dimensionless quantity related to density is the API (American Petroleum Institute) gravity, which is often used to indicate densities of fuel oils and some liquid hazardous wastes. The relationship between the API scale and the specific gravity is:

$$\text{degrees API} = \frac{141.5}{\text{SG}(60/60°\text{F})} - 131.5 \qquad (4.3.7)$$

where SG(60/60°F) = specific gravity of the liquid at 60°F using water at 60°F as the reference

Viscosity

Viscosity is a property associated with a fluid's resistance to flow; more precisely, this property accounts for energy losses, which result from shear stresses that occur between different portions of the fluid, which are moving at different velocities. The *absolute viscosity* (μ) has units of mass per length-time; the fundamental unit is the *poise*, which is defined as 1 g/cm-s. This unit is inconveniently large for many practical purposes and viscosities are frequently given in *centipoises* (0.01 poise), which is abbreviated cP. The viscosity of pure water at 68.6°F is 1.00 cP. In English units, absolute viscosity is expressed either as pounds (mass) per foot second (lb/ft-s) or pounds per foot hour (lb/ft-h). The absolute viscosity depends primarily on temperature and to a lesser degree on pressure. The *kinematic viscosity* (v) is the absolute viscosity divided by the density of the fluid, and is useful in certain fluid flow problems; the units for this quantity are length squared per time, for example, square foot per second (ft²/s) or square meters per hour (m²/h). A kinematic viscosity of 1 cm²/s is called a *stoke*. For pure water at 70°F, $v = 0.983$ cS (centistokes). Because fluid viscosity changes rapidly with temperature, a numerical value of viscosity has no significance unless the temperature is specified.

Liquid viscosity is usually measured by the amount of time it takes for a given volume of liquid to flow through an orifice. The *Saybolt Universal viscometer* is the most widely used device in the United States for the determination of the viscosity of fuel oils and liquid wastes. It should be stressed that Saybolt viscosities, which are expressed in *Saybolt seconds* (SSU), are not even approximately proportional to absolute viscosities except in the range above 200 SSU; hence, converting units from Saybolt seconds to other units requires the use of special conversion tables.[1] As the time of flow decreases, the deviation becomes more marked. In any event, viscosity is an important property because of potential flow problems with liquid wastes and/or fuel oil, particularly with liquid injection incinerators.

The viscosities of air at atmospheric pressure and water are presented in Tables 4.3.2 and 4.3.3, respectively, as functions of temperature. Viscosities of other substances are available in the literature.[1]

Heat Capacity

The *heat capacity* of a substance is defined as the quantity of heat required to raise the temperature of that substance by 1°; the *specific heat capacity* is the heat capacity on a unit mass basis. The term *specific heat* is frequently used in place of *specific heat capacity*. This is not strictly correct, because traditionally *specific heat* has been defined as the ratio of the heat capacity of a substance to the heat capacity of water. However, since the specific heat of water is approximately 1 cal/g-°C or 1 Btu/lb-°F,

TABLE 4.3.2. Viscosity of Air at 1 Atmosphere[a]

$T(°C)$	Viscosity, Micropoise (μP)
0	170.8
18	182.7
40	190.4
54	195.8
74	210.2
229	263.8

[a] $1 \text{ P} = 100 \text{ cP} = 10^6 \text{ }\mu P$; $1 \text{ cp} = 6.72 \times 10^{-4}$ lb/ft-s.

TABLE 4.3.3. Viscosity of Water

$T(°C)$	Viscosity, Centipoise (cP)
0	1.792
5	1.519
10	1.308
15	1.140
20	1.000
25	0.894
30	0.801
35	0.723
40	0.656
50	0.594
60	0.469
70	0.406
80	0.357
90	0.317
100	0.284

the term *specific heat* has come to imply heat capacity per unit mass. For gases, the addition of heat to cause the 1° temperature rise may be accomplished either at constant pressure or at constant volume. Since the amounts of heat necessary are different for the two cases, subscripts are used to identify which heat capacity is being used—c_P, for constant pressure and c_V, for constant volume. For liquids and solids, this distinction does not have to be made since there is little difference between the two. Values of specific heats are available in the literature.[1]

Heat capacities are often used on a *molar* basis instead of a *mass* basis, in which case the units become cal/gmol-°C or Btu/lbmol-°F. To distinguish between the two bases, upper-case letters (C_P, C_V) will be used in this text to represent the *molar*-based heat capacities, and lower-case letters (c_P, c_V) for the *mass*-based heat capacities or specific heats.

Heat capacities are functions of both the temperature and pressure, although the effect of pressure is generally small and is neglected in almost all engineering calculations. The effect of temperature on C_P can be described by.

$$C_P = \alpha + \beta T + \gamma T^2 \tag{4.3.8}$$

or

$$C_P = a + bT + cT^{-2} \tag{4.3.9}$$

Average or *mean* heat capacity data over specific temperature ranges are also available. Values for α, β, γ, and a, b, c, as well as average heat capacity information are provided in tabular form in Chapter 5. These will find extensive application in Chapters 5 and 6, as well as some of the later chapters. Properties such as *enthalpy* and *heat of vaporization* are also discussed in Chapter 5.

Thermal Conductivity

Experience has shown that when a temperature difference exists across a solid body, energy in the form of heat will transfer from the high temperature region to the low temperature region until thermal equilibrium (same temperature) is reached. This mode of heat transfer in which vibrating molecules pass along kinetic energy through the solid is called *conduction*. Liquids and gases may also transport heat in this fashion. (For a more comprehensive discussion of heat transfer mechanisms, the reader is directed to Chapter 9.) The property of *thermal conductivity* provides a measure of how fast (or how easily) heat flows through a substance. It is defined as the amount of heat that flows in unit time through a unit surface area of unit thickness as a result of a unit difference in temperature. Typical units for conductivity are Btu-ft/h-ft^2-°F or Btu/h-ft-°F. Thermal conductivities for some liquids and gases are given in Table 4.3.4, and for some insulating solids in Table 8.2.1.

With regard to hazardous waste incineration systems, this particular property finds application in designing heat exchangers and waste heat boilers (see Chapter 8), as well as in selecting insulating material for the devices within the system, particularly the incinerator.

The Reynolds Number

The Reynolds number, Re, is a dimensionless number that indicates whether a fluid flowing is in the *laminar* or *turbulent* mode. *Laminar* flow is characteristic of fluids flowing slowly enough so that there are no eddies (whirlpools) or *macroscopic* mixing of different portions of the fluid. (*Note:* In any fluid, there is always *molecular* mixing due to the thermal activity of the molecules; this is distinct from *macroscopic* mixing due to the swirling motion of different portions of the fluid.) In laminar flow, a fluid can be imagined to flow like a deck of cards, with adjacent layers sliding past one another. *Trubulent* flow is characterized by eddies and macroscopic currents. In practice, moving gases are generally in the turbulent region. For flow in a pipe, a Reynolds number above 2100 is an indication of turbulent flow.

The Reynolds number is dependent on the fluid velocity, density, viscosity, and

TABLE 4.3.4. Thermal Conductivities of Gases and Liquids

Gases and Vapors (1 atm)	Thermal Conductivity k^a	
	32°F	212°F
Acetone	0.0057	0.0099
Acetylene	0.0108	0.0172
Air	0.0140	0.0184
Ammonia	0.0126	0.0192
Benzene	0.0052	0.0103
Carbon dioxide	0.0084	0.0128
Carbon monoxide	0.0134	0.0176
Carbon tetrachloride		0.0052
Chlorine	0.0043	
Ethane	0.0106	0.0175
Ethyl alcohol		0.0124
Ethyl ether	0.0077	0.0131
Ethylene	0.0101	0.0161
Helium	0.0818	0.0988
Hydrogen	0.0966	0.1240
Methane	0.0176	0.0255
Methyl alcohol	0.0083	0.0128
Nitrogen	0.0139	0.0181
Nitrous oxide	0.0088	0.0138
Oxygen	0.0142	0.0188
Propane	0.0087	0.0151
Sulfur dioxide	0.0050	0.0069
Water vapor		0.0136

Liquids	Temperature (°F)	k^a
Acetic acid	68	0.909
Acetone	86	0.102
Ammonia (anhydrous)	5–86	0.29
Aniline	32–68	0.100
Benzene	86	0.092
n-Butyl alcohol	86	0.097
Carbon bisulfide	86	0.093
Carbon tetrachloride	32	0.107
Chlorobenzene	50	0.083
Ethyl acetate	68	0.101
Ethyl alcohol (absolute)	68	0.105
Ethyl ether	86	0.080
Ethylene glycol	32	0.153
Gasoline	86	0.078
Glycerine	68	0.164
n-Heptane	86	0.081
Kerosene	68	0.086
Methyl alcohol	68	0.124
Nitrobenzene	86	0.095
n-Octane	86	0.083
Sulfur dioxide	5	0.128
Sulfuric acid (90%)	86	0.21
Toluene	86	0.086
Trichloroethylene	122	0.080
o-Xylene	68	0.090

^{a} Units of k: Btu/h-ft-°F.

some *length* characteristic of the system or conduit; for pipes, this characteristic length is the inside diameter.

$$\text{Re} = v D \rho / \mu = v D / \nu \qquad (4.3.10)$$

where Re = Reynolds number
$\quad D$ = inside diameter of the pipe (ft)
$\quad v$ = fluid velocity (ft/s)
$\quad \rho$ = fluid density (lb/ft^3)
$\quad \mu$ = fluid viscosity (lb/ft-s)
$\quad \nu$ = fluid kinematic viscosity (ft^2/s)

Any consistent set of units may be used with. (4.3.10)

pH

An important chemical property of an aqueous solution is its pH. The pH measures the acidity or basicity of the solution. In a neutral solution, such as pure water, the hydrogen (H$^+$) and hydroxyl (OH$^-$) ion concentrations are equal. At ordinary temperatures, this concentration is

$$C_{H^+} = C_{OH^-} = 10^{-7} \text{g-ion/L} \qquad (4.3.11)$$

where $\quad C_{H^+}$ = hydrogen ion concentration
$\quad C_{OH^-}$ = hydroxyl ion concentration

The unit *g-ion* stands for gram-ion, which represents an Avogadro number of ions. In all aqueous solutions, whether neutral, basic, or acidic, a chemical equilibrium or balance is established between these two concentrations, so that

$$K_{eq} = C_{H^+} C_{OH^-} = 10^{-14} \qquad (4.3.12)$$

where K_{eq} = equilibrium constant

The numerical value for K_{eq} given in Eq. (4.3.12) holds for room temperature and only when the concentrations are expressed in gram-ion per liter (g-ion/L). In acid solutions, C_{H^+} is $> C_{OH^-}$; in basic solutions, C_{OH^-} predominates. The pH is a direct measure of the hydrogen ion concentration and is defined by

$$\text{pH} = -\log C_{H^+} \qquad (4.3.13)$$

Thus, an acidic solution is characterized by a pH below 7 (the lower the pH, the higher the acidity); a basic solution, by a pH above 7; and a neutral solution by a pH of 7. It should be pointed out that Eq. (4.3.13) is not the exact definition of pH, but is a close approximation to it. Strictly speaking, the *activity* of the hydrogen ion, a_{H^+}, and not the ion concentration belongs in Eq. (4.3.13). For a discussion of chemical activities, the reader is directed to the literature.[2]

Vapor Pressure

This section concludes with a discussion of *vapor pressure*, which is an important property of liquids, and, to a much lesser extent, of solids. If a liquid is allowed to evaporate in a confined space, the pressure in the vapor space increases as the amount of vapor increases. If there is sufficient liquid present, a point is eventually reached at which the pressure in the vapor space is exactly equal to the pressure exerted by the liquid at its own surface. At this point, a dynamic equilibrium exists in which vaporization and condensation take place at equal rates and the pressure in the vapor space remains constant. The pressure exerted at equilibrium is called the *vapor pressure* of the liquid. The magnitude of this pressure for a given liquid depends on the temperature, but not on the amount of liquid present. Solids, like liquids, also exert a vapor pressure. Evaporation of solids (called *sublimation*) is noticeable only for those with appreciable vapor pressures.

4.4 THE IDEAL GAS LAW

Observations based on physical experimentation often can be synthesized into simple mathematical equations called *laws*. These laws are never perfect and hence are only an approximate representation of reality. The *ideal gas law* (IGL) was derived from experiments in which the effects of pressure and temperature on gaseous volumes were measured over moderate temperature and pressure ranges. This law works well in the pressure and temperature ranges that were used in taking the data; extrapolations outside of the ranges have been found to work well in some cases and poorly in others. As a general rule, this law works best when the molecules of the gas are far apart, that is, when the pressure is low and the temperature is high. Under these conditions, the gas is said to behave *ideally*, that is, its behavior is a close approximation to the so-called *perfect* or *ideal gas*, a hypothetical entity that obeys the ideal gas law perfectly. For engineering calculations, and specifically for hazardous waste incinerator problems, the ideal gas law is almost always assumed to be valid, since it generally works well (usually within a few percent of the correct result) up to the highest pressures and down to the lowest temperatures used in hazardous waste incinerator applications.

The two precursors of the ideal gas law were *Boyle's* and *Charles'* laws. Boyle found that the volume of a given mass of gas is inversely proportional to the absolute pressure if the temperature is kept constant:

$$P_1 V_1 = P_2 V_2 \qquad (4.4.1)$$

where V_1 = volume of gas at absolute pressure P_1 and temperature T
V_2 = volume of gas at absolute pressure P_2 and temperature T

Charles found that the volume of a given mass of gas varies directly with the absolute temperature at constant pressure.

$$\frac{V_1}{T_1} = \frac{V_2}{T_2} \qquad (4.4.2)$$

where V_1 = volume of gas at pressure P and absolute temperature T_1
   V_2 = volume of gas at pressure P and absolute temperature T_2

Boyle's and Charles' laws may be combined into a single equation in which neither temperature nor pressure need be held constant:

$$\frac{P_1 V_1}{T_1} = \frac{P_2 V_2}{T_2} \qquad (4.4.3)$$

For Eq. (4.4.3) to hold, the mass of gas must be constant as the conditions change from (P_1, T_1) to (P_2, T_2). This equation indicates that for a given mass of a specific gas, PV/T has a constant value. Since, at the same temperature and pressure, volume and mass must be directly proportional, this statement may be extended to

$$\frac{PV}{mT} = C \qquad (4.4.4)$$

where m = mass of a specific gas
   C = constant that depends on the gas

Moreover, experiments with different gases showed that Eq. (4.4.4) could be expressed in a far more generalized form. If the number of moles (n) is used in place of the mass (m), the constant is the same for all gases.

$$\frac{PV}{nT} = R \qquad (4.4.5)$$

where R = universal gas constant

Equation (4.4.5) is called the *ideal gas law*. Numerically, the value of R depends on the units used for $P, V, T,$ and n (see Table 4.4.1). In this text, all gases are assumed to approximate ideal gas behavior. As is generally the case in engineering practice, the ideal gas law is assumed to be valid for all illustrative and assigned problems. If a case is encountered in practice where the gas behaves in very nonideal fashion, for example, a high molecular weight gas (such as a chlorinated organic) under high pressures, one of the many *real gas* correlations found in the literature[3] should be used.

   Other useful forms of the ideal gas law are shown in Eq. (4.4.6) and (4.4.7). Equation (4.4.6) applies to gas flow rather than to gas confined in a container.

$$PQ = \dot{n}RT \qquad (4.4.6)$$

where Q = gas volumetric flow rate (ft^3/h)
   P = absolute pressure (psia)
   $\dot{n}$ = molar flow rate (lbmol/h)
   T = absolute temperature (°R)
   R = 10.73 psia-ft^3/lbmol-°R

TABLE 4.4.1. Values of *R* in Various Units

R	Temperature Scale	Units of *V*	Units of *n*	Units of *P*	Units of *PV* (energy)
10.73	°R	ft^3	lbmol	psia	—
0.7302	°R	ft^3	lbmol	atm	—
21.85	°R	ft^3	lbmol	in. Hg	—
555.0	°R	ft^3	lbmol	mm Hg	—
297.0	°R	ft^3	lbmol	in. H$_2$O	—
0.7398	°R	ft^3	lbmol	bar	—
1545.0	°R	ft^3	lbmol	psfa	—
24.75	°R	ft^3	lbmol	ft H$_2$O	—
1.9872	°R	—	lbmol	—	Btu
0.0007805	°R	—	lbmol	—	hp-h
0.0005819	°R	—	lbmol	—	kW-h
500.7	°R	—	lbmol	—	cal
1.314	K	ft^3	lbmol	atm	—
998.9	K	ft^3	lbmol	mm Hg	—
19.32	K	ft^3	lbmol	psia	—
62.361	K	L	gmol	mm Hg	—
0.08205	K	L	gmol	atm	—
0.08314	K	L	gmol	bar	—
8314	K	L	gmol	Pa	—
8.314	K	m^3	gmol	Pa	—
82.057	K	cm^3	gmol	atm	—
1.9872	K	—	gmol	—	cal
8.3144	K	—	gmol	—	J

Equation (4.4.7) combines *n* and *V* from Eq. (4.4.5) to express the law in terms of density.

$$PM = \rho RT \qquad\qquad (4.4.7)$$

where M = molecular weight of gas (lb/lbmol)
   ρ = density of gas (lb/ft^3)

Volumetric flow rates are often given not at the actual conditions of pressure and temperature, but at arbitrarily chosen standard conditions (STP—*standard temperature and pressure*). To distinguish between flow rates based on the two conditions, the letters a and s are often used as part of the unit. The units acfm and scfm stand for actual cubic feet per minute and standard cubic feet per minute, respectively. The ideal gas law can be used to convert from standard to actual conditions, but, since there are many *standard conditions* in use, the STP being used must be known. Standard conditions most often used are shown in Table 4.4.2. The reader is cautioned on the incorrect use of acfm and/or scfm. The use of *standard* conditions is a convenience; when predicting the performance of or designing equipment, the *actual* conditions must be employed. Designs based on standard conditions can lead

TABLE 4.4.2. Common Standard Conditions

System	Temperature	Pressure	Molar Volume
SI	273 K	101.3 kPa	22.4 m^3/kmol
Universal scientific	0°C	760 mm Hg	22.4 L/gmol
Natural gas industry	60°F	14.7 psia	379 ft^3/lbmol
American engineering	32°F	1 atm	359 ft^3/lbmol
Hazardous waste	60°F	1 atm	379 ft^3/lbmol
incinerator industry	70°F	1 atm	387 ft^3/lbmol

to disastrous results, with the unit usually underdesigned. For example, for a flue gas stream at 2140°F, the ratio of acfm to scfm (standard temperature = 60°F) for an incinerator application is *5.0*. Equation (4.4.8), which is a form of Charles' Law, can be used to correct flow rates from standard to actual conditions.

$$Q_a = Q_s(T_a/T_s) \tag{4.4.8}$$

where Q_a = volumetric flow rate at actual conditions (ft^3/h)
  Q_s = volumetric flow rate at standard conditions (ft^3/h)
  T_a = actual absolute temperature (°R)
  T_s = standard absolute temperature (°R)

The reader is again reminded that *absolute* temperatures and pressures must be employed in all ideal gas law calculations.

Partial Pressure and Partial Volume

In engineering practice, mixtures of gases are more often encountered than single or pure gases. The ideal gas law is based on the *number* of molecules present in the gas volume; the *kind* of molecules is not a significant factor, only the *number*. This law applies equally well to mixtures and pure gases alike. Dalton and Amagat both applied the ideal gas law to mixtures of gases. Since pressure is caused by gas molecules colliding with the walls of the container, it seems reasonable that the total pressure of a gas mixture is made up of pressure contributions due to each of the component gases. These pressure contributions are called *partial pressures*. Dalton defined the *partial pressure* of a component as the pressure that would be exerted if the same mass of the component gas occupied the same total volume alone at the same temperature as the mixture. The sum of these partial pressures equals the total pressure:

$$P = P_a + P_b = P_c + \cdots + P_n = \sum_{i=1}^{n} P_i \tag{4.4.9}$$

where P = total pressure
  n = number of components
  P_i = partial pressure of component i

Equation (4.4.9) is known as *Dalton's Law*. Applying the ideal gas law to one component (a) only,

$$P_a V = n_a RT \qquad (4.4.10)$$

where n_a = number of moles of component a

Eliminating R, T, and V between Eq. (4.4.5) and (4.4.10) yields

$$\frac{P_a}{P} = \frac{n_a}{n} = y_a$$

or

$$P_a = y_a P \qquad (4.4.11)$$

where y_a = mole fraction of component a

Amagat's Law is similar to *Dalton's*. Instead of considering the total pressure to be made up of partial pressures where each component occupies the total container volume, Amagat considered total volume to be made up of *partial volumes* in which each component is exerting the total pressure. The definition of the *partial volume* is therefore the volume occupied by a component gas alone at the same temperature and pressure as the mixture.

$$V = V_a + V_b + V_c + \cdots + V_n = \sum_{i=1}^{n} V_i \qquad (4.4.12)$$

Applying Eq. (4.4.5), as before,

$$\frac{V_a}{V} = \frac{n_a}{n} = y_a \qquad (4.4.13)$$

where V_a = partial volume of component a

It is common in the hazardous waste incineration industry to describe low concentrations of pollutants in gaseous mixtures in parts per million (ppm) by volume. Since partial volumes are proportional to mole fractions, it is only necessary to multiply the mole fraction of the pollutant by 1 million to obtain the concentration in parts per million. [For liquids and solids, parts per million (ppm) is also used to express concentration, although it is usually on a *mass* basis rather than a *volume* basis. The terms ppmv and ppmw are sometimes used to distinguish between the volume and mass bases.]

4.5   PHASE EQUILIBRIUM

The term *phase*, for a pure substance, indicates a *state of matter*, that is, solid, liquid, or gas. For mixtures, however, a more stringent connotation must be used, since a

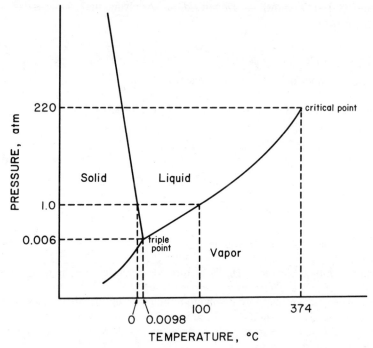

Figure 4.5.1. Phase diagram for water.

totally liquid or solid system may contain more than one phase (a mixture of oil and water, for example). A *phase* is characterized by uniformity or *homogeneity*; the same composition and properties must exist throughout the phase region. At most temperatures and pressures, a pure substance normally exists as a single phase. At certain temperatures and pressures, two or perhaps even three phases can coexist in equilibrium. This is shown on the phase diagram for water (see Fig. 4.5.1). Regarding the interpretation of this diagram, the following points should be noted:

- The line between the gas and liquid phase regions is the *boiling point* line and represents equilibrium between the gas and liquid.
- The boiling point of a liquid is the temperature at which its vapor pressure is equal to the *external* pressure. The temperature at which the vapor pressure is equal to 1 atm is the *normal* boiling point.
- The line between the solid and gas phase regions is the *sublimation point* line and represents equilibrium between the solid and gas.
- The line between the solid and liquid phase regions is the *melting point* or *freezing point* line and represents equilibrium between the liquid and solid.
- The point at which all three equilibrium lines meet (i.e., the one pressure and temperature where solid, liquid, and gas phases can all coexist) is the *triple point*.
- The liquid–gas equilibrium line is bounded on one end by the triple point and the

other end by the *critical* point. The *critical temperature* (the temperature coordinate of the critical point) is defined as the temperature above which a gas or vapor cannot be liquefied by the application of pressure alone.

The term *vapor*, strictly speaking, is used only for a *condensable* gas, that is, a gas below its *critical* temperature, and should not be applied to a noncondensable gas. It should also be pointed out that the phase diagram for water (Fig. 4.5.1) differs from that of other substances in one respect—the freezing point line is *negatively* sloped; for other materials, the slope of this line is *positive*. This is a consequence of the fact that liquid water is denser than ice and isothermal (constant temperature) compression of the liquid can result in a transformation to the solid.

Incineration calculations rarely involve single (pure) components, however. Phase equilibria for multicomponent systems are considerably more complex, mainly because of the addition of *composition* variables; for example, in a *ternary* or three-component system, the mole fractions of two of the components are pertinent variables along with temperature and pressure. In a single-component system, dynamic equilibrium between two phases is achieved when the rate of molecular transfer from one phase to the second equals that in the opposite direction. In multicomponent systems, the equilibrium requirement is more stringent—the rate of transfer of *each component* must be the same in both directions.

In hazardous waste incineration applications, the most important equilibrium phase relationship is that between liquid and vapor. *Raoult's and Henry's* Laws theoretically describe liquid–vapor behavior and under certain conditions are applicable in practice. *Raoult's Law* is sometimes useful for mixtures of components of similar structure and states that the partial pressure of any component in the vapor is equal to the product of the vapor pressure of the *pure* component and the mole fraction of that component in the liquid, that is,

$$P_i = P_i' x_i \qquad (4.5.1)$$

where P_i = partial pressure of component i in the vapor
P_i' = vapor pressure of pure i at the same temperature
x_i = mole fraction of component i in the liquid

This expression may be applied to all components. If the gas phase is ideal, Eq. (4.5.1) becomes

$$y_i = (P_i'/P)x_i \qquad (4.5.2)$$

where y_i = mole fraction component i in the vapor
P = total system pressure

Thus, the mole fraction of water vapor in a gas that is saturated, that is, in equilibrium contact with pure water ($x = 1.0$), is given simply by the ratio of the vapor pressure of water at the system temperature divided by the system pressure. This finds application in many quench and absorber calculations (see Part III).

Unfortunately, relatively few mixtures follow Raoult's Law. *Henry's Law* is a more empirical relationship used for representing data on many systems.

$$P_i = H_i x_i \tag{4.5.3}$$

where H_i = Henry's law constant for component i (in units of pressure)

If the gas behaves ideally, Eq. (4.5.3) may be written as

$$y_i = m_i x_i \tag{4.5.4}$$

where m_i = constant (dimensionless)

This is a more convenient form of Henry's Law to work with. The constant m_i (or H_i) has been determined experimentally for a large number of compounds and is usually valid at low concentrations.

In other applications, mixtures of condensable vapors and noncondensable gases musι ɒe handled. A common example is water vapor and air; a mixture of organic vapors and air is another such example that often appears in air pollution problems. Condensers can be used to control organic emissions to the atmosphere by lowering the temperature of the gaseous stream, although an increase in pressure will produce the same result. The calculation for this is accomplished using the phase equilibrium constant (K_i). This constant has been referred to in industry as a *componential split factor* since it provides the ratio of the mole fractions of a component in two equilibrium phases. The defining equation is

$$K_i = y_i / x_i \tag{4.5.5}$$

where K_i = phase equilibrium constant for component i (dimensionless)

This equilibrium constant is a function of the temperature, pressure, the other components in the system, and the mole fractions of these components. However, as a first approximation, K_i is generally treated as a function only of the temperature and pressure. For this condition, K_i may be approximated by

$$K_i = P_i' / P \tag{4.5.6}$$

Another phase equilibrium application involves the *psychrometric* or *humidity* chart (see Fig. 4.5.2). A humidity chart is used to determine the properties of moist air and to calculate moisture content in air. The ordinate of the chart is the absolute humidity $\mathscr{H}$, which is defined as the mass of water vapor per mass of bone-dry air. (Some charts base the ordinate on moles instead of mass.)

Based on this definition, Eq. (4.5.7) gives $\mathscr{H}$ in terms of moles and also in terms of partial pressure.

$$\mathscr{H} = \frac{18 n_{H_2O}}{29(n_T - n_{H_2O})} = \frac{18 P_{H_2O}}{29(P - P_{H_2O})} \tag{4.5.7}$$

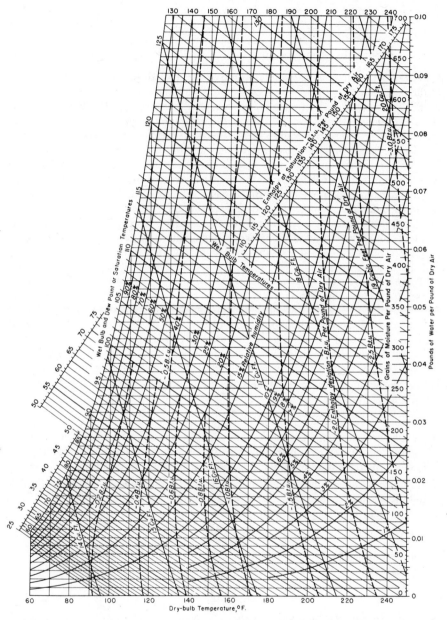

Figure 4.5.2. Psychrometric chart. (From R. Perry and D. Green, *Perry's Chemical Engineering Handbook* 6th ed., © 1984, McGraw-Hill, New York; reproduced with permission of McGraw-Hill.)

where n_{H_2O} = number of moles of water vapor
$\quad n_T$ = total number of moles in gas
$\quad P_{H_2O}$ = partial pressure of water vapor
$\quad P$ = total system pressure

Curves showing the *relative humidity* (ratio of the mass of the water vapor in the air to the maximum mass of water vapor the air could hold at that temperature, that is, if the air were saturated) of humid air also appear on the charts. The curve for 100% relative humidity is also referred to as the *saturation curve*. The abscissa of the humidity chart is air temperature, also known as the *dry-bulb* temperature (T_{DB}). The *wet-bulb* temperature (T_{WB}) is another measure of humidity; it is the temperature at which a thermometer with a wet wick wrapped around the bulb stabilizes. As water evaporates from the wick to the ambient air, the bulb is cooled; the rate of cooling depends on how humid the air is. No evaporation occurs if the air is saturated with water, hence T_{WB} and T_{DB} are the same. The lower the humidity, the greater the difference between these two temperatures. On the psychrometric chart, constant wet-bulb temperature lines are straight with negative slopes. The value of T_{WB} corresponds to the value of the abscissa at the point of intersection of this line with the saturation curve.

The *humid volume* is the volume of wet air per mass of dry air and is linearly related to the humidity. (In Fig. 4.5.2, this quantity is used as an alternate ordinate. Note the straight parallel lines labeled with units of *cu ft*.) The product of the humid volume and the absolute humidity gives the volume of the moist air. The *humid enthalpy* (also called *humid heat*) is the enthalpy of the moist air on a bone-dry air basis. The term *enthalpy* is a measure of the energy content of the mixture and is defined in Chapter 5. The enthalpy for *saturated* air can be read from the chart by extending the approximate wet-bulb temperature line upwards to the diagonal scale labeled *Enthalpy of Saturation*.

The following are some helpful points on the use of psychrometric charts.

• Heating or cooling at temperatures above the *dew point* (temperature at which the vapor begins to condense) corresponds to a horizontal movement on the chart. As long as no condensation occurs, the absolute humidity stays constant.

• If the air is cooled, the system follows the appropriate horizontal line to the left until it reaches the saturation curve and follows it thereafter.

• In problems involving use of the humidity chart, it is convenient to choose the *mass of dry air* as a basis, since the chart uses this basis.

Steam tables, charts, and diagrams are often employed in hazardous waste incineration calculations. A somewhat abbreviated steam table is given in Table 4.5.1; more comprehensive versions are available in the literature.

4.6   ILLUSTRATIVE EXAMPLES

Example 4.6.1.   Convert units of acceleration in cm/s^2 to miles/yr^2.

SOLUTION

$$\frac{1\,cm}{s^2} \times \frac{3600^2\,s^2}{1\,h^2} \times \frac{24^2\,h^2}{1\,day^2} \times \frac{365^2\,day^2}{1\,yr^2} \times \frac{1\,in.}{2.54\,cm} \times \frac{1\,ft}{12\,in.} \times \frac{1\,mile}{5280\,ft} = 6.18 \times 10^9 \text{ miles/yr}^2$$

Table 4.5.1. Properties of Saturated Steam and Water[a]

Temp. T, °F	Vapor press. P_A lb$_f$/in.²	Specific vol., ft³/lb Liquid v_x	Sat. vapor v_y	Enthalpy, Btu/lb Liquid h_x	Vaporization λ	Sat. vapor h_y
32	0.08854	0.01602	3,306	0.00	1075.8	1075.8
35	0.09995	0.01602	2,947	3.02	1074.1	1077.1
40	0.12170	0.01602	2,444	8.05	1071.3	1079.3
45	0.14752	0.01602	2,036.4	13.06	1068.4	1081.5
50	0.17811	0.01603	1,703.2	18.07	1065.6	1083.7
55	0.2141	0.01603	1,430.7	23.07	1062.7	1085.8
60	0.2563	0.01604	1,206.7	28.06	1059.9	1088.0
65	0.3056	0.01605	1,021.4	33.05	1057.1	1090.2
70	0.3631	0.01606	867.9	38.04	1054.3	1092.3
75	0.4298	0.01607	740.0	43.03	1051.5	1094.5
80	0.5069	0.01608	633.1	48.02	1048.6	1096.6
85	0.5959	0.01609	543.5	53.00	1045.8	1098.8
90	0.6982	0.01610	468.0	57.99	1042.9	1100.9
95	0.8153	0.01612	404.3	62.98	1040.1	1103.1
100	0.9492	0.01613	350.4	67.97	1037.2	1105.2
110	1.2748	0.01617	265.4	77.94	1031.6	1109.5
120	1.6924	0.01620	203.27	87.92	1025.8	1113.7
130	2.2225	0.01625	157.34	97.90	1020.0	1117.9
140	2.8886	0.01629	123.01	107.89	1014.1	1122.0
150	3.718	0.01634	97.07	117.89	1008.2	1126.1
160	4.741	0.01639	77.29	127.89	1002.3	1130.2
170	5.992	0.01645	62.06	137.90	996.3	1134.2
180	7.510	0.01651	50.23	147.92	990.2	1138.1
190	9.339	0.01657	40.96	157.95	984.1	1142.0
200	11.526	0.01663	33.64	167.99	977.9	1145.9
210	14.123	0.01670	27.82	178.05	971.6	1149.7
212	14.696	0.01672	26.80	180.07	970.3	1150.4
220	17.186	0.01677	23.15	188.13	965.2	1153.4
230	20.780	0.01684	19.382	198.23	958.8	1157.0
240	24.969	0.01692	16.323	208.34	952.2	1160.5
250	29.825	0.01700	13.821	218.48	945.5	1164.0
260	35.429	0.01709	11.763	228.64	938.7	1167.3
270	41.858	0.01717	10.061	238.84	931.8	1170.6
280	49.203	0.01726	8.645	249.06	924.7	1173.8
290	57.556	0.01735	7.461	259.31	917.5	1176.8
300	67.013	0.01745	6.466	269.59	910.1	1179.7
310	77.68	0.01755	5.626	279.92	902.6	1182.5
320	89.66	0.01765	4.914	290.28	894.9	1185.2
330	103.06	0.01776	4.307	300.68	887.0	1187.7
340	118.01	0.01787	3.788	311.13	879.0	1190.1
350	134.63	0.01799	3.342	321.63	870.7	1192.3
360	153.04	0.01811	2.957	332.18	862.2	1194.4
370	173.37	0.01823	2.625	342.79	853.5	1196.3
380	195.77	0.01836	2.335	353.45	844.6	1198.1
390	220.37	0.01850	2.0836	364.17	835.4	1199.6
400	247.31	0.01864	1.8633	374.97	826.0	1201.0
410	276.75	0.01878	1.6700	385.83	816.3	1202.1
420	308.83	0.01894	1.5000	396.77	806.3	1203.1
430	343.72	0.01910	1.3499	407.79	796.0	1203.8
440	381.59	0.01926	1.2171	418.90	785.4	1204.3
450	422.6	0.0194	1.0993	430.1	774.5	1204.6

[a] From J.H. Keenan and F.G. Keyes, *Thermodynamic Properties of Steam*, John Wiley & Sons, New York, 1936.

Thus, $1.0\,\text{cm/s}^2$ is equal to 6.18×10^9 miles/yr^2.

Example 4.6.2.   If a 55-gal tank contains 20.0 lb of water, (1) how many lbmoles of water does it contain; (2) how many gmoles does it contain; and (3) how many molecules does it contain?

SOLUTION. The molecular weight of the water is

$$\text{MW} = (2)(1.008) + (15.999) = 18.015\,\text{g/gmol}$$

$$= 18.015\,\text{lb/lbmol}$$

(1) $20.0\,\text{lb} \times \dfrac{\text{lbmol}}{18.015\,\text{lb}} = 1.11\,\text{lbmol water}$

(2) $20.0\,\text{lb} \times \dfrac{453.593\,\text{g}}{1\,\text{lb}} \times \dfrac{\text{gmol}}{18.015\,\text{g}} = 503.6\,\text{gmol water}$

(3) $503.6\,\text{gmol} \times \dfrac{6.023 \times 10^{23}\,\text{molecules}}{1\,\text{gmol}} = 3.033 \times 10^{26}\,\text{molecules}.$

Example 4.6.3.   Perform the following temperature conversions:

1. Convert 55°F to (a) Rankine, (b) Celsius, and (c) kelvin
2. Convert 55°C to (a) Fahrenheit, (b) Rankine and (c) kelvin

SOLUTION

1. (a) $°R = °F + 460 = 55 + 460 = 515$
   (b) $°C = \frac{5}{9}(°F - 32) = \frac{5}{9}(55 - 32) = 12.8$
   (c) $K = \frac{5}{9}(°F + 460) = \frac{5}{9}(55 - 460) = 286$
2. (a) $°F = 1.8(°C) + 32 = 1.8(55) + 32 = 131$
   (b) $°R = 1.8(°C) + 492 = 1.8(55) + 492 = 591$
   (c) $K = °C + 273 = 55 + 273 = 328$

Example 4.6.4.   Consider the following pressure calculations.

1. A liquid weighing 100 lb held in a cylindrical column with a base area of 3 in.^{2} exerts how much pressure at the base in lb_f/ft^2?
2. If the pressure is 35 psig (pounds per square inch gauge), what is the absolute pressure?

SOLUTION

1. $F = mg/g_c = 100\,\text{lb}(1\,\text{lb}_f/\text{lb})$

$$= 100\,\text{lb}_f$$

Note: g_c is a conversion factor equal to $32.16\,\text{lb-ft/lb}_f\text{-s}^2$. g is the gravitational acceleration, which is equal, or close to, $32.16\,\text{ft/s}^2$ on the earth's surface.

$$P = F/\text{area} = 100 \, \text{lb}_f/3 \, \text{in.}^2$$

$$= 33.33 \, \text{lb}_f/\text{in.}^2$$

$$= 4800 \, \text{lb}_f/\text{ft}^2$$

2.
$$P_{\text{absolute}} = P_{\text{gauge}} + P_{\text{atmosphere}} = 35 + 14.7$$

$$= 49.7 \, \text{psia}$$

Example 4.6.5

1. What is the density of air at 75°F and 14.7 psia? The molecular weight of air is 29.
2. Calculate the volume (in ft^3) of 1.0 lbmol of any ideal gas at 60°F and 14.7 psia.
3. Calculate the density of a gas (MW = 29) in g/cm^3 at 20°C and 1.2 atm using the ideal gas law.
4. Data from an incinerator indicate a volumetric flow rate of 10,000 scfm (60°F, 1 atm). If the operating temperature and pressure of the unit are 1100°F and 1 atm, respectively, calculate the actual flow rate in actual cubic feet per minute.
5. The exhaust gas flow rate from a facility is 1000 scfm. All of the gas is vented through a small stack that has an inlet area of 1.0 ft^2. The exhaust gas temperature is 300°F. What is the velocity of the gas through the stack inlet in feet per second? Assume standard conditions to be 70°F and 1.0 atm. Neglect the pressure drop across the stack.

SOLUTION. This example is solved using the ideal gas law.

1.
$$PV = nRT = \frac{m}{M} RT$$

$$\rho = \frac{PM}{RT} = \frac{(14.7 \, \text{psia})(29 \, \text{lb/lbmol})}{(10.73 \, \text{ft}^3\text{-psi/lbmol-°R})(75 + 460)}$$

$$= 0.0743 \, \text{lb/ft}^3$$

2. Solve the ideal gas law for V and calculate the volume.

$$V = \frac{nRT}{P} = \frac{(1)(10.73)(60 + 460)}{(14.7)}$$

$$= 379 \, \text{ft}^3$$

This result is an important number to remember in combustion calculations—1 lbmol of any (ideal) gas at 60°F and 1 atm occupies 379 ft^3.

3. Calculate the density of the gas again using the ideal gas law.

$$PV = nRT = \frac{m}{M} RT$$

$$\frac{m}{V} = \rho = \frac{PM}{RT} = \frac{(1.2)(29)}{(82.06)(20 + 273)}$$

$$= 0.00145 \, \text{g/cm}^3$$

The effects of pressure, temperature, and molecular weight on density can be obtained directly from the ideal gas law equation. Increasing the pressure and molecular weight increases the density; increasing the temperature decreases the density.

4. Since the pressure remains constant, calculate the actual cubic feet per minute using Charles' Law.

$$Q_a = Q_s \left(\frac{T_a}{T_s}\right) = 10,000 \left(\frac{1100 + 460}{60 + 460}\right)$$

$$= 30,000 \, \text{acfm}$$

The reader is again cautioned on the use of acfm and/or scfm. Predicting the performance of and designing equipment should always be based on *actual* conditions. Designs based on standard conditions can lead to disastrous results, with the unit underdesigned. The reader is also reminded that absolute temperatures and pressures must be employed in all ideal gas law calculations.

5. Calculate the actual flow rate, in acfm, using the Charles' law.

$$Q_a = Q_s \left(\frac{T_a}{T_s}\right) = 1000 \left(\frac{460 + 300}{460 + 70}\right)$$

$$= 1434 \, \text{acfm}$$

Note that since the gas in vented through the stack to the atmosphere, the pressure is 1.0 atm. Calculate the velocity of the gas.

$$v = \frac{Q_a}{A} = \frac{1434}{1.0}$$

$$= 1434 \, \text{ft/min}$$

Example 4.6.6. The exhaust to the atmosphere from an incinerator has a CO concentration of 0.15 mm Hg partial pressure. Calculate the parts per million of CO in the exhaust.

SOLUTION. First calculate the mole fraction (y). By definition,

$$y = P_{co}/P$$

Since the exhaust is discharged to the atmosphere, the atmospheric pressure (760 mm Hg) is the total pressure (P).

$$y = (0.15)/(760) = 1.97 \times 10^{-4}$$

$$\text{ppm} = (y)(10^6) = (1.97 \times 10^{-4})(10^6)$$

$$= 197 \, \text{ppm}$$

Note: Since the concentration of *gases* is involved, it is understood that ppm is on a volume basis.

Example 4.6.7. What is the kinematic viscosity of a gas, the specific gravity and absolute viscosity of which are 0.8 and 0.02 cP, respectively?

SOLUTION

$$\frac{0.02\,\text{cP}}{1} \times \frac{6.720 \times 10^{-4}\,\text{lb/ft-s}}{1\,\text{cP}} = 1.344 \times 10^{-5}\,\text{lb/ft-s}$$

$$\rho = (\text{SG})(\rho_{\text{ref}}) = (0.8)(62.43\,\text{lb/ft}^3) = 49.94\,\text{lb/ft}^3$$

$$\nu = \mu/\rho = (1.344 \times 10^{-5}\,\text{lb/ft-s})/(49.94\,\text{lb/ft}^3)$$

$$= 2.691 \times 10^{-7}\,\text{ft}^2/\text{s}$$

Example 4.6.8. If the heat capacity of liquid water is 1.0 Btu/lb-°F at 17°C, what is the change in temperature if 50 Btu of heat is added to 3 lb of water at constant pressure, assuming no loss of heat?

SOLUTION. From the definition of heat capacity (see Chapter 5 for additional details),

$$Q = mc_P \Delta T$$

where Q = amount of heat (Btu)
m = mass (lb)

$$\Delta T = Q/mc_P = (50\,\text{Btu})/(3\,\text{lb})(1\,\text{Btu/lb-°F})$$

$$= 16.67°\text{F}$$

Example 4.6.9. What is the heat flux (Btu/h-ft^2) through a copper plate 0.3 in. thick if one side is at 100°C and the other side is 400°C? The thermal conductivity of copper is 214 Btu/h-ft-°F.

SOLUTION. From the difinition of thermal conductivity (see Chapter 8 for additional details),

$$\frac{q}{A} = \frac{k\Delta T}{x} = (214\,\text{Btu/h-ft-°F})[(400 - 100)(1.8)]°\text{F}/(0.3/12)\,\text{ft}$$

$$= 4.62 \times 10^6\,\text{Btu/h-ft}^2$$

where q/A = heat flux
x = thickness

Note: The direction of heat transfer is from the high temperature region to the low temperature region.

Example 4.6.10. Calculate the Reynolds number for a liquid waste flowing through a 5-in. diameter pipe at 10 fps (feet per second) with a density of 50 lb/ft^3 and a viscosity of 0.65 cP? Is the flow turbulent or laminar?

SOLUTION. By definition

$$Re = \rho v D / \mu$$

Substitution yields

$$Re = \frac{50\,lb}{ft^3} \times \frac{10\,ft}{s} \times \frac{5\,in.}{1} \times \frac{1\,ft}{12\,in.} \times \frac{1}{0.65\,cP} \times \frac{1\,cP}{6.720 \times 10^{-4}\,lb/ft\text{-}s}$$

$$= (50\,lb/ft^3)(10\,ft/s)(5/12\,ft)/(0.65 \times 6.72 \times 10^{-4}\,lb/ft\text{-}s)$$

$$= 477,000$$

The Reynolds number is > 2100; therefore, this flow is turbulent.

Example 4.6.11.   Perform the following ideal gas law calculations.

1. What is the final volumetric flow rate of a gas that is heated from 100 to 300°F if its initial flow is 3500 scfm (60°F, 1 atm)?
2. What is the volumetric flow rate of the gas (100°F, 1 atm) if it is compressed isothermally (constant temperature) to 3 atm?
3. What is the final volumetric flow rate if the gas is compressed to 3 atm and heated from 100 to 300°F with an initial volumetric flow rate of 3500 scfm (60, 1 atm)?

SOLUTION

1. Converting the initial standard conditions to actual conditions using Charles' Law,

$$Q_a = Q_s(T_a/T_s) = \frac{(3500\,scfm)(100 + 460)}{(60 + 460)}$$

$$= 3769\,acfm$$

Using Charles' Law again to obtain the final volumetric flow rate,

$$Q_{a2} = Q_{a1}\left(\frac{T_2}{T_1}\right) = \frac{(3769)(760)}{(560)}$$

$$= 5115\,acfm$$

2. Using Boyle's law,

$$Q_{a2} = Q_{a1}\left(\frac{P_1}{P_2}\right) = \frac{(1)(3769)}{(3)}$$

$$= 1256\,acfm$$

3. Using the combined gas law,

$$Q_{a2} = Q_{a1}\left(\frac{P_1}{P_2}\right)\left(\frac{T_2}{T_1}\right) = \frac{(1)(3769)(760)}{(560)(3)}$$

$$= 1705\,acfm$$

123

Example 4.6.12. Refer to the psychrometric chart to answer the following:

1. List key properties for humid air at a dry-bulb temperature of 160°F and a wet-bulb temperature of 100°F.
2. A stream of moist air is cooled and humidified adiabatically from T_{DB} of 100°F and T_{WB} of 70°F to a T_{DB} of 80°F. How much moisture is added per pound of dry air?

SOLUTION

1. If the air were to be cooled until the moisture just begins to condense, the dew point would be reached. This is represented by a horizontal line at constant humidity intersecting the saturation curve at a dew point of 87.5°F.

   The *relative humidity* is approximately 14% (interpolating between the 15 and 10% relative humidity lines). The absolute humidity is the horizontal line extended to the right intersecting the ordinate at a humidity of 0.0285 lb H_2O/lb dry air.

   The *humid volume* is approximately 16.3 ft³ moist air/lb dry air (interpolating between 16 and 17 ft³ moist air volume.)

   The *enthalpy* for saturated air at a T_{WB} of 100°F is 71.8 Btu/lb dry air. For the unsaturated air, the enthalpy deviation is −1.0 Btu/lb dry air; therefore, the actual enthalpy for the moist air at a T_{WB} of 100°F and a T_{DB} of 160°F is 70.8 Btu/lb dry air.

2. Adiabatic cooling follows the wet-bulb temperature line upwards (toward the saturation curve). The difference in the final and initial humidities is the added moisture.

	$\mathscr{H}$
Initial	0.0090
Final	0.0133

$\Delta \mathscr{H} = 0.0133 - 0.0090 = 4.3 \times 10^{-3}$ lb H_2O/lb dry air

Example 4.6.13. A flue gas is discharged at 120°F from an HCl absorber in a HWI facility in which carbon tetrachloride (CCl_4) is being incinerated. If 9000 lb/h (MW = 30) of gas enters the absorber essentially dry (negligible water) at 560°F, calculate the moisture content, the mass flow rate, and the volumetric flow rate of the discharge gas. The discharge gas from the absorber may safely be assumed to be saturated with water vapor.

SOLUTION. From Fig. 4.5.2, the discharge humidity of the flue gas is approximately

$$\mathscr{H}_{out} = 0.0814 \text{ lb } H_2O/\text{lb bone dry air}$$

This represents the moisture content of the gas at outlet conditions in lb H_2O/lb dry air. If the gas is assumed to have the properties of air, the discharge water vapor rate is

$$w_{H_2O} = (0.0814)(9000)$$

$$= 733 \, \text{lb/h}$$

The total flow rate leaving the absorber is

$$w_{total} = 733 + 9000$$

$$= 9733 \, \text{lb/h}$$

The volumetric (or molar) flow rate can only be calculated if the molecular weight of the gas is known. The average molecular weight of the discharge flue gas must first be calculated from the mole fraction of the flue gas (fg) and water vapor (wv).

$$y_{fg} = \frac{9000/30}{(9000/30) + (733/18)} = 0.88$$

$$y_{wv} = \frac{733/18}{(733/18) + (9000/30)} = 0.12$$

$$\overline{MW} = (0.88)(30) + (0.12)(18) = 28.6 \, \text{lb}$$

The ideal gas law is employed to calculate the volumetric flow rate, Q_a

$$PQ_a = \frac{w}{\overline{MW}} RT$$

$$Q_a = \frac{9733}{28.6} \frac{(0.73)(460 + 140)}{1.0}$$

$$= 1.49 \times 10^5 \, \text{ft}^3/\text{h}$$

PROBLEMS

Use the correct number of significant figures and scientific notation whenever possible.

1. Convert (a) 8.03 yr to s, (b) 150 miles/h to yd/h, (c) 100.0 m/s^2 to ft/min^2, and (d) 0.03 g/cm^3 to lb/ft^3.

2. (a) What is the molecular weight of nitrobenzene ($C_6H_5O_2N$)?
   (b) How many moles are there in 50.0 g of nitrobenzene?
   (c) If the specific gravity is 1.203, what is the density in g/cm^3?
   (d) What is the volume occupied by 50.0 g of nitrobenzene in cm^3, in ft^3, and in in.^{3}?
   (e) If the nitrobenzene is held in a cylindrical container with a base of 1 in. in diameter, what is the pressure at the base? What is it in gauge pressure?
   (f) How many molecules are contained in 50.0 g of nitrobenzene?

3. Convert the following temperature (a) 20°C to °F, K, °R and (b) 20°F to °C, K, °R.

4. **(a)** What is the density of air at 115°F and 2 atm?
   (b) What volume does 100 g of air occupy at the condition of (a)?
   (c) What is the new volume if the temperature is changed to 200°F and the pressure is 1 atm?

5. A liquid with a viscosity of 0.78 cP and a density of 1.50 g/cm^3 flows through a 1-in. diameter pipe at 20 cm/s. Calculate the Reynolds number. What region (laminar or turbulent) is it in?

6. A heat pump takes in 3500 gpm of water at a temperature of 38°F and discharges back to the lake at 36.2°F. How many Btu are removed from the water per day? (C_P for H_2O = 75.4 J/gmol-°C, ρ = 62.4 lb/ft^3)

7. One wall of an oven has a 3-in. insulation cover. The temperature on the inside of the wall is at 400°F; the temperature on the outside is at 25°C. What is the heat flux (heat flow rate per unit area) across the wall if the insulation is made of glass wool (k = 0.022 Btu/h-ft-°F)?

8. A container holds a gaseous mixture comprised of 30% CO_2, 5% CO, 5% H_2O, 50% N_2 and 10% O_2, by volume. What is the partial pressure of each component if the total pressure is 2 atm? What are their partial volumes if the total volume is 10 ft^3? What are their concentrations in ppm (parts per million)?

9. The exhaust to the atmosphere from an incinerator shows a CO concentration with a partial pressure of 0.19 mmHg. What is this concentration in ppm?

10. The inlet gas to a spray tower is at 1600°F. It is piped through a 3.0-ft inside diameter duct at 25 ft/s to the spray tower. The scrubber cools the gas to 500°F. In order to maintain the velocity of 25 ft/s, what size duct would be required at the outlet of the unit? Neglect the pressure across the spray tower and any moisture considerations.

11. **(a)** What is the density of air at 60°F and 14.7 psia?
    (b) Calculate the volume (in ft^3) of 1.0 lbmol of any ideal gas at 77°F and 14.7 psia.
    (c) Calculate the density of a gas (MW = 32) in g/cm^3 at 20°C and 1.0 atm using the ideal gas law.
    (d) Data from an incinerator indicate a volumetric flow rate of 10,000 scfm (60°F, 1 atm). If the operating temperature and pressure of the unit are 1950°F and 1 atm, respectively, calculate the actual flow rate in cubic feet per minute.

REFERENCES

1. R.C. Weast, ed., *CRC Handbook of Chemistry and Physics*, 51st ed., The Chemical Rubber Company, Cleveland, OH, 1971.

2. S. Maron and C. Prutton, *Principles of Physical Chemistry*, 4th ed., Macmillan, New York, 1970.

3. J.M. Smith and H.C. Van Ness, *Introduction to Chemical Engineering Thermodynamics*, 2nd ed., McGraw-Hill, New York, 1959.

4. R. Perry and D. Green, *Perry's Chemical Engineering Handlook*, 6th ed., McGraw-Hill, New York, 1984, p. 12-5.

5

Stoichiometric and Thermodynamic Considerations

Partial contribution: Mitchell Summerfield

5.1 INTRODUCTION

In order to understand completely the design as well as the operation and performance of hazardous waste incinerators, it is necessary first to understand the chemical theory underlying this technology. How can one predict what products will be emitted from effluent streams? At what temperature must the incinerator be operated to ensure compliance with the *four nines* (99.99% destruction and/or removal efficiency)? How much energy in the form of heat is given off during combustion? Is it economically feasible to recover this heat? Is the waste feed high enough in heating value, or must additional fuel be added to assist in the combustion process? If so, how much fuel must be added? The answers to these questions are rooted in the various theories of thermodynamics, thermochemistry, and chemical reaction equilibrium. Answers will be provided in the following sections of this chapter and Chapter 6.

5.2 THE CONSERVATION LAWS

Conservation of Mass

The *conservation law* for mass can be applied to any process or system. The general form of this law is given by Eq. (5.2.1).

$$\text{mass in} - \text{mass out} + \text{mass generated} = \text{mass accumulated}$$

or on a time rate basis by

$$\text{rate of mass in} - \text{rate of mass out} + \text{rate of mass generated} = \text{rate of mass accumulated}$$

$$(5.2.1)$$

This equation may be applied either to the total mass involved or to a particular species, on either a mole or mass basis. In incineration processes, it is often necessary to obtain quantitative relationships by writing mass balances on the various elements in the system. This law can be applied to steady-state or unsteady-state (transient) processes and to batch or continuous systems. In order to isolate a system for study, it is separated from the surroundings by a boundary or envelope. This boundary may be real (e.g., the walls of an incinerator) or imaginary. Mass crossing the boundary and entering the system is part of the *mass in* term in Eq. (5.2.1), while that crossing the boundary and leaving the system is part of the *mass out* term. Equation (5.2.1) may be written for any compound whose quantity is not changed by chemical reaction, and for any chemical element whether or not it has participated in a chemical reaction. It may be written for one piece of equipment, around several pieces of equipment, or around an entire process. It may be used to calculate an unknown quantity directly, to check the validity of experimental data, or to express one or more of the independent relationships among the unknown quantities in a particular problem situation.

A *steady-state* process is one in which there is no change in conditions (pressure, temperature, composition, etc.) or rates of flow with time at any given point in the system. The accumulation term in Eq. (5.2.1) is then zero. (If there is no chemical or nuclear reaction, the generation term is also zero.) All other processes are *unsteady state*. In a *batch* process, a given quantity of reactants is placed in a container, and by chemical and/or physical means, a change is made to occur. At the end of the process, the container (or adjacent containers to which material may have been transferred) holds the product or products. In a *continuous* process, reactants are continuously fed to a piece of equipment or to several pieces in series, and products are continuously removed from one or more points. A continuous process may or may not be steady-state. A coal-fired power plant, for example, operates continuously. However, because of the wide variation in power demand between peak and slack periods, there is an equally wide variation in the rate at which the coal is fired. For this reason, power plant problems may require the use of average data over long periods of time. Most hazardous waste incinerator operations are assumed to be steady-state and continuous.

As indicated previously, Eq. (5.2.1) may be applied to the total mass of each stream (referred to as an *overall* or *total material balance*) or to the individual component(s) of the streams (referred to as a *componential* or *component material balance*). Often the primary task in preparing a material balance in hazardous waste calculations is to develop the quantitative relationships among the streams. The primary factors, therefore, are those that *tie* the streams together. An element, compound, or unreactive mass (ash, for example) that enters or leaves in a single stream or passes through a process unchanged is so convenient for this purpose that it may be considered a *key* to the calculations. If sufficient data are given about this component, it can be used in a component balance to determine the total masses of the entering and exiting streams. Such a component is sometimes referred to as a *key component*. Since a key component does not react in a process, it must retain its identity as it passes through the process. Obviously, except for nuclear reactions, elements may always be used as key components because they do not change identity even though they may undergo a chemical reaction. Thus CO (carbon monoxide) may be used as a key component only when it does not react, but C (carbon) may always be used as a key component. A component that enters the system in only one

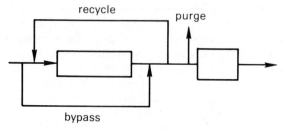

Figure 5.2.1. Recycle, bypass, and purge.

stream and leaves in only one stream is usually the most convenient choice for a key component.

Four important processing concepts are bypass, recycle, purge, and makeup. With *bypass*, part of the inlet stream is diverted around the equipment to rejoin the (main) stream after the unit (see Fig. 5.2.1). This stream effectively moves in parallel with the stream passing through the equipment. In *recycle*, part of the product stream is sent back to mix with the feed. If a small quantity of nonreactive material is present in the feed to a process that includes recycle, it may be necessary to remove the nonreactive material in a *purge* stream to prevent its building up above a maximum tolerable value. This can also occur in a process without recycle; if a nonreactive material is added in the feed and not totally removed in the products, it will accumulate until purged. The purging process is sometimes referred to as *blowdown*. *Makeup*, as its name implies, involves adding or making up part of a stream that has been removed from a process. Makeup may be throught of, in a final sense, as the opposite of purge and/or blowdown.

Conservation of Energy

A presentation of the conservation law for energy would be incomplete without a brief review of some introductory thermodynamic principles. *Thermodynamics* is defined as that science that deals with the relationships among the various forms of energy. A system may possess energy due to its

1. temperature,
2. velocity,
3. position,
4. molecular structure,
5. surface, and so on.

The energies corresponding to these states are

1. internal,
2. kinetic,
3. potential,
4. chemical, and
5. surface, and so on.

Engineering themodynamics is founded on three basic laws. Energy, like mass and momentum, is conserved. Application of the conservation law for energy gives rise to the first law of thermodynamics. This law, in steady-state equation form for batch and flow processes, is presented here.

For *batch* processes:

$$\Delta E = Q - W \tag{5.2.2}$$

For *flow* processes:

$$\Delta H = Q - W_s \tag{5.2.3}$$

where potential, kinetic, and other energy effects have been neglected and

Q = energy in the form of heat transferred across the boundaries of the system,

W = energy in the form of work transferred across the boundaries of the system,

W_s = energy in the form of mechanical work transferred across the boundaries of the system,

E = internal energy of the system,

H = enthalpy of the system, and

$\Delta E, \Delta H$ = changes in the internal energy and enthalpy, respectively, during the process.

The internal energy and enthalpy in Eq. (5.2.2) and (5.2.3), as well as the other equations in this section may be on a *mass* basis (i.e., for 1 g or 1 lb of material), on a *mole* basis (i.e., for 1 gmol or 1 lbmol of material), or represent the total internal energy and enthalpy of the entire system. As long as these equations are dimensionally consistent, it makes no difference. For the sake of clarity, the same convention that was used for the heat capacities will be employed throughout this text—uppercase letters (e.g., H, E, C_p) represent properties on a mole basis, while lower-case letters (e.g., h, e, c_P) represent properties on a *mass* basis. Properties for the entire system will rarely be used and therefore require no special symbols.

Perhaps the most important thermodynamic function the engineer works with is the *enthalpy*. The enthalpy is defined by the equation

$$H = E + PV$$

where P = pressure of the system
V = volume of the system

The terms E and H (and free energy G, to be discussed later) are *state* or *point* functions. By fixing a certain number of variables on which the function depends, this automatically fixes the numerical value of the function; that is, it is single valued (see Section 4.5). For example, fixing the temperature and pressure of a one-component single-phase system immediately specifies the enthalpy and internal energy.

The change in enthalpy as it undergoes a change in state from (T_1, P_1) to (T_2, P_2) is given by

$$\Delta H = H_2 - H_1 \tag{5.2.4}$$

Note that H and ΔH are independent of the path. This is a characteristic of all *state* or *point* functions, that is, the state of the system is independent of the path by which the state is reached. The terms Q, W, and W_s in Eqs. (5.2.2) and (5.2.3) are *path* functions; their values depend on the path used between the two states; unless a process or change of state is occurring, path functions have no value.

For a mathematical representation of this thermodynamic point function, the following can be written:

$$H = H(T, P)$$

By the rules of partial differentiation, a differential change in H is given by

$$dH = \left(\frac{\partial H}{\partial T}\right)_P dT + \left(\frac{\partial H}{\partial P}\right)_T dP \tag{5.2.5}$$

The term $(\partial H/\partial P)_T$ is assumed to be negligible in most engineering applications. It is exactly zero for an ideal gas and is small for solids and liquids, and gases near ambient conditions. The term $(\partial H/\partial T)_P$ is defined as the heat capacity at constant pressure (see Chapter 4 for additional details):

$$C_P = \left(\frac{\partial H}{\partial T}\right)_P \tag{5.2.6}$$

Equation (5.2.5) may now be written as

$$dH = C_P dT \tag{5.2.7}$$

If average molar heat capacity data are available, this equation may be integrated to yield

$$\Delta H = \overline{C_P} \Delta T \tag{5.2.8}$$

where $\overline{C_p}$ = average value of C_P in the temperature range ΔT

Average molar heat capacity data are provided in Table 5.2.1. Thus, the calculation of enthalpy change(s) associated with a temperature change may be accomplished though the application of either Eq. (5.2.4) or Eq. (5.2.8). Actual enthalpy data as provided in Table 5.2.2 must be available for use of Eq. (5.2.4). If the heat capacity is a function of temperature, for example,

$$C_P = \alpha + \beta T + \gamma T^2 \tag{5.2.9}$$

TABLE 5.2.1. Mean Molar Heat Capacities of Gases for the Temperature Range 0°F to T^a

$$C_P = \text{Btu/lbmol-°F.}$$

T (°F)	N_2	O_2	H_2O	CO_2	H_2	CO	CH_4	SO_2	NH_3	HCl	NO
0	6.94	6.92	7.93	8.50	6.86	6.92	8.25	9.9	8.80	6.92	7.1
200	6.96	7.03	8.04	9.00	6.89	6.96	8.42	10.0	8.85	6.96	7.2
400	6.98	7.14	8.13	9.52	6.93	7.00	9.33	10.3	9.05	7.01	7.2
600	7.02	7.26	8.25	9.97	6.95	7.05	10.00	10.6	9.40	7.05	7.3
800	7.08	7.39	8.39	10.37	6.97	7.13	10.72	10.9	9.75	7.10	7.3
1000	7.15	7.51	8.54	10.72	6.98	7.21	11.45	11.2	10.06	7.15	7.4
1200	7.23	7.62	8.69	11.02	7.01	7.30	12.13	11.4	10.43	7.19	7.5
1400	7.31	7.71	8.85	11.29	7.03	7.38	12.78	11.7	10.77	7.24	7.6
1600	7.39	7.80	9.01	11.53	7.07	7.47	13.38	11.8		7.29	7.7
1800	7.46	7.88	9.17	11.75	7.10	7.55		12.0		7.33	7.7
2000	7.53	7.96	9.33	11.94	7.15	7.62		12.1		7.38	7.8
2200	7.60	8.02	9.48	12.12	7.20	7.68		12.2		7.43	7.8
2400	7.66	8.08	9.64	12.28	7.24	7.75		12.3		7.47	7.9
2600	7.72	8.14	9.79	12.42	7.28	7.80		12.4		7.52	8.0
2800	7.78	8.19	9.93	12.55	7.33	7.86		12.5		7.57	8.0
3000	7.83	8.24	10.07	12.67	7.38	7.91		12.5		7.61	8.1
3200	7.87	8.29	10.20	12.79	7.43	7.95					
3400	7.92	8.34	10.32	12.89	7.48	8.00					
3600	7.96	8.38	10.44	12.98	7.53	8.04					
3800	8.00	8.42	10.56	13.08	7.57	8.08					
4000	8.04	8.46	10.67	13.16	7.62	8.11					
4200	8.07	8.50	10.78	13.23	7.66	8.14					
4400	8.10	8.54	10.88	13.31	7.70	8.18					
4600	8.13	8.58	10.97	13.38	7.75	8.20					
4800	8.16	8.62	11.08	13.44	7.79	8.23					

[a] From E.T. Williams, and R.C. Johnson, *Stoichiometry for Chemical Engineers*, McGraw-Hill, New York, 1958.

Equation (5.2.7) may be integrated directly between some reference or standard temperature (T_0) and the final temperature (T_1).

$$\Delta H = H_1 - H_0$$

$$= \alpha(T_1 - T_0) + \frac{\beta}{2}(T_1^2 - T_0^2) + \frac{\gamma}{3}(T_1^3 - T_0^3) \tag{5.2.10}$$

Equation (5.2.7) may also be integrated if the heat capacity as a function of temperature is of the form

$$C_P = a + bT + cT^{-2} \tag{5.2.11}$$

The enthalpy change is then given by

TABLE 5.2.2. Molar Enthalpies of Combustion Gases[a] in Btu/lbmol

T (°F)	N_2	Air (MW 28.97)	CO_2	H_2O
32	0	0	0	0
60	194.9	194.6	243.1	224.2
77	312.2	312.7	392.2	360.5
100	473.3	472.7	597.9	545.3
200	1,170	1,170	1,527	1,353
300	1,868	1,870	2,509	2,171
400	2,570	2,576	3,537	3,001
500	3,277	3,289	4,607	3,842
600	3,991	4,010	5,714	4,700
700	4,713	4,740	6,855	5,572
800	5,443	5,479	8,026	6,460
900	6,182	6,227	9,224	7,364
1,000	6,929	6,984	10,447	8,284
1,200	8,452	8,524	12,960	10,176
1,500	10,799	10,895	16,860	13,140
2,000	14,840	14,970	23,630	18,380
2,500	19,020	19,170	30,620	23,950
3,000	23,280	23,460	37,750	29,780

[a] K.A. Kobe, and E.G. Long, "Thermochemistry for the Petroleum Industry," *Pet. Refiner*, **28** (11), 127–132, (1949).

$$\Delta H = a(T_1 - T_0) + \frac{b}{2}(T_1^2 - T_0^2) - c\left(\frac{1}{T_1} - \frac{1}{T_0}\right) \qquad (5.2.12)$$

Tabulated values of α, β, γ and a, b, c for a host of compounds (including some chlorinated organics) are given in Tables 5.2.3 and 5.2.4, respectively. The reader is cautioned that the use of Eqs. (5.2.9)–(5.2.12) required the use of absolute temperature in kelvins.

Most hazardous waste incineration facilities operate in a steady-state flow mode with no significant mechanical or shaft work added (or withdrawn) from the system. For this condition, Eq. (5.2.3) reduces to

$$Q = \Delta H \qquad (5.2.13)$$

This equation is routinely used in many hazardous waste incineration calculations. If a unit or system is operated adiabatically, $Q = 0$ and Eq. (5.2.13) becomes

$$\Delta H = 0 \qquad (5.2.14)$$

Although the topics of material and energy balances have been covered separately in this section, it should be emphasized that this segregation does not exist in reality. Incineration is invariably accompanied by heat effects, and one must work with both energy and material balances simultaneously.

TABLE 5.2.3. Molar Heat Capacities[a,b]

Compound	Formula	α	$\beta \times 10^3$	$\gamma \times 10^6$
	Normal Paraffins (gases)			
Methane	CH_4	3.381	18.044	−4.300
Ethane	C_2H_6	2.247	38.201	−11.049
Propane	C_3H_8	2.410	57.195	−17.533
n-Butane	C_4H_{10}	3.844	73.350	−22.655
n-Pentane	C_5H_{12}	4.895	90.113	−28.039
n-Hexane	C_6H_{14}	6.011	106.746	−33.363
n-Heptane	C_7H_{16}	7.094	123.447	−38.719
n-Octane	C_8H_{18}	8.163	140.217	−44.127
Increment per C atom above C_8	—	1.097	16.667	−5.338
	Normal Monoolefins (gases) (1-alkenes)			
Ethylene	C_2H_4	2.830	28.601	−8.726
Propylene	C_3H_6	3.253	45.116	−13.740
1-Butene	C_4H_8	3.909	62.848	−19.617
1-Pentene	C_5H_{10}	5.347	78.990	−24.733
1-Hexene	C_6H_{12}	6.399	95.752	−30.116
1-Heptene	C_7H_{14}	7.488	112.440	−35.462
1-Octene	C_8H_{16}	8.592	129.076	−40.775
Increment per C atom above C_8	—	1.097	16.667	−5.338
	Miscellaneous Gases			
Acetaldehyde[c]	C_2H_4O	3.364	35.722	−12.236
Acetylene	C_2H_2	7.331	12.622	−3.889
Ammonia	NH_3	6.086	8.812	−1.506
Benzene	C_6H_6	−0.409	77.621	−26.429
1,3-Butadiene	C_4H_6	5.432	53.224	−17.649
Carbon dioxide	CO_2	6.214	10.396	−3.545
Carbon monoxide	CO	6.420	1.665	−0.196
Chlorine	Cl_2	7.576	2.424	−0.965
Cyclohexane	C_6H_{12}	−7.701	125.675	−41.584
Ethyl alcohol	C_3H_6O	6.990	39.741	−11.926
Hydrogen	H_2	6.947	−0.200	0.481
Hydrogen chloride	HCl	6.732	0.433	0.370
Hydrogen sulfide	H_2S	6.662	5.134	−0.854
Methyl alcohol	CH_4O	4.394	24.274	−6.855
Nitric oxide	NO	7.020	−0.370	2.546
Nitrogen	N_2	6.524	1.250	−0.001
Oxygen	O_2	6.148	3.102	−0.923
Sulfur dioxide	SO_2	7.116	9.512	3.511
Sulfur trioxide	SO_3	6.077	23.537	−0.687
Toluene	C_7H_8	0.576	93.493	−31.227
Water	H_2O	7.256	2.298	0.283

TABLE 5.2.3. (*Continued*)

Compound	Formula	α	$\beta \times 10^3$	$\gamma \times 10^6$
	Clorinated Organic Liquids			
Methyl chloride	CH_3Cl	8.994	10.280	-21.218
Dichloromethane	CH_2Cl_2	13.672	7.616	-33.272
Chloroform	$CHCl_3$	18.453	4.932	-37.923
Carbon tetrachloride	CCl_4	23.777	2.224	-39.558
Ethyl chloride	C_2H_5Cl	15.267	17.060	-48.798
1,1-Dichloroethane	$C_2H_4Cl_2$	20.014	14.315	-54.917
1,1,2,2-Tetrachloroethane	$C_2H_2Cl_4$	28.397	10.338	-66.739
n-Propyl chloride	C_3H_7Cl	20.703	24.070	-69.351
1,3-Dichloropropane	$C_3H_6Cl_2$	24.840	21.810	-68.214
n-Butyl chloride	C_4H_9Cl	26.258	31.208	-89.482
1-Chloropentane	$C_5H_{11}Cl$	31.812	38.345	-109.612
1-Chloroethylene	C_2H_3Cl	14.058	10.939	-40.586
trans-1,2-Dichloroethylene	$C_2H_6Cl_2$	18.050	8.845	-42.981
Trichloroethylene	C_2HCl_3	22.410	6.394	-46.257
Tetrachloroethylene	C_2Cl_4	26.595	4.106	-46.208
3-Chloro-1-propene	C_3H_5Cl	18.728	18.514	-56.495
Chlorobenzene	C_6H_5Cl	28.558	25.786	-116.150
p-Dichlorobenzene	$C_6H_5Cl_2$	33.265	23.025	-116.848
Hexachlorobenzene	C_6Cl_6	50.759	13.286	-115.620

[a] Selected mainly from values given by H.M. Spencer, J. Justice, and G. Flanagan, *J. Am. Chem. Soc.*, **56**, 2311, 1934; **64**, 2511, 1942; **67**, 1859, 1945; *Ind. Eng. Chem.* **40**, 2152, 1948; also from personal notes: L. Theodore and J. Reynolds, 1986.

[b] Constants for the equation $C_P = \alpha + \beta T + \gamma T^2$, where T is in K and C_P is in Btu/lbmol-°F or cal/gmol-°C.

[c] Applicable range 298–1000 K.

5.3 STOICHIOMETRY

The term *stoichiometry* has come to mean different things to different people. In a loose sense, stoichiometry involves the balancing of an equation for a chemical reaction that provides a quantitative relationship among the reactants and products. In the simplest stoichiometric situation, exact quantities of pure reactants are available, and these quantities react completely to give the desired product(s). In an industrial process, the reactants usually are not pure, one reactant is usually in excess of what is needed for the reaction, and the desired reaction may not go to completion because of a host of other considerations. For example, complete combustion of pure hydrocarbons yields carbon dioxide and water as the reaction products. Consider the combustion of methane in oxygen:

$$CH_4 + O_2 \longrightarrow CO_2 + H_2O \tag{5.3.1}$$

In order to balance this reaction, *two* molecules of oxygen are needed. This requires that there be *four* oxygen atoms on the right side of the reaction. This is satisfied by introducing *two* molecules of water as product. The final balanced reaction becomes

TABLE 5.2.4. Molar Heat Capacities^{a,b}

Compound	Formula	Temperature Range (K)	a	$b \times 10^3$	$c \times 10^{-5}$
Inorganic Gases					
Ammonia	NH_3	298–1800	7.11	6.00	−0.37
Bromine	Br_2	298–3000	8.92	0.12	−0.30
Carbon monoxide	CO	298–2500	6.79	0.98	−0.11
Carbon dioxide	CO_2	298–2500	10.57	2.10	−2.06
Carbon disulfide	CS_2	298–1800	12.45	1.60	−1.80
Chlorine	Cl_2	298–3000	8.85	0.16	−0.68
Hydrogen	H_2	298–3000	6.52	0.78	+0.12
Hydrogen sulfide	H_2S	298–2300	7.81	2.96	−0.46
Hydrogen chloride	HCl	298–2000	6.27	1.24	+0.30
Hydrogen cyanide	HCN	298–2500	9.41	2.70	−1.44
Nitrogen	N_2	298–3000	6.83	0.90	−0.12
Nitrous oxide	N_2O	298–2000	10.92	2.06	−2.04
Nitric oxide	NO	298–2500	7.03	0.92	−0.14
Nitrogen dioxide	NO_2	298–2000	10.07	2.28	−1.67
Nitrogen tetroxide	N_2O_4	298–1000	20.05	9.50	−3.56
Oxygen	O_2	298–3000	7.16	1.00	−0.40
Sulfur dioxide	SO_2	298–2000	11.04	1.88	−1.84
Sulfur trioxide	SO_3	298–1500	13.90	6.10	−3.22
Water	H_2O	298–2750	7.30	2.46	0.00
Chlorinated Organic Liquids					
Methyl chloride	CH_3Cl	298–1800	3.44	23.52	−0.81
Dichloromethane	CH_2Cl_2	298–1800	5.34	27.00	−1.16
Chloroform	$CHCl_3$	298–1800	8.91	27.18	−1.34
Carbon tetrachloride	CCl_4	298–1800	1.39	24.97	−1.36
Ethyl chloride	C_2H_5Cl	298–1800	2.69	46.81	−1.81
1,1-Dichloroethane	$C_2H_4Cl_2$	298–1800	5.99	47.31	−2.00
1,1,2,2-Tetrachloroethane	$C_2H_2Cl_4$	298–1800	1.15	49.83	−2.38
n-Propyl chloride	C_3H_7Cl	298–1800	2.88	66.18	−2.56
1,3-Dichloropropane	$C_3H_6Cl_2$	298–1800	7.35	63.07	−2.51
n-Butyl chloride	C_4H_9Cl	298–1800	3.22	85.65	−3.32
1-Chloropentane	$C_5H_{11}Cl$	298–1800	3.57	10.51	−4.07
1-Chloroethylene	C_2H_3Cl	298–1800	3.70	35.30	−1.48
trans-1,2-Dichloroethylene	$C_2H_6Cl_2$	298–1800	7.13	34.46	−1.55
Trichloroethylene	C_2HCl_3	298–1800	10.76	33.59	−1.64
Tetrachloroethylene	C_2Cl_4	298–1800	14.99	31.13	−1.63
3-Chloro-1-propene	C_3H_5Cl	298–1800	4.26	52.61	−2.07
Chlorobenzene	C_6H_5Cl	298–1800	−1.12	95.63	−4.24
p-Dichlorobenzene	$C_6H_5Cl_2$	298–1800	3.49	92.98	−4.24
Hexachlorobenzene	C_6Cl_6	298–1800	21.53	81.64	−4.13

^{a} Selected from K. Kelley, *U.S. Bur. Mines Bull.* **584** (1960); also from personal notes, L. Theodore and J. Reynolds, 1986.
^{b} Constants for the equation $C_P = a + bT + cT^{-2}$, where T is in K and C_P is in Btu/lbmol-°F or cal/gmol-°C.

$$CH_4 + 2O_2 \longrightarrow CO_2 + 2H_2O \tag{5.3.2}$$

Thus, *two* molecules (or moles) of oxygen are required to completely combust *one* molecule (or mole) of methane to yield one molecule (or mole) of carbon dioxide and two molecules (or moles) of water. Note that the numbers of carbon, oxygen, and hydrogen atoms on the right-hand side of this reaction are equal to those on the left-hand side. The reader should verify that the total mass (obtained by multiplying the number of each molecule by its molecular weight and summing) on each side of the reaction is the same. [The term *moles* here may refer to either gram-moles (gmol) or pound-moles (lbmol); it makes no difference.]

The terms used to describe a reaction that does not involve stoichiometric ratios of reactants must be carefully defined in order to avoid confusion. If the reactants are not present in formula or stoichiometric ratio, one reactant is said to be *limiting*; the others are said to be *in excess*. Consider the reaction

$$CO + \tfrac{1}{2}O_2 \longrightarrow CO_2 \tag{5.3.3}$$

If the starting amounts are 1 mol of CO and 3 mol of oxygen, CO is the *limiting* reactant, with O_2 present in excess. There are 2.5 mol of excess O_2, because only 0.5 mol is required to combine with the CO. Thus there is 500% excess oxygen present. The percentage of excess must be defined in relation to the amount of the reactant necessary to react completely with the limiting reactant. Thus, if for some reason only part of the CO actually reacts, this does not alter the fact that the oxygen is in excess by 500%. However, there are often several possible products. For instance, the reactions

$$C + O_2 \longrightarrow CO_2$$

and

$$C + \tfrac{1}{2}O_2 \longrightarrow CO \tag{5.3.4}$$

can occur simultaneously. In this case, if there are 3 mol of oxygen present per mole of carbon, the oxygen is in excess. The extent of this excess, however, cannot be definitely fixed. It is customary to choose one product (e.g., the desired one) and specify the excess reactant in terms of this product. For this case, there is 200% excess oxygen for the reaction going to CO_2, and there is 500% excess oxygen for the reaction going to CO. The discussion on excess oxygen can be extended to excess air using the same approach. Stoichiometric or theoretical oxygen (or air) is defined as 0% excess oxygen (or air). This is an important concept since hazardous waste incinerators operate with excess air. For example, approximately 25 to 50% excess air is employed with liquid injection incinerators. Minimum excess air requirements for rotary kilns are approximately 100% if only bulk solids are burned and 150% if containerized wastes are incinerated.

Consider the combustion of 1 mol of ethane. The reaction for complete combustion may be written

$$C_2H_6 + 3\tfrac{1}{2}O_2 \longrightarrow 2CO_2 + 3H_2O \tag{5.3.5}$$

Thus, 2 mol of CO_2 and 3 mol of H_2O will be formed from the complete combustion of 1 mol of C_2H_6. The oxygen required is 3.5 mol. If 60% excess is used, this means an additional 2.1 mol, or a total of 5.6 mol of oxygen is required. Equation (5.3.5) describes a gas phase reaction. In accordance with Charles' Law, one may also interpret this equation as follows: when 1 ft^3 of C_2H_6 reacts with 3.5 ft^3 of O_2, 2 ft^3 of CO_2 and 3 ft^3 of H_2O will form. Similar balanced stoichiometric reactions may be written for other hydrocarbons and organics to determine oxygen (or air) requirements and the products of combustion. However, this somewhat tedious calculation may be bypassed by use of Table 5.3.1. The reader should carefully review this table before proceeding to the illustrative examples at the end of this chapter.

This discussion may now be extended to reaction systems that involve the combustion of carbon and/or carbonaceous hazardous wastes and fuels. To simplify matters, it will be assumed that both the air and waste–fuel mixture are dry. Throughout this text, air is assumed to contain 21% O_2 by volume. This is perhaps a bit high by a few hundredths of a percent. The remaining 79% is assumed to be inert and to consist of nitrogen (and a trace of the noble gases). The usual incineration products of hydrocarbons are CO_2 and H_2O. The combustion products of other organics may contain additional compounds; for example, an organic chloride will also produce chlorine and/or hydrochloric acid. Combustion products from certain fuels will yield sulfur dioxide and nitrogen. So-called complete combustion of an organic or a fuel involves conversion of all the elemental carbon to carbon dioxide, hydrogen to water, sulfur to sulfur dioxide, and nitrogen to its elemental form of N_2. Thus, the *theoretical* or *stoichiometric* oxygen described for a combustion reaction is the amount of oxygen required to burn all the carbon to carbon dioxide, all the hydrogen to water, and so on; excess oxygen (or excess air) is the oxygen furnished in excess of the theoretical oxygen required for combustion. Since combustion calculations assume dry air to contain 21% oxygen and 79% nitrogen on a mole or volume basis, 4.76 mol of air consists of 1.0 mol of oxygen and 3.76 mol of nitrogen. The temperature of the products that results from complete combustion under adiabatic conditions is defined as the *adiabatic flame temperature*; this will be discussed in more detail in Section 5.4.

For the combustion of 1.0 lbmol of chlorobenzene (C_6H_5Cl) in stoichiometric oxygen, the balanced reaction is

$$C_6H_5Cl + 7O_2 \longrightarrow 6CO_2 + 2H_2O + HCl \qquad (5.3.6)$$

Since air, not oxygen, is employed in incineration processes, this reaction with air becomes

$$C_6H_5Cl + 7O_2 + 26.3N_2 \longrightarrow 6CO_2 + 2H_2O + HCl + 26.3N_2 \qquad (5.3.7)$$

where the nitrogen in the air has been retained on both sides of the equation since it does not participate in the combustion reaction. The moles and masses involved in this reaction, based on the stoichiometric combustion of 1.0 lbmol of C_6H_5Cl, are given here:

TABLE 5.3.1. Combustion Constants[a]

Compound	Density lb/ft³ᶜ	Specific Volume ft³/lbᶜ	Heat of Combustion (Btu/ft³)ᶜ Gross (high)	Net (low)	Heat of Combustion (Btu/lb) Gross (high)	Net (low)	For 100% Total Air (mole/mole or ft³/ft³ of combustible) Required for Combustion O₂	N₂	Air	Flue Products CO₂	H₂O	N₂	For 100% Total Air (lb/lb of combustible) Required for Combustion O₂	N₂	Air	Flue Products CO₂	H₂O	N₂	Flammability Limits (% by volume) Lower	Upper
Carbon, Cᵇ	—	—	—	—	14,093	14,093	1.0	3.76	4.76	1.0	—	3.76	2.66	8.86	11.53	3.66	—	8.86	—	—
Hydrogen, H₂	0.0053	187.723	325	275	61,100	51,623	0.5	1.88	2.38	—	1.0	1.88	7.94	26.41	34.34	—	8.94	26.41	4.00	74.20
Oxygen, O₂	0.0846	11.819	—	—	—	—	—	—	—	—	—	—	—	—	—	—	—	—	—	—
Nitrogen (atm), N₂	0.0744	13.443	—	—	—	—	—	—	—	—	—	—	—	—	—	—	—	—	—	—
Carbon monoxide, CO	0.0740	13.506	322	322	4,347	4,347	0.5	1.88	2.38	1.0	—	1.88	0.57	1.90	2.47	1.57	—	1.90	12.50	74.20
Carbon dioxide, CO₂	0.1170	8.548	—	—	—	—	—	—	—	—	—	—	—	—	—	—	—	—	—	—
Paraffin Series																				
Methane, CH₄	0.0424	23.565	1013	913	23,879	21,520	2.0	7.53	9.53	1.0	2.0	7.53	3.99	13.28	17.27	2.74	2.25	13.28	5.00	15.00
Ethane, C₂H₆	0.0803	12.455	1792	1641	22,320	20,432	3.5	13.18	16.68	2.0	3.0	13.18	3.73	12.39	16.12	2.93	1.80	12.39	3.00	12.50
Propane, C₃H₈	0.1196	8.365	2590	2385	21,661	19,944	5.0	18.82	23.82	3.0	4.0	18.82	3.63	12.07	15.70	2.99	1.68	12.07	2.12	9.35
n-Butane, C₄H₁₀	0.1582	6.321	3370	3113	21,308	19,680	6.5	24.47	30.97	4.0	5.0	24.47	3.58	11.91	15.49	3.03	1.55	11.91	1.86	8.41
Isobutane, C₄H₁₀	0.1582	6.321	3363	3105	21,257	19,629	6.5	24.47	30.97	4.0	5.0	24.47	3.58	11.91	15.49	3.03	1.55	11.91	1.80	8.44
n-Pentane, C₅H₁₂	0.1904	5.252	4016	3709	21,091	19,517	8.0	30.11	38.11	5.0	6.0	30.11	3.55	11.81	15.35	3.05	1.50	11.81	1.80	—
Isopentane, C₅H₁₂	0.1904	5.252	4008	3716	21,052	19,478	8.0	30.11	38.11	5.0	6.0	30.11	3.55	11.81	15.35	3.05	1.50	11.81	—	—
Neopentane, C₅H₁₂	0.1904	5.252	3993	3693	20,970	19,396	8.0	30.11	38.11	5.0	6.0	30.11	3.55	11.81	15.35	3.05	1.50	11.81	—	—
n-Hexane, C₆H₁₄	0.2274	4.398	4762	4412	20,940	19,403	9.5	35.76	45.26	6.0	7.0	35.76	3.53	11.74	15.27	3.06	1.46	11.74	1.18	7.40

Olefin Series

Ethylene, C_2H_4	0.0746	13.412	1614	1513	21,644	20,295	3.0	2.0	2.0	11.29	3.42	11.39	14.81	3.14	1.29	11.39	2.75	28.60
Propylene, C_3H_6	0.1110	9.007	2336	2186	21,041	19,691	4.5	3.0	3.0	16.94	3.42	11.39	14.81	3.14	1.29	11.39	2.00	11.10
1-Butene, C_4H_8	0.1480	6.756	3084	2885	20,840	19,496	6.0	4.0	4.0	22.59	3.42	11.39	14.81	3.14	1.29	11.39	1.75	9.70
Isobutene, C_4H_8	0.1480	6.756	3068	2869	20,730	19,382	6.0	4.0	4.0	22.59	3.42	11.39	14.81	3.14	1.29	11.39	—	—
1-Pentene, C_5H_{10}	0.1852	5.400	3836	3586	20,712	19,363	7.5	5.0	5.0	28.23	3.42	11.39	14.81	3.14	1.29	11.39	—	—

Aromatic Series

Benzene, C_6H_6	0.2060	4.852	3751	3601	18,210	17,480	7.5	6.0	3.0	28.23	3.07	10.22	13.50	3.38	0.69	10.22	1.40	7.10
Toluene, C_7H_8	0.2431	4.113	4484	4284	18,440	17,620	9.0	7.0	4.0	33.88	3.13	10.40	13.53	3.34	0.78	10.40	1.27	6.75
Xylene, C_8H_{10}	0.2803	3.567	5230	4980	18,650	17,760	10.5	8.0	5.0	39.52	3.17	10.53	13.70	3.32	0.85	10.53	1.00	6.00

Miscellaneous Gases

Acetylene, C_2H_2	0.0697	14.344	1499	1448	21,500	20,776	2.5	2.0	1.0	9.41	3.07	10.22	13.30	3.38	0.69	10.22	—	—
Napthalene, $C_{10}H_8$	0.3384	2.955	5854	5654	17,298	16,708	12.0	10.0	4.0	45.17	3.00	9.97	12.96	3.43	0.56	9.97	—	—
Methyl alcohol, CH_3OH	0.0846	11.820	868	768	10,259	9,078	1.5	1.0	2.0	5.65	1.50	4.98	6.48	1.37	1.13	4.98	6.72	36.50
Ethyl alcohol, C_2H_5OH	0.1216	8.221	1600	1451	13,161	11,929	3.0	2.0	3.0	11.29	2.08	6.93	9.02	1.92	1.17	6.93	3.28	18.95
Ammonia, NH_3	0.0456	21.914	441	365	9,668	8,001	0.75	1.5	1.5	3.32	1.41	4.69	6.10	—	1.59	4.69	15.50	27.00
Sulfur, S[b]	—	—	—	—	3,983	3,983	1.0 (SO2)	1.0	—	3.76	1.00	3.29	4.29	2.00 (SO2)	0.53	3.29	—	—
Hydrogen sulfide, H_2S	0.0911	10.979	647	596	7,100	6,545	1.5	1.0	1.0	5.65	1.41	4.69	6.10	1.88	0.53	4.69	4.30	45.50
Sulfur dioxide, SO_2	0.1733	5.770	—	—	—	—	—	—	—	—	—	—	—	—	—	—	—	—
Water vapor, H_2O	0.0476	21.017	—	—	—	—	—	—	—	—	—	—	—	—	—	—	—	—
Air	0.0766	13.063	—	—	—	—	—	—	—	—	—	—	—	—	—	—	—	—
Gasoline	—	—	—	—	—	—	—	—	—	—	—	—	—	—	—	—	1.40	7.60

[a] Adapted from "Fuel Flue Gases," *Combustion Flame and Explosions of Gases*, American Gas Association, New York, NY, 1951.

[b] Carbon and sulfur are considered as gases for molal calculations only.

[c] All gas volumes corrected to 60°F and 30 in. Hg dry.

$$C_6H_5Cl + 7O_2 + 26.3N_2 \longrightarrow 6CO_2 + 2H_2O + HCl + 26.3N_2$$

moles	1	7	26.3	6	2	1	26.3	
MW	112.5	32	28	44	18	36.5	28	
mass	112.5	224	736.4	264	36	36.5	736.4	(5.3.8)

initial mass = 1072.9; initial number of moles = 34.3

final mass   = 1072.9; final number of moles   = 35.3

Note that, in accordance with the conservation law for mass, the initial and final masses balance. The number of moles, as is typically the case in incinerator calculations, do not balance. The concentrations of the various species may also be calculated. For example,

$$\%CO_2 \text{ by weight} = (264/1072.9)\ 100\% = 24.61\%$$

$$\%CO_2 \text{ by mol (or volume)} = (6/35.3)\ 100\% = 17.0\%$$

$$\%CO_2 \text{ by weight (dry basis)} = (264/1036.9)\ 100\% = 25.46\%$$

$$\%CO_2 \text{ by mol (dry basis)} = (6/33.3)\ 100\% = 18.0\%$$

The air requirement for this reaction is 33.3 lbmol. This is stoichiometric or 0% excess air (EA). For 100% EA (100% above stoichiometric), one would use (33.3) (2.0) or 66.6 lbmol of air. For 50% EA one would use (33.3)(1.5) or 50 lbmol of air; for this condition 16.7 lbmol excess or additional air is employed.

If 100% excess air is employed in the incineration of 1.0 lbmol of C_6H_5Cl, the combustion reaction would become

$$C_6H_5Cl + 14O_2 + 52.6N_2 \longrightarrow 6CO_2 + 2H_2O + HCl + 7O_2 + 52.6N_2 \qquad (5.3.9)$$

The reader is left with the exercise of verifying these results:

$$\text{Total final number of moles} = 68.6$$

$$\%O_2 \text{ by mol (or volume)} = (7/68.6)\ 100\% = 10.2\%$$

$$\%HCl \text{ by mol} = (1/68.6)\ 100\% = 1.46\%$$

$$\%H_2O \text{ by mol} = (2/68.6)\ 100\% = 2.92\%$$

Since incinerators are operated at essentially atmospheric pressure, then

$$\text{partial pressure } O_2 = 0.102 \text{ atm}$$

$$\text{partial pressure } HCl = 0.0146 \text{ atm}$$

$$\text{partial pressure } H_2O = 0.0292 \text{ atm}$$

If the C_6H_5Cl waste contains 0.5% sulfur (S) by mass, then

$$\text{weight of } S = (0.005)(112.5) = 0.5625 \text{ lbs}$$

$$\text{number of lbmol of } S = 0.5625/32 = 0.0176 \text{ lbmol}$$

$$\text{number of lbmol of } SO_2 \text{ formed} = \text{number of lbmol of } S = 0.0176 \text{ lbmol}$$

For this condition (approximately)

$$\% SO_2 \text{ by mol} = (0.0176/68.6) \; 100\% = 0.0257\%$$

$$\text{partial pressure } SO_2 = 2.57 \times 10^{-4} \text{ atm}$$

5.4 THERMOCHEMISTRY

Consider now the energy effects associated with a chemical reaction. To introduce this subject, the reader is reminded that the engineer and applied scientist are rarely concerned with the *magnitude* or *amount* of the energy in a system; the primary concern is with *changes* in the amount of energy. In measuring energy changes for systems, the enthalpy (H) has been found to be the most convenient term to work with. There are many different types of enthalpy effects; these include:

Sensible (temperature).
Latent (phase).
Reaction (chemical).
Dilution (with water), for example, HCl with H_2O.
Solution (nonaqueous), for example, HCl with a solvent other than H_2O.

The *sensible* enthalpy change was reviewed in Section 5.2. The *latent* enthalpy change finds application in hazardous waste incineration calculations in determining the heat (enthalpy) of condensation or vaporization of water. Steam tables (or the equivalent) are usually employed for this determination. The *dilution* and *solution* enthalpy effects are often significant in some industrial absorber calculations, but may safely be neglected in hazardous waste incineration calculations. The *heat of reaction* is defined as the enthalpy change of a system undergoing chemical reaction. If the reactants and products are at the same temperature and in their standard states, the heat of reaction is termed the *standard heat of reaction*. For engineering purposes, the *standard state* of a chemical may be taken as the *pure* chemical at 1-atm pressure. A superscript zero is often employed to identify a standard heat of reaction, for example, $\Delta H°$. A T subscript ($\Delta H_T°$) is sometimes used to indicate the temperature; standard heat of reaction data are meaningless unless the temperatures are specified. $\Delta H_{298}°$ data (i.e., for 298 K or 25°C) for many reactions are available in the literature.[1,2]

As described earlier, the first law of thermodynamics provides that, in a steady flow process with no mechanical work,

$$Q = \Delta H \tag{5.4.1}$$

Since enthalpy is a point function, it is independent of the path for any process. If heat of reaction determinations are made in a flow reactor, the energy (in the form of heat) transferred across the reactor boundary or surface is exactly equal to the heat of reaction. This is not the case for batch or nonflow systems. For this reason, *heat of reaction* is a misleading term. More recently it has been referred to as the *enthalpy of reaction*, although a more accurate term would be *enthalpy change* of reaction.

The *heat of formation* (ΔH_f) is defined as the enthalpy change occurring during a chemical reaction where 1 mol of a product is formed from its elements. The *standard* heat of formation (ΔH_f°) is applied to formation reactions that occur at constant temperature with each element and the product in its standard state.

Consider the formation reaction for CO_2 at standard conditions at 25°C.

$$C + O_2 \xrightarrow{\Delta H_{f\,298}^\circ} CO_2$$
$$\text{Enthalpy: } 0 \quad 0 \qquad H_{CO_2}$$

(5.4.2)

Once again, this reaction reads; "1 mol of carbon and 1 mol of oxygen react to form 1 mol of carbon dioxide." The enthalpies of reactants and products are printed below the symbols in the reaction, and the enthalpy change for the reaction, ΔH_f°, placed above the arrow. Note that the enthalpies of elements in their standard states (pure, 1 atm) at 25°C have arbitrarily been set to zero. The enthalpy change accompanying this reaction is the *standard heat of formation* and is given by

$$\Delta H_{f\,298}^\circ = H_{CO_2} - (H_C^0 + H_{O_2}^0) = H_{CO_2}$$

(5.4.3)

The *standard heat of combustion* at temperature T is defined as the enthalpy change during a chemical reaction where 1 mol of material is burned in oxygen, where all reactants and products are in their standard states. This quantity finds extensive application in calculating enthalpy changes for incineration reactions, and is often given in the literature for 60°F (16°C). Although much of the literature data on standard heats of reaction are given for 25°C (76°F), there is little sensible enthalpy change between these two temperatures and the two sets of data may be considered compatible.

Note that this *formation* reaction for CO_2 is a *combustion* reaction; the heat of formation, in this case, is therefore equal to the heat of combustion. Since a combustion reaction is one type of chemical reaction, the development to follow will concentrate on chemical reactions in general.

Chemical (stoichiometric) equations may be combined by addition or subtraction. The standard heat of reaction (ΔH°) associated with each equation may likewise be combined to give the standard heat of reaction associated with the resulting chemical equation. This is possible, once again, because enthalpy is a point function, and these changes are independent of path. In particular, formation equations and standard heats of formation may always be combined by addition and subtraction to produce any desired equation and its accompanying standard heat of reaction. This desired equation cannot itself be a formation equation. Thus, the enthalpy change for a chemical reaction is the same whether it takes place in one or several steps. This is referred to as the *law of constant enthalpy summation* and is a direct consequence of the first law of thermodynamics.

Consider the general reaction

$$aA + bB + \cdots \longrightarrow cC + dD + \cdots \qquad (5.4.4)$$

where $\quad$ A, B = formulas for the reactants (r)
$\qquad$ C, D = formulas for the products (p)
$\quad$ a, b, c, d = the stoichiometric coefficients of the balanced reaction

To simplify the presentation that follows, Eq. (5.4.4) is shortened to

$$aA + bB \longrightarrow cC + dD \qquad (5.4.5)$$

(Although this presentation plus those in later sections of this chapter and Chapter 6 will deal with the hypothetical species, A, B, C, etc., application to real systems can be found in the illustrative examples at the end of this and Chapter 6.) This reaction reads: "a mol of A react with b mol of B to form c mol of C and d mol of D." The heat of reaction for this chemical change is given by

$$\Delta H^{\circ} = cH_C^{\circ} + dH_D^{\circ} - aH_A^{\circ} - bH_B^{\circ} \qquad (5.4.6)$$

where $\quad$ H_C° = enthalpy of C in its standard state
$\quad$ C, D, A, B = subscripts indicating chemical species C, D, A, and B

If the temperature is 25°C, the enthalpies of the elements at standard state are, by convention, equal to zero. Therefore,

$$(\Delta H_f^{\circ})_{i\,298} = H_{i\,298}^{\circ} \qquad (5.4.7)$$

Substituting Eq. (5.4.7) for each component in Eq. (5.4.6) yields

$$\Delta H_{298}^{\circ} = c(\Delta H_f^{\circ})_C + d(\Delta H_f^{\circ})_D - a(\Delta H_f^{\circ})_A - b(\Delta H_f^{\circ})_B$$

or

$$\Delta H_{298}^{\circ} = \sum_p n_p(\Delta H_f^{\circ})_p - \sum_r n_r(\Delta H_f^{\circ})_r \qquad (5.4.8)$$

where $\quad$ p = products
$\qquad$ r = reactants
$\quad$ n_p, n_r = coefficients from the chemical equation

The standard heat of a reaction is obtained by taking the difference between the standard heat of formation of the products and that of the reactants. If the standard heat of reaction or formation is negative (*exothermic*), as is the case with most incineration reactions, then energy is liberated as a result of the reaction. Energy is absorbed if ΔH is positive (*endothermic*). Standard heat of formation and standarrd enthalpy of combustion data at 25°C are provided in Table 5.4.1. Both of these heat (or enthalpy) effects find extensive application in incinerator calculations.

The reader is cautioned on the use of the published heat of combustion data in the literature for chlorinated organics. It appears that the laboratory combustion

TABLE 5.4.1. Standard Heats of Formation and Combustion at 25°C in Calories per Gram-Mole[a,b]

Compound	Formula	State	ΔH°_{f298}	$-\Delta H^\circ_{c298}$
Normal Paraffins				
Methane	CH_4	g	−17,889	212,800
Ethane	C_2H_6	g	−20,236	372,820
Propane	C_2H_8	g	−24,820	530,600
n-Butane	C_4H_{10}	g	−30,150	687,640
n-Pentane	C_5H_{12}	g	−35,000	845,160
n-Hexane	C_6H_{14}	g	−39,960	1,002,570
Increment per C atom above C_6	—	g	−4,925	157,440
Normal Monoolefins (1-alkenes)				
Ethylene	C_2H_4	g	12,496	337,230
Propylene	C_3H_6	g	4,879	491,990
1-Butene	C_4H_8	g	−30	649,450
1-Pentene	C_5H_{10}	g	−5,000	806,850
1-Hexene	C_6H_{12}	g	−9,960	964,260
Increment per C atom above C_6	—	g	−4,925	157,440
Miscellaneous Organic Compounds				
Acetaldehyde	C_2H_4O	g	−39,760	
Acetic acid	$C_2H_4O_2$	l	−116,400	
Acetylene	C_2H_2	g	54,194	310,620
Benzene	C_6H_6	g	19,820	789,080
Benzene	C_6H_6	l	11,720	780,980
1,3-Butadiene	C_4H_6	g	26,330	607,490
Cyclohexane	C_6H_{12}	g	−29,430	944,790
Cyclohexane	C_6H_{12}	l	−37,340	936,880
Ethanol	C_2H_6O	g	−56,240	
Ethanol	C_2H_6O	l	−66,356	
Ethylbenzene	C_8H_{10}	g	7,120	1,101,120
Ethylene glycol	$C_2H_6O_2$	l	108,580	
Ethylene oxide	C_2H_4O	g	−12,190	
Methanol	CH_4O	g	−48,100	
Methanol	CH_4O	l	−57,036	
Methylcyclohexane	C_7H_{14}	g	−36,990	1,099,590
Methylcyclohexane	C_7H_{14}	l	−45,450	1,091,130
Styrene	C_8H_8	g	35,220	1,060,900
Toluene	C_7H_8	g	11,950	943,580
Toluene	C_7H_8	l	2,870	934,500
Miscellaneous Inorganic Compounds				
Ammonia	NH_3	g	−11,040	
Calcium carbide	CaC_2	s	−15,000	
Calcium carbonate	$CaCO_3$	s	−288,450	
Calcium chloride	$CaCl_2$	s	−190,000	

TABLE 5.4.1. (*Continued*)

Compound	Formula	State	ΔH°_{f298}	$-\Delta H^\circ_{c298}$
Calcium chloride	$CaCl_2 \cdot 6H_2O$	s	$-623,150$	
Calcium hydroxide	$Ca(OH)_2$	s	$-235,800$	
Calcium oxide	CaO	s	$-151,900$	
Carbon	C	Graphite	—	$94,052$
Carbon dioxide	CO_2	g	$-94,052$	
Carbon monoxide	CO	g	$-26,416$	$67,636$
Hydrochloric acid	HCl	g	$-22,063$	
Hydrogen	H_2	g	—	$68,317$
Hydrogen sulfide	H_2S	g	$-4,815$	
Iron oxide	FeO	s	$-64,300$	
Iron oxide	Fe_3O_4	s	$-267,000$	
Iron oxide	Fe_2O_3	s	$-196,500$	
Iron sulfide	FeS_2	s	$-42,520$	
Lithium chloride	$LiCl$	s	$-97,700$	
Lithium chloride	$LiCl \cdot H_2O$	s	$-170,310$	
Lithium chloride	$LiCl \cdot 2H_2O$	s	$-242,100$	
Lithium chloride	$LiCl \cdot 3H_2O$	s	$-313,500$	
Nitric acid	HNO_3	l	$-41,404$	
Nitrogen oxides	NO	g	$21,600$	
	NO_2	g	$8,041$	
	N_2O	g	$19,490$	
	N_2O_4	g	$2,309$	
Sodium carbonate	Na_2CO_3	s	$-270,300$	
Sodium carbonate	$Na_2CO_3 \cdot 10H_2O$	s	$-975,600$	
Sodium chloride	$NaCl$	s	$-98,232$	
Sodium hydroxide	$NaOH$	s	$101,990$	
Sulfur dioxide	SO_2	g	$-70,960$	
Sulfur trioxide	SO_3	g	$-94,450$	
Sulfur trioxide	SO_3	l	$-104,800$	
Sulfuric acid	H_2SO_4	l	$-193,910$	
Water	H_2O	g	$-57,798$	
Water	H_2O	l	$-68,317$	

Chlorinated Organic Compounds

Compound	Formula	State	ΔH°_{f298}	$-\Delta H^\circ_{c298}$
Methyl chloride	CH_3Cl	l	$-20,630$	
Dichloromethane	CH_2Cl_2	l	$-22,800$	
Chloroform	$CHCl_3$	l	$-24,200$	$36,900$
Carbon tetrachloride	CCl_4	l	$-24,000$	
Ethyl chloride	C_2H_5Cl	l	$-26,700$	
1,1-Dichloroethane	$C_2H_4Cl_2$	l	$-31,050$	
1,1,2,2-Tetrachloroethane	$C_2H_2Cl_4$	l	$-36,500$	
n-Propyl chloride	C_3H_7Cl	l	$-31,100$	
1,3-Dichloropropane	$C_3H_6Cl_2$	l	$-38,600$	
n-Butyl chloride	C_4H_9Cl	l	$-35,200$	
1-Chloropentane	$C_5H_{11}Cl$	l	$-41,800$	
1-Chloroethylene	C_2H_3Cl	l	$-8,400$	
trans-1,2-Dichloroethylene	$C_2H_6Cl_2$	l	$-1,000$	
Trichloroethylene	C_2HCl_3	l	$-1,400$	

TABLE 5.4.1. (*Continued*)

Compound	Formula	State	ΔH°_{f298}	$-\Delta H^\circ_{c298}$
Tetrachloroethylene	C_2Cl_4	l	−3,400	
3-Chloro-1-propene	C_3H_5Cl	l	−150	
Chlorobenzene	C_6H_5Cl	l	12,390	
p-Dichlorobenzene	$C_6H_5Cl_2$	l	5,500	
Hexachlorobenzene	C_6Cl_6	l	−8,100	510,000
Benzylchloride	C_7H_7Cl	l		782,000
1,1,1-trichloro-2,2-bis (p-chlorophenyl)ethane (DDT)	$C_{14}H_9Cl_5$	l		1,600,000

[a] Selected mainly from F.D. Rossini, ed., "Selected Values of Physical and Thermodynamic Properties of Hydrocarbons and Related Compounds," *American Petroleum Institute Research Project 44*, Carnegie Institute of Technology, Pittsburgh, PA, 1953; F.D. Rossini, D.D. Wagman, W.H. Evans, S. Levine, and I. Jaffe, "Selected Values of Chemical Thermodynamic Properties," *Natl. Bur. Stand. Circ. 500*, 1952; also from personal notes, L. Theodore and J. Reynolds, 1985.
[b] For combustion reactions the products are H_2O (l) and CO_2 (g).

tests used to generate these data are conducted in bomb calorimeters using excess oxygen. Burning carbon tetrachloride under these conditions yields carbon dioxide and chlorine oxides. Based on the published heat of combustion data, calculations suggest that chlorine monoxide or chlorine dioxide are the likely products, but there may be an equilibrium mixture of several species. The important lesson to be learned is that the published heat of combustion data for halogenated species may be useless since the products of combustion in an incinerator are acid gas (e.g., HCl), not halogen oxides. The use of heat of formation data is therefore recommended; these can be used to determine the heat of reaction for the burning of a particular waste that produces specific products. This avoids the pitfalls of using published data that may not apply to the reaction occurring in the incinerator.[3]

Other tables of heat of formation, combustion, and reaction are available in the literature (particularly thermodynamics text/reference books) for a wide variety of compounds.[1,2] It is important to note that these are valueless unless the stoichiometric equation, temperature, and the state of the reactants and products are included. However, *heat of reaction* is a term rarely employed in air pollution and/or incinerator calculations. The two terms most often used in this field are the *gross* (or *higher*) *heating value* and the *net* (or *lower*) *heating value*. The former is designated by HHV or HV_G and the latter by NHV or HV_N. The *gross heating value* represents the enthalpy change or heat released when a compound is stoichiometrically combusted (reacted) at a reference temperature with the final (flue) products also at the reference temperature and any water present in the liquid state. Most of these data are available at a reference temperature of 60°F. The *net heating value* is similar to the gross heating value except the water is in the vapor state. The difference (if any) between the two values represents the energy necessary to vaporize any water present. Thus, the standard heat of reaction and the gross and/or net heating values employed in the incinerator industry both represent the same phenomenon. Gross and net heating values for a number of hydrocarbons are presented in Table 5.3.1. In addition, the net heating value may be approximated by a form of Dulong's equation that includes the chlorine content:

$$NHV = 14,000m_C + 45,000(m_H - \tfrac{1}{8}m_O) - 760m_{Cl} + 4500m_S \qquad (5.4.9)$$

where   NHV = net heating value of waste–fuel mixture (Btu/lb)
    m_C = mass fraction of C (carbon) in the waste–fuel mixture
    C, H, O, Cl, S = subcripts indicating carbon, hydrogen, oxygen, chlorine, and sulfur, respectively

Another common term employed in incinerator calculations is the *available heat*, usually designated as HA_T. The available heat at any temperature T is the gross heating value minus the amount of heat ($\Sigma\Delta H$) required to take the product(s) of combustion (flue gas) from the reference temperature to that temperature T. Thus.

$$HA_T = HHV - \Sigma\Delta H$$

If all the heat liberated by the reaction goes into heating up the products of combustion (the flue gas), the temperature achieved is defined as the *flame temperature*. If the combustion process is conducted adiabatically, with no heat transfer to the surroundings, the final temperature achieved by the flue gas is defined as the *adiabatic flame temperature*. If the combustion process is conducted with theoretical or stoichiometric air (0% excess), the resulting temperature is defined as the *theoretical adiabatic flame temperature*.

In order to calculate fuel requirements, operating temperature, and (excess) air requirements for an incineration operation, one must apply the conservation laws for mass and energy in conjunction with thermochemical principles. This is an extremely involved, rigorous calculation. An enthalpy balance is applied around the incinerator following comprehensive overall and componental material balances. The enthalpy balance must account for all temperature changes of both the feed and fuel (reactants) as well as the flue gas (products). Latent (phase) and combustion (reaction) enthalpy effects must also be included in the analysis. Fortunately, algorithms are available to perform these detailed thermodynamic calculations; these can be found in Chapter 6.

Effect of Temperature on Heats of Reaction

The heat of reaction is a function of temperature because the heat capacities of both the reactants and products vary with temperature. Smith and Van Ness[2] have described this effect mathematically in the following manner

$$\Delta H_T^\circ = \Delta H_{298}^\circ + \int_{298}^{T} \Delta C_P dT \qquad (5.4.10)$$

where T is the absolute temperature (K), and

$$\Delta C_P = \sum_{products} nC_P - \sum_{reactants} nC_P$$

If the heat capacity for each product and reactant is expressed by

$$C_P = \alpha + \beta T + \gamma T^2$$

then

$$\Delta C_P = \Delta\alpha + (\Delta\beta)\,T + (\Delta\gamma)\,T^2 \qquad (5.4.11)$$

where

$$\Delta\alpha = \sum_{\text{products}} n\alpha - \sum_{\text{reactants}} n\alpha$$

with similar definitions for $\Delta\beta$ and $\Delta\gamma$. Combining Eq. (5.4.10) and (5.4.11), the following expression for the standard heat of reaction at temperature T is obtained:

$$\Delta H_T^\circ = \Delta H_{298}^\circ + \int_{298}^{T} [\Delta\alpha + (\Delta\beta)\,T + (\Delta\gamma)\,T^2]\,dT$$

or

$$\Delta H_T^\circ = \Delta H_{298}^\circ + \Delta\alpha(T - 298) + \tfrac{1}{2}\Delta\beta(T^2 - 298^2) + \tfrac{1}{3}\Delta\gamma(T^3 - 298^3) \qquad (5.4.12)$$

If all the constant terms in this equation are collected and lumped together into a constant designated ΔH_0, the result is

$$\Delta H_T^\circ = \Delta H_0 + \Delta\alpha T + \tfrac{1}{2}(\Delta\beta)\,T^2 + \tfrac{1}{3}(\Delta\gamma)\,T^3 \qquad (5.4.13)$$

where ΔH_T° = standard heat of reaction at temperature T
ΔH_0 = constant = $\Delta H_{298}^\circ - 298\Delta\alpha - \tfrac{1}{2}(298)^2\Delta\beta - \tfrac{1}{3}(298)^3\Delta\gamma$

Note: the use of the equations in this subsection requires that the temperature T be expressed in kelvins.

If the standard heat of reaction is known at a single temperature, for example, 25°C, ΔH_T° can be calculated. The constant ΔH_0 may then be calculated directly from Eq. (5.4.13).

The reader should attempt to rederive the equivalent of Eq. (5.4.12) and (5.4.13) if heat capacity variation with temperature is given by

$$C_P = a + bT + cT^{-2}$$

One application of this calculation is that associated with the adiabatic flame temperature (to be discussed in Chapter 6). It represents the maximum temperature the products of combustion (flue) can achieve if the reaction is conducted adiabatically. For this condition, all the energy liberated on combustion at or near standard conditions (ΔH_c°) appears as sensible heat in heating up the flue products, ΔH_p, that is,

$$\Delta H_c^\circ + \Delta H_p = 0 \qquad\qquad (5.4.14)$$

where ΔH_c° = standard heat of combustion at 25°C

ΔH_p = enthalpy change of the products as the temperature increases from 25°C to the theoretical flame temperature

The right term of Eq. (5.4.10), in conjunction with Eq. (5.4.11), should be applied to the product flue gases in evaluating ΔH_p. Details of this calculation are presented in the illustrative examples provided in Section 5.7.

5.5 CHEMICAL REACTION EQUILIBRIUM

With regard to chemical reactions, there are two important questions that are of concern to the engineer: (1) how *far* will the reaction go; and (2) how *fast* will the reaction go? Chemical thermodynamics provides the answer to the first question; however, it tells nothing about the second. Reactor rates fall within the domain of chemical kinetics and will be treated in Section 5.6. To illustrate the difference and importance of both questions on an engineering analysis of a chemical reaction, consider the following process. Substance A, which costs 1 cent/ton, can be converted to B, which costs 1 million $/lb, by the reaction A $\leftrightarrow$ B. Chemical thermodynamics will provide information on the maximum amount of B that can be formed. If 99.99% of A can be converted to B, the reaction would then appear to be economically feasible, from a *thermodynamic* point of view. However, a *kinetic* analysis might indicate that the reaction is so slow that, for all practical purposes, its rate is vanishingly small. For example, it might take 10^6 yr to obtain a $10^{-6}\%$ conversion of A. The reaction is then economically unfeasible. Thus, it can be seen that both equilibrium and kinetic effects must be considered in an overall engineering analysis of a chemical reaction. The same principle applies to control of gaseous pollutants by combustion and incineration of hazardous wastes.

A rigorous, detailed presentation of this topic is beyond the scope of this text. However, a superficial treatment is presented in the hope that it may provide at least a qualitative introduction to chemical reaction equilibrium. This topic does find application in incinerator calculations involving SO_3 and Cl_2 discharges and will hopefully explain, in part, the role of reaction equilibria in these calculations.

Chemical reaction equilibrium calculations are structured around another thermodynamic term referred to as *free energy*. This so-called *free energy* (*G*) is a thermodynamic property that cannot be easily defined without some basic grounding in thermodynamics. No attempt will be made to define it here and the interested reader is again directed to the literature[2] for further development of this topic. In the opinion of the authors, the fact that standard free energy data are available in the literature and can be used to calculate chemical equilibrium constants warrants the inclusion of this property in the discussion to follow. Free energy has the same units as enthalpy and internal energy and may be used on a mole, or total mass basis. The upper-case *G* signifies a mole basis.

Consider the equilibrium reaction

$$a\mathrm{A} + b\mathrm{B} = c\mathrm{C} + d\mathrm{D} \qquad\qquad (5.5.1)$$

where A, B, C, D = chemical formulas of the reactant and product species
     a, b, c, d = stoichiometric coefficients

and the = sign is a reminder that the reacting system is at equilibrium. For this reaction (as with enthalpy),

$$\Delta G^\circ = c G_C^\circ + d G_D^\circ - a G_A^\circ - b G_B^\circ \tag{5.5.2}$$

ΔG° represents the free energy change for this reaction when reactants and products are in their standard states. Note that ΔG° may be calculated and obtained in a manner similar to ΔH°. Note further that, at 25°C, the free energy of an element in its standard state is arbitrarily set at zero. Therefore,

$$(\Delta G_f^\circ)_{i\,298} = G_{i\,298}^\circ$$

so that

$$\Delta G_{298}^\circ = c(\Delta G_f^\circ)_C + d(\Delta G_f^\circ)_D - a(\Delta G_f^\circ)_A - b(\Delta G_f^\circ)_B \tag{5.5.3}$$

According to Eq. (5.5.3), at 25°C, the standard free energy of reaction ΔG° may be calculated from standard free energy of formation data. Some of this information is presented in Table 5.5.1. The following equation is used to calculate the chemical reaction equilibrium constant K at a temperature T.

$$\Delta G_T^\circ = -RT \ln K \tag{5.5.4}$$

TABLE 5.5.1. Standard Free Energy of Formation at 25°C in Calories per Gram-Mole[a]

Compound	Formula	State	$\Delta G_{f\,298}^\circ$
Normal Paraffins			
Methane	CH_4	g	−12,140
Ethane	C_2H_6	g	−7,860
Propane	C_3H_8	g	−5,614
n-Butane	C_4H_{10}	g	−4,100
n-Pentane	C_5H_{12}	g	−2,000
n-Hexane	C_6H_{14}	g	−70
n-Heptane	C_7H_{16}	g	1,920
n-Octane	C_8H_{18}	g	3,920
Increment per C atom above C_8		g	2,010
Normal Monoolefins (1-alkenes)			
Ethylene	C_2H_4	g	16,282
Propylene	C_3H_6	g	14,990
1-Butene	C_4H_8	g	17.090
1-Pentene	C_5H_{10}	g	18,960
1-Hexene	C_6H_{12}	g	20,940
Increment per C atom above C_6		g	2,010

TABLE 5.5.1. (*Continued*)

Compound	Formula	State	ΔG°_{f298}
Miscellaneous Organic Compounds			
Acetaldehyde	C_2H_4O	g	−31,960
Acetic acid	$C_2H_4O_2$	l	−93,800
Acetylene	C_2H_2	g	50,000
Benzene	C_6H_6	g	30,989
Benzene	C_6H_6	l	29,756
1,3-Butadiene	C_4H_6	g	36,010
Cyclohexane	C_6H_{12}	g	7,590
Cyclohexane	C_6H_{12}	l	6,370
Ethanol	C_2H_6O	g	−40,130
Ethanol	C_2H_6O	l	−41,650
Ethylbenzene	C_8H_{10}	g	31,208
Ethylene glycol	$C_2H_6O_2$	l	−77,120
Ethylene oxide	C_2H_4O	g	−2,790
Methanol	CH_4O	g	−38,810
Methanol	CH_4O	l	−39,850
Methylcyclohexane	C_6H_{14}	g	6,520
Methylcyclohexane	C_6H_{14}	l	4,860
Styrene	C_8H_8	g	51,100
Toluene	C_7H_8	g	29,228
Toluene	C_7H_8	l	27,282
Miscellaneous Inorganic Compounds			
Ammonia	NH_3	g	−3,976
Ammonia	NH_3	aq	−6,370
Calcium carbide	CaC_2	s	−16,200
Calcium carbonate	$CaCO_3$	s	−269,780
Calcium chloride	$CaCl_2$	s	−179,300
Calcium chloride	$CaCl_2$	aq	−194,880
Calcium hydroxide	$Ca(OH)_2$	s	−214,330
Calcium hydroxide	$Ca(OH)_2$	aq	−207,370
Calcium oxide	CaO	s	−144,400
Carbon dioxide	CO_2	g	−94,258
Carbon monoxide	CO	g	−32,781
Hydrochloric acid	HCl	g	−22,778
Hydrogen sulfide	H_2S	g	−7,892
Iron oxide	Fe_3O_4	s	−242,400
Iron oxide	Fe_2O_3	s	−177,100
Iron sulfide	FeS_2	s	−39,840
Nitric acid	HNO_3	l	−19,100
Nitric acid	HNO_3	aq	−26,410
Nitrogen oxides	NO	g	20,690
	NO_2	g	12,265
	N_2O	g	24,933
	N_2O_4	g	23,395
Sodium carbonate	Na_2CO_3	s	−250,400
Sodium chloride	$NaCl$	s	−91,785
Sodium chloride	$NaCl$	aq	−93,939
Sodium hydroxide	$NaOH$	s	−90,600

TABLE 5.5.1. (*Continued*)

Compound	Formula	State	$\Delta G^{\circ}_{f\,298}$
Sodium hydroxide	NaOH	aq	$-100{,}184$
Sulfur dioxide	SO_2	g	$-71{,}790$
Sulfur trioxide	SO_3	g	$-88{,}520$
Sulfuric acid	H_2SO_4	aq	$-177{,}340$
Water	H_2O	g	$-54{,}635$
Water	H_2O	l	$-56{,}690$
Chlorinated Organics			
Methyl chloride	CH_3Cl	l	$-15{,}030$
Dichloromethane	CH_2Cl_2	l	$-16{,}460$
Chloroform	$CHCl_3$	l	$-16{,}380$
Carbon tetrachloride	CCl_4	l	$-13{,}920$
Ethyl chloride	C_2H_5Cl	l	$-14{,}340$
1,1-Dichloroethane	$C_2H_4Cl_2$	l	$-17{,}470$
1,1,2,2-Tetrachloroethane	$C_2H_2Cl_4$	l	$-20{,}480$
n-Propyl chloride	C_3H_7Cl	l	$-12{,}110$
1,3-Dichloropropane	$C_3H_6Cl_2$	l	$-19{,}740$
n-Butyl chloride	C_4H_9Cl	l	$-9{,}270$
1-Chloropentane	$C_5H_{11}Cl$	l	$-8{,}940$
1-Chloroethylene	C_2H_3Cl	l	$-12{,}310$
trans-1,2-Dichloroethylene	$C_2H_6Cl_2$	l	$-6{,}350$
Trichloroethylene	C_2HCl_3	l	$-4{,}750$
Tetrachloroethylene	C_2Cl_4	l	$-4{,}900$
3-Chloro-1-propene	C_3H_5Cl	l	$10{,}420$
Chlorobenzene	C_6H_5Cl	l	$23{,}700$
p-Dichlorobenzene	$C_6H_5Cl_2$	l	$18{,}440$
Hexachlorobenzene	C_6Cl_6	l	$10{,}560$

[a] Selected mainly from F.D. Rossini, ed., "Selected Values of Physical and Thermodynamic Properties of Hydrocarbons and Related Compounds," *American Petroleum Institute Research Project 44*, Carnegie Institute of Technology, Pittsburgh, PA, 1953; F.D. Rossini, D.D. Wagman, W.H. Evans, S. Levine, and I. Jaffe, "Selected Values of Chemical Thermodynamic Properties," *Natl. Bur. Stand. Circ.* **500**, 1952; and also from personal notes, L. Theodore and J. Reynolds, 1986.

The value of this equilibrium constant depends on the temperature at which the equilibrium is established. The effect of temperature on K must now be examined.

Effect of Temperature on the Equilibrium Constant

The dependence of ΔG° on temperature is given by[2]

$$\frac{d(\Delta G^{\circ}/RT)}{dT} = \frac{-\Delta H^{\circ}}{RT^2} \tag{5.5.5}$$

According to Eq. (5.5.4)

$$\frac{\Delta G°}{RT} = -\ln K$$

Substitution yields

$$\frac{d\ln K}{dT} = \frac{\Delta H°}{RT^2} \tag{5.5.6}$$

Equation (5.5.6) shows the effect of temperature on the equilibrium constant, and hence on the equilibrium yield. If the reaction is exothermic, $\Delta H°$ is negative and the equilibrium constant decreases with an increase in temperature; for an endothermic reaction, the equilibrium constant increases with an increase in temperature.

If the term $\Delta H°$, which is the standard enthalpy change (heat of reaction), is assumed to be constant with temperature, Eq. (5.5.6) can be integrated between the temperatures T and T_1 to give

$$\ln \frac{K}{K_1} = -\frac{\Delta H°}{R}\left(\frac{1}{T} - \frac{1}{T_1}\right) \tag{5.5.7}$$

This approximate equation may be used to determine the equilibrium constant at a temperature T from the known value at some other temperature T_1 if the difference between the two temperatures is small. If the standard heat of reaction is known as a function of temperature, however, Eq. (5.5.6) can be integrated rigorously.

$$\ln K = \frac{1}{R}\int \frac{\Delta H°}{T^2}\,dT + I \tag{5.5.8}$$

where I is a constant of integration

In the previous section, it was shown that, if the molar heat capacity for each chemical species taking part in the reaction is known and can be expressed as a power series in T (kelvins),

$$C_P = \alpha + \beta T + \gamma T^2$$

then $\Delta H°$ at a given temperature T becomes

$$\Delta H°_T = \Delta H_0 + \Delta\alpha T + \frac{\Delta\beta T^2}{2} + \frac{\Delta\gamma T^3}{3} \tag{5.4.13}$$

In this equation, the constant ΔH_0 can be calculated from the standard heat of reaction at 25°C [see Eq. (5.4.13)]. With ΔH_0 determined, $\Delta H°_T$ can be substituted into Eq. (5.5.8) and integrated to obtain

$$\ln K = -\frac{\Delta H_0}{RT} + \frac{\Delta\alpha}{R}\ln T + \frac{\Delta\beta}{2R}T + \frac{\Delta\gamma}{6R}T^2 + I \tag{5.5.9}$$

Here the constant I may be evaluated from a knowledge of the equilibrium constant at one temperature. This is usually obtained from standard free energy of formation data at 25°C. Equation (5.5.4) is employed to obtain K at this temperature. A similar equation for ΔG_T° may be obtained by combining Eqs. (5.5.4) and (5.5.9).

$$\Delta G_T^{\circ} = \Delta H_0 - (\Delta \alpha) T \ln T - \left(\frac{\Delta \beta}{2}\right) T^2 - \left(\frac{\Delta \gamma}{6}\right) T^3 - IRT \qquad (5.5.10)$$

The reader is left the exercise of developing companion equations to Eqs. (5.5.9) and (5.5.10) if the heat capacity variation with temperature is given by

$$C_P = a + bT + cT^{-2}$$

The problem that remains is to relate K to understandable physical quantities. For gas phase reactions, as in an incinerator operation, the term K for Eq. (5.5.4) may be approximately represented in terms of the partial pressures of the components involved. This functional relationship is given in Eqs. (5.5.11) and (5.5.12).

$$K = K_P \qquad (5.5.11)$$

where K is an equilibrium constant based on partial pressures.

$$K_P = \frac{P_C^c P_D^d}{P_A^a P_B^b} \qquad (5.5.12)$$

where P_A = partial pressure of component A, etc.

This definition of K_P obviously applies to the reaction of.Eq. (5.5.1). Assuming a K value is available or calculable, this equation may be used to determine the partial pressures of the participating components at equilibrium. It is important to note that the component partial pressures (P_i) are *equilibrium* values. For product gases, P_i usually represents the *maximum* values that can ultimately be achieved; for reactant gases, P_i represents *minimum* values. For liquid phase reactions, K is approximately given by

$$K = K_C \qquad (5.5.13)$$

where $\qquad K_C = C_C^c C_D^d / C_A^a C_B^b$
$\qquad\qquad C_C$ = concentration of component C (gmol/L)
$\qquad$ C, D, A, B = subscripts indicating chemical species, C, D, A, and B

5.6   CHEMICAL KINETICS

Chemical kinetics involves the study of reaction rates and the variables that affect these rates. It is a topic that is critical for the analysis of incinerator/combustion systems. The objective of this section is to develop a working understanding of this subject in order to apply chemical kinetics principles to hazardous waste incineration applications. The topic is treated from an engineering point of view, that is, in terms of physically measurable quantities. The rate of a chemical reaction can be described n any of several different ways. The most commonly used definition involves the time rate of change in the amount of one of the components participating in the reaction; this rate is usually based on some arbitrary factor related to the reacting system size or geometry, such as volume, mass, and interfacial area. The definition shown in Eq. (5.6.1), which applies to homogeneous reactions (all reactants and products are in the same phase), is a convenient one from an engineering point of view and is adopted for use in this text.

$$R_A = \frac{1}{V}\left(\frac{dn_A}{d\theta}\right)$$

(5.6.1)

where R_A = reaction rate based on component A .
V = volume of reacting system
n_A = number of moles of A at time θ
θ = time

If the volume term is constant, one may write Eq. (5.6.1) as

$$R_A = \frac{d(n_A/V)}{d\theta}$$

$$= \frac{dC_A}{d\theta}$$

(5.6.2)

where C_A = molar concentration of A at time θ

Equation (5.6.2) states that the reaction rate is equal to the rate of change in concentration of one of the components with respect to time. When rates are expressed in terms of concentration changes, the assumption of constant volume is implied. The units of the rate become lbmol/h-ft^3 and gmol/s-L in the engineering and metric systems, respectively. When the reaction is nonhomogeneous, the geometric factor is usually some parameter other than the reaction volume. For fluid–fluid surface reactions, for example, the geometric factor is usually the inter-facial area, in which case the units of the reaction rate are lbmol/h-ft^2. For fluid–solid reaction systems, the factor is often the mass of the solid; for example, in gas-phase catalytic combustion reactions, the factor is the mass of the catalyst and the units of the rate are pound moles per hour pound (lbmol/h-lb).

Based on Eq. (5.6.2), the reaction rate is positive if species A is being formed since C_A increases with time, and negative if A is reacting since C_A decreases with time. The rate is zero if the system is at chemical equilibrium (see Section 5.5).

An equation expressing the rate in terms of measurable and/or desirable quantities may now be developed. Based on experimental evidence, the rate of reaction is a function of the concentration of the components present in the reaction mixture (this includes reacting and inert species), temperature, pressure, catalyst variables, or, in equation form,

$$R_A = R_A \,(C_i, \ P, \ T, \text{ catalyst variables})$$

(5.6.3)

where C_i = molar concentration of each component present in the reaction mixture

Equation (5.6.3) may be condensed into

$$R_A = \pm k_A f(C_i) \tag{5.6.4}$$

where k_A incorporates all the variables in Eq. (5.6.3) other than the concentration variable. The $\pm$ notation is included to indicate whether component A is being consumed $(-)$ or produced $(+)$. The k_A may be regarded as a constant of proportionality; it is termed the *specific reaction rate* or, more commonly, the *reaction velocity constant*. Although this "constant" is independent of concentration, it is a function of the other variables. This approach has, in a sense, isolated one of the variables. The reaction velocity constant, like the rate of reaction, must be based on one of the components of the reacting system, almost always the same component on which the reaction rate is based. This term is very definitely influenced by temperature and catalyst activity. For the present, however, it is assumed that k is solely a function of temperature. Thus,

$$R_A = \pm k_A(T) f(C_i) \tag{5.6.5}$$

The effect of temperature on k is generally represented by the Arrhenius equation:

$$k = A e^{-E/RT} \tag{5.6.6}$$

where A = frequency factor, same units as k
   E = activation energy, same units as RT
   R = universal gas constant
   T = absolute temperature

The values of A and E are usually obtained experimentally.

The functional relationship for the molar concentration, $f(C_i)$, can be obtained from the stoichiometric reaction equation, which assumes elementary or power law kinetics to apply. Consider again the reaction

$$aA + bB \longrightarrow cC + dD \tag{5.6.7}$$

where the single arrow represents an *irreversible* reaction, that is, if stoichiometric amounts of A and B are initially present, the reaction will proceed to the products until all the A and B have reacted. The rate of this reaction is given by

$$R_A = -k_A C_A^a C_B^b \tag{5.6.8}$$

where the negative sign is introduced to account for the the fact that A is being consumed. The product concentrations do not affect the rate. The term $f(C_i)$ is then the product of the molar concentrations of the reactants, each raised to the power of its stoichiometric coefficient.

The *order* of this reaction *with respect to a particular species* is given by the exponent of that concentration term appearing in the rate expression. The reaction of Eq. (5.6.8) is, therefore, of order a with respect to A, and of order b with respect to B. The overall order n, usually referred to as the *order of the reaction*, is the sum of the individual orders, that is,

$$n = a + b$$

This analysis has assumed elementary law kinetics to apply. This is frequently an idealization of more complex theory and should therefore be looked upon as a general and first introduction to the rate laws.

All real and naturally occurring reactions are reversible. A *reversible* reaction is one in which the products also react to form the reactants. Unlike irreversible reactions that proceed to the right until completion, reversible reactions achieve an equilibrium state after an infinite period of time. Reactants and products are still present in the system. At this equilibrium state the net reaction rate is zero. Consider the following reversible reaction

$$a\text{A} + b\text{B} \longleftrightarrow c\text{C} + d\text{D} \tag{5.6.9}$$

where the double headed arrow is a reminder that the reaction is reversible; the notation $=$ is employed if the reaction system is at equilibrium. The rate of this reaction is given by

$$R_\text{A} = \underbrace{-k_\text{A} C_\text{A}^a C_\text{B}^b}_{\substack{\text{forward}\\ \text{reaction}}} + \underbrace{k_\text{A}' C_\text{C}^c C_\text{D}^d}_{\substack{\text{reverse}\\ \text{reaction}}} \tag{5.6.10}$$

where k_A = reaction velocity constant for the forward reaction
$\qquad k_\text{A}'$ = reaction velocity constant for the reverse reaction

The orders of the forward and reverse reactions are $(a + b)$ and $(c + d)$, respectively. For equilibrium systems, the forward reaction rate is equal to the reverse rate and Eq. (5.6.10) becomes

$$R_\text{A} = 0 \tag{5.6.11}$$

so that

$$k_\text{A} C_\text{A}^a C_\text{B}^b = k_\text{A}' C_\text{C}^c C_\text{D}^d \tag{5.6.12}$$

If an equilibrium constant K_C is defined by the ratio of the forward to the reverse reaction velocity constant

$$K_\text{C} = \frac{k_\text{A}}{k_\text{A}'} \tag{5.6.13}$$

then

$$K_\text{C} = \frac{C_\text{C}^c C_\text{C}^d}{C_\text{A}^a C_\text{B}^b} \tag{5.6.14}$$

This is the same as Eq. (5.5.13). Thus, if k_A is given and K_C is known, then k'_A can be calculated. Note that K_C is approximately the true equilibrium constant K referred to in Section 5.5.

Gas phase reactions often have rates expressed in a different form. For the nth-order irreversible reaction

$$aA + bB \longrightarrow products$$

where $n = a + b$, it was previously stated that

$$R_A = \frac{dC_a}{d\theta} = -k_A C_A^a C_B^b \tag{5.6.15}$$

This equation may also be written in terms of the partial pressures of A and B. Assuming ideal gas conditions permits the following substitution:

$$C_i = \frac{P_i}{RT} \tag{5.6.16}$$

where P_i = partial pressure of component i
   R = universal gas constant
   T = absolute temperature

Substituting Eq. (5.6.16) for each reactant into the right-hand side of Eq. (5.6.15) gives

$$R_A = \frac{dC_A}{d\theta} = -k_{PA} P_A^a P_B^b \tag{5.6.17}$$

where k_{PA} = reaction velocity constant for A based on partial pressure. Further substitution of Eq. (5.6.16) into Eq. (5.6.17) yields

$$R_A^* = \frac{dP_A}{d\theta} = -k_{PA}^* P_A^a P_B^b \tag{5.6.18}$$

where $R_A^* = R_A (RT)$
   k_{PA}^* = reaction velocity constant of A based on R_A^*

The three reaction velocity constants defined by Eq. (5.6.15), (5.6.17), and (5.6.18) are related as follows:

$$k_{Pi} = \frac{k_i}{(RT)^n}$$

$$k_{Pi}^* = \frac{k_i}{(RT)^{n-1}}$$

$$k_{Pi} = \frac{k_{Pi}^*}{RT} \tag{5.6.19}$$

These results are now applied to different kinetic systems. Some of the more common and elementary chemical reactions are presented here for constant volume systems.

1. *Zero-Order, Irreversible Reaction*

$$A \longrightarrow products$$

$$R_A = dC_A/d\theta = -k_A$$

This is an idealized reaction system. It has been shown that all real *zero-order* reactions can be represented by this rate equation only at high concentrations.

2. *First-Order, Irreversible Reaction*

$$A \longrightarrow products$$

$$R_A = dC_A/d\theta = -k_A C_A$$

3. *Second-Order, Irreversible Reaction*

(a)
$$A + B \longrightarrow products$$

$$R_A = dC_A/d\theta = -k_A C_A C_B$$

(b)
$$2A \longrightarrow products$$

$$R_A = dC_A/d\theta = -k_A C_A^2$$

4. *Third-Order, Irreversible Reaction*

(a)
$$3A \longrightarrow products$$

$$R_A = dC_A/d\theta = -k_A C_A^3$$

(b)
$$2A + B \longrightarrow products$$

$$R_A = dC_A/d\theta = -k_A C_A^2 C_B$$

(c)
$$A + B + C \longrightarrow products$$

$$R_A = dC_A/d\theta = -k_A C_A C_B C_C$$

5. *Fractional- or Higher-Order, Irreversible Reaction*

$$nA \longrightarrow products$$

$$R_A = dC_A/d\theta = -k_A C_A^n$$

For many complex reactions, the order is often a large integer or a fraction.

6. *First-Order, Reversible Reaction*

$$A \longleftrightarrow B$$

$$R_A = dC_A/d\theta = -k_A C_A + k'_A C_B$$
$$= -k_A[C_A - (C_B/K_A)]$$

where $K_A = k_A/k'_A$

7. *Second-order, Reversible Reaction*

(a) $$A \longleftrightarrow 2B$$

$$R_A = dC_A/d\theta = -k_A C_A + k'_A C_B^2$$
$$= -k_A[C_A - (C_B^2/K_A)]$$

(b) $$A \longleftrightarrow B + C$$

$$R_A = dC_A/d\theta = -k_A C_A + k'_A C_B C_C$$
$$= -k_A[C_A - (C_B C_C/K_A)]$$

(c) $$A + B \longleftrightarrow C$$

$$R_A = dC_A/d\theta = -k_A C_A C_B + k'_A C_C$$
$$= -k_A[(C_A C_B - (C_C/K_A)]$$

(d) $$2A \longleftrightarrow B$$

$$R_A = dC_A/d\theta = -k_A C_A^2 + k'_A C_B$$
$$= -k_A[C_A^2 - (C_B/K_A)]$$

(e) $$A + B \longleftrightarrow C + D$$

$$R_A = dC_A/d\theta = -k_A C_A C_B + k'_A C_C C_D$$
$$= -k_A[C_A C_B - (C_C C_D/K_A)]$$

8. Similar equations can be written for *third-and higher-order reversible reactions.*

9. *Simultaneous Irreversible Reactions*

$$A \xrightarrow{k_{A_1}} \text{products}$$

$$A + B \xrightarrow{k_{A_2}} \text{products}$$

$$3A \xrightarrow{k_{A_3}} \text{products}$$

$$R_A = dC_A/d\theta = -k_{A_1} C_A - k_{A_2} C_A C_B - k_{A_3} C_A^3$$
$$R_B = dC_B/d\theta = -k_{A_2} C_A C_B$$

10. *Consecutive Irreversible Reactions (two first-order reactions)*

$$A \xrightarrow{k_A} B$$

$$B \xrightarrow{k_B} C$$

$$R_A = dC_A/d\theta = -k_A C_A$$
$$R_B = dC_B/d\theta = k_A C_A - k_B C_B$$
$$R_C = dC_C/d\theta = k_B C_B$$

In recent times, investigators in the hazardous waste incineration field have taken the path of least resistance in analyzing incineration reactions and assumed the reactions to approach first-order behavior. Available experimental data are employed to generate *average* or *best fit* values of k for first-order models. The rate equation

$$dC/d\theta = -kC \tag{5.6.20}$$

can then be integrated to provide a simple quantitative relationship between concentration and time, as shown in Eq. (5.6.21).

$$C = C_0 e^{-k\theta} \tag{5.6.21}$$

where C_0 = initial concentration

The extent of the reaction or the destruction efficiency (DRE) may then be calculated.

Experimentally determined reaction velocity constants for incinerator reactors are based on *thermal destruction*, as opposed to *flame mode destruction*. When a waste passes through the flame front, some of it is destroyed by flame zone radicals at a rate much faster than the rate of waste destruction in the *thermal destruction* zone beyond the flame zone. Most of the waste is consumed in the thermal destruction zone. Rate constants for specific compounds and waste samples are obtained by a variety of experimental methods. Some of the most common are the TDAS (Thermal Destruction Analytical System) method of the University of Dayton Research Institute and thermogravimetric analysis.[4]

5.7 ILLUSTRATIVE EXAMPLES

Example 5.7.1. Waste material is fed into a hazardous waste incinerator at a rate of 10,000 lb/h in the presence of 20,000 lb/h of air. Due to the low heating value of the waste, 2000 lb/h of methane is added to assist in the combustion of the waste. At what rate (lb/h) do the product gases exit the incinerator?

SOLUTION. Apply the conservation law for mass to the incinerator. Assume steady-state conditions to apply.

$$\text{rate of mass in } (w_{in}) = \text{rate of mass out } (w_{out})$$

$$w_{in} = (10{,}000 + 20{,}000 + 2000)$$

$$= 32{,}000 \, lb/h$$

Therefore, $w_{out} = 32{,}000 \, lb/h$

Example 5.7.2. C_6H_5Cl is fed into a hazardous waste incinerator at a rate of 5000 scfm (60°F, 1 atm) and is combusted in the presence of air fed at a rate of 3000 scfm (60°F, 1 atm). Both streams enter the incinerator at 70°F. The products are cooled from 2000°F and exit a cooler at 180°F. At what rate (lb/h) do the products exit the cooler? The molecular weight of C_6H_5Cl is 112.5; the molecular weight of air is 29.

SOLUTION. First convert scfm to acfm using Charles' Law.

$$5000 \, scfm \times \frac{(460 + 70)}{(460 + 60)} = 5096 \, acfm \text{ of } C_6H_5Cl$$

$$3000 \, scfm \times \frac{(460 + 70)}{(460 + 60)} = 3058 \, acfm \text{ of air}$$

One lbmol of any ideal gas occupies 387 ft³ at 70°F and 1 atm. Therefore, the molar flow rate ($\dot{n}$) may be calculated by dividing these results by 387.

$$\dot{n}(C_6H_5Cl) = \frac{5096}{387}$$

$$= 13.17 \, lbmol/min$$

$$\dot{n}(air) = \frac{3058}{387}$$

$$= 7.90 \, lbmol/min$$

The mass flow rate is obtained by multiplying these results by the molecular weight.

$$w(C_6H_5Cl) = (13.17)(112.5)(60)$$

$$= 88{,}898 \, lb/h$$

$$w(air) = (7.90)(29)(60)$$

$$= 13{,}746 \, lb/h$$

Note: One can calculate these results by simply applying the conservation of mass law. Since 1 lbmol of any ideal gas occupies 379 ft³ at 60°F and 1 atm, $\dot{n}$ can be calculated directly, and w by multiplying $\dot{n}$ by the molecular weight.

$$w_{in}(C_6H_5Cl) = \frac{(5000)(112.5)(60)}{(379)}$$

$$= 88{,}898 \text{ lb/h}$$

$$w_{in}(air) = \frac{(3000)(29)(60)}{(379)}$$

$$= 13{,}746 \text{ lb/h}$$

Since mass is conserved, w_{in} is equal to w_{out}.

$$w_{in,\,total} = 88{,}898 + 13{,}746$$

$$= 102{,}644 \text{ lb/h}$$

Example 5.7.3. A flue gas from an incinerator enters a waste heat boiler at 1800°F at a rate of 50,000 lb/h. It transfers 2.3×10^7 Btu/h of heat adiabatically. Calculate the outlet temperature of the gas stream. Assume the average heat capacity of the flue gas to be 0.36 Btu/lb-°F.

SOLUTION. Since average heat capacity information is given, an equation based on Eq. (5.2.8) may be employed.

$$q = w\bar{c}_P \Delta T$$

$$\Delta T = \frac{q}{w\bar{c}_P} = \frac{(2.3 \times 10^7)}{(50{,}000)(0.36)}$$

$$= 1278°F = (1800 - T_2)$$

The outlet temperature (T_2) is then 522°F.

Example 5.7.4. Propane is combusted in the presence of stoichiometric oxygen at 1500°F to form carbon dioxide and water as the combustion products. Balance the combustion reaction

$$C_3H_8 + O_2 \longrightarrow CO_2 + H_2O$$

SOLUTION. Since there are three carbon atoms on the left-hand side of the reaction, it is balanced with $3CO_2$ on the right-hand side:

$$C_3H_8 + O_2 \longrightarrow 3CO_2 + H_2O$$

Since 8 hydrogen atoms appear on the left-hand side, this is balanced by 4 mol of H_2O:

$$C_3H_8 + O_2 \longrightarrow 3CO_2 + 4H_2O$$

There are 10 oxygen atoms on the right-hand side of the reaction. This is balanced by 5 oxygen molecules on the left-hand side to yield the final balanced equation for the combustion of propane:

$$C_3H_8 + 5O_2 \longrightarrow 3CO_2 + 4H_2O$$

Note: This result could have been obtained directly from Table 5.3.1.

Example 5.7.5.   The offensive odor of butanol (butyl alcohol) can be removed from stack gases by its complete combustion to CO_2 and H_2O. It is of interest that the incomplete combustion of butanol can actually result in a more serious environmental and odor pollution problem than the original one. Write the reactions showing the formation of the two intermediate malodorous products formed if butanol undergoes incomplete combustion. The three possible products of combustion are listed here.

1. Butylaldehyde (C_4H_8O)
2. Butyric acid (C_3H_7COOH)
3. Carbon dioxide (CO_2) and water (H_2O)

SOLUTION.   For butylaldehyde:

$$C_4H_9OH + \tfrac{1}{2}O_2 \longrightarrow C_4H_8O + H_2O$$

For butyric acid:

$$C_4H_9OH + O_2 \longrightarrow C_3H_7COOH + H_2O$$

For "complete" combustion:

$$C_4H_9OH + 6O_2 \longrightarrow 4CO_2 + 5H_2O$$

This is one of the main disadvantages of waste management control by incineration. As shown, it is possible to create a worse pollution problem if the combustion is incomplete. These products of incomplete combustion are defined as *PICs*. Incomplete combustion can arise because of low temperatures, insufficient oxygen, incomplete mixing, incomplete retention time, or a combination of these.

Example 5.7.6.   The combustion reaction for C_6H_5Cl is

$$C_6H_5Cl + O_2 \longrightarrow CO_2 + H_2O + HCl$$

1. Balance the equation.
2. 2000 lb/h of C_6H_5Cl is fed into an incinerator with 50% excess air. What quantities of combustion products are produced (lb/h)?
The molecular weight of C_6H_5Cl is 112.5. The molecular weight of air is 29.

SOLUTION

1. The balanced reaction is

$$C_6H_5Cl + 7O_2 \longrightarrow 6CO_2 + 2H_2O + HCl$$

2. This reaction for air becomes

$$C_6H_5Cl + 7O_2 + 26.3N_2 \longrightarrow 6CO_2 + 2H_2O + HCl + 26.3N_2$$

This appears as Eq. (5.3.7) in the text.
   For 50% excess air,

$$\frac{\text{lbmol of } O_2}{\text{lbmol } C_6H_5Cl} = 7 \times 1.5$$

$$= 10.5$$

$$\frac{\text{lbmol of } N_2}{\text{lbmol } C_6H_5Cl} = 26.3 \times 1.5$$

$$= 39.45$$

The balanced reaction for 50% excess air is

$$C_6H_5Cl + 10.5O_2 + 39.45N_2 \longrightarrow 6CO_2 + 2H_2O + HCl + 3.5O_2 + 39.45N_2$$

Convert the mass flow rate of C_6H_5Cl to a molar flow rate.

$$\dot{n}(C_6H_5Cl) = \frac{2000}{112.5}$$

$$= 17.78 \, \text{lbmol/h}$$

$$w(CO_2) = \dot{n}(C_6H_5Cl)(6)(MW) = (17.78)(6)(44)$$

$$= 4694 \, \text{lb/h}$$

$$w(H_2O) = (17.78)(2)(18)$$

$$= 640 \, \text{lb/h}$$

$$w(HCl) = (17.78)(1)(36.46)$$

$$= 648 \, \text{lb/h}$$

$$w(O_2) = (17.78)(3.5)(32)$$

$$= 1991 \, \text{lb/h}$$

$$w(N_2) = (17.78)(39.45)(28)$$

$$= 19640 \, \text{lb/h}$$

The reader is left the exercise of calculating the mass fractions of the flue components and the total mass flow rate.

Note: On combusting C_6H_5Cl, the resulting equilibrium conditions will not permit all of the Cl to go to HCl. The amount of Cl_2 formed (which will be small) is a function of temperature, excess air, and the H/Cl ratio. This effect is treated in a later example.

Example 5.7.7. Verify that the enthalpy of combustion (Table 5.4.1) may be calculated from enthalpy of formation data (also in Table 5.4.1). Use *n*-hexane as an example.

SOLUTION. From Table 5.4.1,

$$\Delta H_c^\circ(\text{n-hexane}) = -1{,}002{,}570 \, \text{cal/gmol}$$

First, write the combustion reaction.

$$\text{C}_6\text{H}_{14} + 9.5\text{O}_2 \longrightarrow 6\text{CO}_2 + 7\text{H}_2\text{O(l)}$$

From Table 5.4.1, one obtains

$$\text{C}_6\text{H}_{14} \, (g); \quad \Delta H_f^\circ = -39{,}960 \, \text{cal/gmol}$$
$$\text{CO}_2 \, (g); \quad \Delta H_f^\circ = -94{,}052 \, \text{cal/gmol}$$
$$\text{H}_2\text{O} \, (l); \quad \Delta H_f^\circ = -68{,}317 \, \text{cal/gmol}$$

$$\Delta H_c^\circ = \Sigma \Delta H_{f,p}^\circ - \Sigma \Delta H_{f,r}^\circ = 6(-94{,}052) + 7(-68{,}317) - (-39{,}960)$$
$$= -1{,}002{,}571 \, \text{cal/gmol}$$

Example 5.7.8. Using standard heat of combustion data, calculate the standard heat of formation of $\text{C}_{14}\text{H}_9\text{Cl}_5$. (The organic and generic names for this compound are 1,1,1-trichloro-2,2-bis(*p*-chlorophenyl)ethane and DDT, respectively. DDT is the abbreviation for "dichlorodiphenyltrichloroethene", another name sometimes used for this compound.)

SOLUTION. The standard heat of combustion for this organic is obtained directly from Table 5.4.1, noting that the H_2O and HCl formed are in the liquid and gaseous states, respectively.

$$\Delta H_c^\circ = -1600 \, \text{kcal/gmol}$$

First write a balanced stoichiometric equation for this combustion reaction.

$$\text{C}_{14}\text{H}_9\text{Cl}_5 + 15\text{O}_2 \longrightarrow 14\text{CO}_2 + 2\text{H}_2\text{O(l)} + 5\text{HCl(g)}$$

For this reaction,

$$\Delta H_c^\circ = 14\Delta H_{f,\text{CO}_2}^\circ + 2\Delta H_{f,\text{H}_2\text{O(l)}}^\circ + 5\Delta H_{f,\text{HCl(g)}}^\circ - \Delta H_{f,\text{C}_{14}\text{H}_9\text{Cl}_5}^\circ$$

From Table 5.4.1,

$$\Delta H_{f,\text{CO}_2}^\circ = -94.052 \, \text{kcal/gmol}$$
$$\Delta H_{f,\text{H}_2\text{O(l)}}^\circ = -68.317 \, \text{kcal/gmol}$$
$$\Delta H_{f,\text{HCl(g)}}^\circ = -22.063 \, \text{kcal/gmol}$$

Solving this equation for $\Delta H^\circ_{f, C_{14}H_9Cl_5}$ yields

$$\Delta H^\circ_{f, C_{14}H_9Cl_5} = 36.32 \text{ kcal/gmol}$$

The reader is again reminded that the heat of formation of elements in their standard states is zero. In addition, care should be exercised in using heat of combustion data for highly chlorinated organics.

Example 5.7.9. Calculate the theoretical adiabatic flame temperature of C_6H_5Cl. Assume that the heat capacity variation with temperature takes the form

$$C_P = a + bT + cT^{-2}$$

SOLUTION. The theoretical adiabatic flame temperature for an organic compound has been previously defined as the maximum temperature the flue products of combustion will achieve if the reactants of combustion are at ambient conditions. This temperature is achieved with adiabatic (no heat lost to the surroundings) operation and with theoretical (stoichiometric or 0% excess) air.

The standard heat of combustion for chlorobenzene is obtained from the heats of formation data in Table 5.4.1. Since

$$C_6H_5Cl + 7O_2 \longrightarrow 6CO_2 + 2H_2O(g) + HCl(g)$$

$$\Delta H^\circ_c = (6)(-94,052) + (2)(-57,789) + (-22,063) - 12,390$$

$$= -714,361 \text{ cal/gmol}$$

Note: This value is the net or lower heating value since the water is in the gaseous state.

This stoichiometric reaction is now written for combustion in air. We first note that there are 7.0(79/21) or 26.33 lbmol of nitrogen present in the theoretical combustion air.

$$C_6H_5Cl + 7O_2 + 26.33N_2 \longrightarrow 6CO_2 + 2H_2O(g) + HCl(g) + 26.33N_2$$

The heat capacity for the flue gas products in the form

$$C_P = a + bT + cT^{-2}$$

are read from Table 5.2.4.

$$C_{P,CO_2} = 10.57 + 2.10 \times 10^{-3} T - 2.06 \times 10^5 T^{-2}$$
$$C_{P,H_2O} = 7.30 + 2.46 \times 10^{-3} T + 0.0 \times 10^5 T^{-2}$$
$$C_{P,HCl} = 6.27 \times 1.24 \times 10^{-3} T + 0.3 \times 10^5 T^{-2}$$
$$C_{P,N_2} = 6.83 + 0.90 \times 10^{-3} T - 0.12 \times 10^5 T^{-2}$$

However,

$$\Delta C_P = 6C_{P,CO_2} + 2C_{P,H_2O} + C_{P,HCl} + 26.33C_{P,N_2}$$

Thus,

$$\begin{aligned}
\Delta C_P = {} & 6(10.57 + 2.10 \times 10^{-3}\,T - 2.06 \times 10^5\,T^{-2}) \\
& + 2(7.30 + 2.46 \times 10^{-3}\,T - 0.0 \times 10^5\,T^{-2}) \\
& + 1(6.27 + 1.24 \times 10^{-3}\,T + 0.30 \times 10^5\,T^{-2}) \\
& + 26.33(6.83 + 0.90 \times 10^{-3}\,T - 0.12 \times 10^5\,T^{-2})
\end{aligned}$$

If this is expressed in the form $\Delta C_P = \Delta a + \Delta bT + \Delta cT^{-2}$, then

$$\Delta C_P = 264.12 + 0.0425T - 1.522 \times 10^6\,T^{-2} \text{ cal/gmol-K or Btu/lbmol-°R}$$

Equation (5.4.14) applies in calculating the adiabatic flame temperature. The energy liberated on combustion appears as sensible energy in heating the flue (product) gas. The sum of these two effects is zero if the operation is conducted adiabatically. This is treated in more detail in Chapter 6.

$$\Delta H_c^\circ + \Delta H_p = \Delta H = 0$$

Since 25°C = 298 K, the enthalpy change associated with heating the flue products is given by

$$\Delta H_p = \int_{298}^{T_2} \Delta C_P\, dT; \quad T_2 = \text{theoretical adiabatic temperature (K)}$$

Substituting ΔC_P obtained previously and integrating yields

$$\begin{aligned}
\Delta H_p &= \Delta a(T_2 - 298) + (\Delta b/2)(T_2^2 - 298^2) - \Delta c\left(\frac{1}{T_2} - \frac{1}{298}\right) \\
&= 264.12(T_2 - 298) + (0.0425/2)(T_2^2 - 298^2) + 1.522 \times 10^6\left(\frac{1}{T_2} - \frac{1}{298}\right) \\
&= -\Delta H_c^\circ \\
&= 714{,}361 \text{ cal/gmol}
\end{aligned}$$

This equation may now be rewritten in the form

$$800{,}063 = 264.12\,T_2 + 0.02125\,T_2^2 + (1.522 \times 10^6)/T_2$$

This is a nonlinear cubic equation. It may be solved by any one of several analytical or numerical methods. The final result can be obtained by a crude trial and error procedure. However, this trial and error process can be simplified by recognizing that on the right-hand side, the last term is an order of magnitude smaller than the first two terms. Therefore, an excellent first guess can be obtained by solving the equation

$$0.02125\,T_2^2 + 264.12\,T_2 - 800{,}063 = 0$$

$$T_2 = \frac{-264.12 \pm \sqrt{(264.12)^2 + (4)(0.02125)(800{,}063)}}{(2)(0.02125)}$$

$$T_2 = 2519\,\text{K} = 4534°\text{R} = 4074°\text{F}$$

Final theoretical adiabatic flame temperature is 4074°F.

Example 5.7.10. Calculate the theoretical adiabatic flame temperature for DDT. Perform the calculations using SI (cgs) units.

SOLUTION. The standard enthalpy (heat) of combustion is again obtained from Table 5.4.1.

$$\Delta H_c^\circ = -1600\,\text{kcal/gmol}$$

As noted earlier, these combustion values assume that the water and hydrogen chloride are in the liquid and gaseous state, respectively. The enthalpy of vaporization of water at 60°F is 10.60 kcal/gmol of H_2O vaporized. The balanced stoichiometric combustion equation is

$$C_{14}H_9Cl_5 + 15O_2 + 56.43N_2 \longrightarrow 14CO_2 + 2H_2O + 5HCl + 56.43N_2$$

With water in the vapor state, the standard heat of combustion (the net heating value) is

$$\Delta H_c^\circ = -1600 + 2(10.60)$$
$$= -1578.8 \approx -1580\,\text{kcal/gmol}$$

(A value of -1580 kcal/gmol will be used in this and subsequent problems for the net heating value of DDT at 60°F.)

The describing equation for the adiabatic flame temperature is

$$\Delta H_c^\circ + \Delta H_p = \Delta H = 0$$

Heat capacity data of the form

$$C_P = a + bT + \frac{c}{T^2}$$

are again obtained from Table 5.2.4.

The term ΔC_P is

$$\Delta C_P = 14C_{P,\,CO_2} + 2C_{P,\,H_2O} + 5C_{P,\,HCl} + 56.43C_{P,\,N_2}$$

Substituting values for $C_{P,\,i}$ yields

$$\Delta C_P = 579.35 + 9.13 \times 10^{-2}\,T - 3.41 \times 10^6\,T^{-2}$$

The sensible enthalpy term is

$$\Delta H_p = \int_{298}^{T_2} C_P\,dT$$

$$= 579.35(T_2 - 298) + \frac{9.13 \times 10^{-2}}{2}(T_2^2 - 298^2) + 3.41 \times 10^6\left(\frac{1}{T_2} - \frac{1}{298}\right)$$

Setting the right-hand side of this equation equal to the net heating value yields

$$1580 \times 10^3 = 579.35(T_2 - 298) + \frac{9.13 \times 10^{-2}}{2}(T_2^2 - 298^2) + 3.41 \times 10^6\left(\frac{1}{T_2} - \frac{1}{298}\right)$$

A trial-and-error solution yields

$$T_2 = 2541\,\text{K} = 2268°\text{C} = 4574°\text{R} = 4114°\text{F}$$

The theoretical adiabatic flame temperature for DDT is therefore 2541 K.

The reader should note that this is a *theoretical* thermochemical calculation. For practical purposes, 3400°F is about the maximum flame temperature that a refractory-lined incinerator can withstand.

Example 5.7.11. Calculate the chemical reaction equilibrium constant K for reactions that have $\Delta G°$ values of $+20.0$ and -20.0 kcal/gmol at 70°F.

SOLUTION. From Eq. (5.5.4)

$$\ln K = -\frac{\Delta G_T°}{RT}$$

With $T = 70°\text{F} = 294\,\text{K}$
   For $\Delta G° = 20.0$ kcal/gmol

$$\ln K = -\frac{(20,000)}{(1.99)(294)}$$

$$K = 1.4251 \times 10^{-15}$$

   For $\Delta G° = -20.0$ kcal/gmol

$$\ln K = \frac{(20,000)}{(1.99)(294)}$$

$$K = 7.017 \times 10^{14}$$

The reader should note the effect of $\Delta G°$ has on the numerical value of K.

Example 5.7.12. The equilibrium constant K_P for the reaction (at 1 atm)

$$2HCl(g) + 0.5O_2 \longrightarrow Cl_2 + H_2O(g)$$

may be expressed in the form [5]

$$K_P = A e^{B/T}$$

where $A = 0.229 \times 10^{-3}$
$B = 7340$
$T = $ absolute temperature (K)

As part of a hazardous waste incineration project, Dr. Theodore is required to develop a more rigorous equation for K_P as a function of the absolute temperature T. Tied up by other nonessential departmental assignments, he has chosen to hire you as a subcontractor on the project to obtain this information.

SOLUTION. First calculate ΔG_{298}° for the reaction using Eq. (5.5.3)

$$\Delta G_{298}^\circ = 1\Delta G_{f, H_2O}^\circ - 2\Delta G_{f, HCl}^\circ$$

From Table 5.5.1, $\Delta G_{f, H_2O}^\circ$ and $\Delta G_{f, HCl}^\circ$ can be obtained:

$$\Delta G_{f, H_2O}^\circ = -54{,}635 \, \text{cal/gmol}$$

$$\Delta G_{f, HCl}^\circ = -22{,}778 \, \text{cal/gmol}$$

$$\Delta G_{298}^\circ = (1)(-54{,}635) - (2)(-22{,}778)$$

$$= -9079 \, \text{cal/gmol}$$

Similarly, ΔH_{298}° is obtained from Table 5.4.1:

$$\Delta H_{298}^\circ = -13{,}672 \, \text{cal/gmol}$$

According to Eq. (5.5.9),

$$\ln K = \frac{-\Delta H_0}{RT} + \frac{\Delta \alpha}{R} \ln T + \frac{\Delta \beta}{2R} T + \frac{\Delta \gamma}{6R} T^2 + I$$

or, its equivalent form for $C_P = a + bT + cT^{-2}$,

$$\ln K = \frac{-\Delta H_0}{RT} + \frac{\Delta a}{R} \ln T + \frac{\Delta b}{2R} T + \frac{\Delta c}{2R} T^{-2} + I$$

Next, ΔH_0 and I must be determined. ΔH_0 is found by a procedure similar to that used in Section 5.4. From Eq. (5.4.10),

$$\Delta H_T^\circ = \Delta H_{298}^\circ + \int_{298}^{T} \Delta C_P \, dT \qquad (5.4.10)$$

The combustion reaction is

$$2\,HCl(g) + 0.5\,O_2(g) \longrightarrow H_2O(g) + Cl_2(g)$$

The heat capacities for the flue gas products in the form

$$C_P = a + bT + cT^{-2}$$

are read from Table 5.2.4.

$$\Delta a = (7.30 + 8.85) - [(2)(6.27) + (0.5)(7.16)]$$

$$= 0.03$$

$$\Delta b = (2.46 \times 10^{-3} + 0.16 \times 10^{-3}) - [(2)(1.24 \times 10^{-3}) + (0.5)(1.0 \times 10^{-3})]$$

$$= -3.6 \times 10^{-4}$$

$$\Delta c = (0.0 - 0.68 \times 10^5) - [(2)(0.30 \times 10^5) + (0.5)(-0.4 \times 10^5)]$$

$$= -1.08 \times 10^5$$

Equation (5.4.10) becomes

$$\Delta H_T^\circ = \Delta H_{298}^\circ + \int_{298}^{T} [\Delta a + (\Delta b)\,T + (\Delta c)\,T^{-2}]\,dT$$

or

$$\Delta H_T^\circ = H_{298}^\circ + \Delta a\,(T - 298) + \frac{1}{2}\Delta b\,(T^2 - (298)^2) - \Delta c\left(\frac{1}{T} - \frac{1}{298}\right)$$

Combining the constant terms into ΔH_0 (as in Section 5.4) yields the following:

$$\Delta H_T^\circ = \Delta H_0 + (\Delta a)\,T + \tfrac{1}{2}(\Delta b)\,T^2 - (\Delta c)\,T^{-1}$$

where

$$\Delta H_0 = \Delta H_{298}^\circ - 298\Delta a - \frac{1}{2}(298)^2\,\Delta b + \frac{1}{298}\Delta c$$

$$= -13{,}672 - 298(0.03) - \frac{1}{2}(298)^2(-3.6 \times 10^{-4}) + \frac{1}{298}(-1.08 \times 10^5)$$

$$= -14{,}027\,\text{cal/gmol}$$

From Eq. (5.5.4),

$$\ln K = \frac{-\Delta G_{298}^\circ}{RT} = \frac{9079}{(1.99)(298)}$$

$$= 15.31$$

Therefore,

$$15.31 = \frac{14027}{(1.99)(298)} + \frac{0.03}{1.99}(\ln 298) + \frac{-3.6 \times 10^{-4}}{(2)(1.99)}(298) + \frac{-1.08 \times 10^5}{(2)(1.99)}(298)^{-2} + I$$

Solving for I,

$$I = -8.09$$

The final form of the equation for K is

$$\ln K = \frac{7048.7}{T} + 0.0151 \ln T - 9.06 \times 10^{-5}\, T - 2.714 \times 10^4\, T^{-2} - 8.09$$

Example 5.7.13. Refer to the previous example. The chemical reaction equilibrium constant based on partial pressure (K_P) was obtained as a function of temperature for the reaction

$$2HCl + 0.5O_2 \longrightarrow Cl_2 + H_2O$$

The final result took the form:

$$\ln K = \frac{7048.7}{T} + 0.0151 \ln T - 9.06 \times 10^{-5}\, T - 2.714 \times 10^4\, T^{-2} - 8.09$$

If the initial partial pressures of HCl, O_2, Cl_2, and H_2O are 0.146, 0.106, 0.0, and 0.0292, respectively, calculate the equilibrium partial pressure of the Cl_2 if the operating conditions are 1.0 atm and 1250 K.

SOLUTION. By definition,

$$K_P = \frac{P_{Cl_2} P_{H_2O}}{P_{HCl}^2 P_{O_2}^{0.5}}$$

A rigorous procedure to calculate the equilibrium partial pressures is available elsewhere.[3] The following procedure is satisfactory for purposes of engineering calculations. At equilibrium,

$$P_{Cl_2} = P_{Cl_2}(\text{initial}) + x$$
$$= x$$

The term x represents the increase in the partial pressure of the chlorine due to this equilibrium reaction:

$$P_{H_2O} = P_{H_2O}(\text{initial}) + x = 0.0292 + x$$
$$P_{HCl} = P_{HCl}(\text{initial}) - 2x = 0.146 - 2x$$
$$P_{O_2} = P_{O_2}(\text{initial}) - 0.5x = 0.106 - 0.5x$$

K_P can then be expressed as

$$K_P = \frac{(x)(0.0292 + x)}{(0.146 - 2x)^2(0.106 - 0.5x)^{0.5}}$$

Now, calculate K_P at 1250 K using the result from the previous example.

$$\ln K_P = \frac{7048.7}{1250} + 0.015(\ln 1250) - 9.06 \times 10^{-5}(1250) - 2.714 \times 10^4(1250)^{-2} - 8.09$$

$$= -2.475$$

$$K_P = 0.0842$$

Therefore,

$$0.0842 = \frac{(x)(0.0292 + x)}{(0.146 - 2x)^2(0.106 - 0.5x)^{0.5}}$$

Solving for x, which is the equilibrium partial pressure of Cl_2, by trial-and-error calculation yields

$$P_{Cl_2} = x = 0.01050 \, \text{atm}$$

Note: Approximately 1% of the discharge flue gas is chlorine—a rather sizable amount.

Example 5.7.14. Calculate the residence time required to achieve 99.99% destruction efficiency of benzene if an incinerator is operated at a temperature of 980°C. The literature provides values of the Arrhenius constant (A) and the activation energy (E) of $3.3 \times 10^{10} \, \text{s}^{-1}$ and 35,900 cal/gmol, respectively.

SOLUTION. As discussed in the previous section, many chemical reactions have been shown to be satisfactorily described by a first-order reaction, that is,

$$\frac{dC}{d\theta} = -kC$$

where C = concentration of the material undergoing reaction
k = reaction rate constant
θ = time

If the initial concentration is C_0, the solution to this differential equation becomes [see Eq. (5.6.20)]

$$\ln\left(\frac{C}{C_0}\right) = -k\theta$$

The temperature dependence of k is given by the Arrhenius equation (5.6.6):

$$k = A \exp\left(\frac{-E}{RT}\right)$$

where A = frequency factor (units of time^{-1})
$\quad E$ = activation energy (cal/gmol or Btu/lbmol)
$\quad T$ = K or °R
$\quad R$ = 1.987 cal/gmol-K or Btu/lbmol-°R

For 99.99% destruction efficiency,

$$\frac{C}{C_0} = 0.0001$$

From the Arrhenius equation,

$$k = A \exp\left(-\frac{E}{RT}\right) = 3.3 \times 10^{10} \exp\frac{(-35,900)}{(1.987)(980 + 273)}$$

$$= 1.804 \times 10^4\, s^{-1}$$

The required residence time is, therefore,

$$\theta = -\frac{\ln(C/C_0)}{k} = -\frac{(\ln 0.0001)}{(1.804 \times 10^4)}$$

$$= 5.11 \times 10^{-4}\, s$$

PROBLEMS

1. Along with 2000 lb/h of hazardous waste, 5000 lb/h of air is fed into an incinerator. To assist in the complete combustion of the waste 3000 lb/h of propane is added to the incinerator. What is the amount of product in lb/h?

2. Carbon tetrachloride at 7000 scfm (60°F, 1 atm) is combusted at 1250°F in the presence of oxygen fed at the rate of 12,000 scfm. All feeds enter the incinerator at 120°F. At what rate (lb/min) do the products exit from the incinerator? The molecular weight of CCl_4 is 153.82.

3. Complete combustion of carbon disulfide results in combustion products of CO_2 and SO_2 according to the reaction

$$CS_2 + O_2 \longrightarrow CO_2 + SO_2$$

(a) Balance this reaction.
(b) If 500 lb of CS_2 is combusted with 225 lb of oxygen, which is the limiting reactant?
(c) How much of each product is formed (lb)?
MW of CS_2 is 76.14; MW of SO_2 is 64.07; MW of CO_2 is 44.

4. Combustion products at 25,000 lb/h exit an incinerator at 1700°F and then

transfer 6.00×10^6 Btu/h of heat to a waste heat boiler. If the waste heat boiler operates adiabatically, and the heat capacity of the flue products over the appropriate temperature range is 0.25 Btu/lb-°F, at what temperature will the gases leave the boiler?

5. Benzene is incinerated at 2100°F in the presence of 50% excess oxygen. Balance the combustion reaction

$$C_6H_6 + O_2 \longrightarrow CO_2 + H_2O + O_2$$

6. Find the enthalpy of combustion of cyclohexane from enthalpy of formation data. Perform the calculation for both liquid and gaseous cyclohexane.

7. Calculate the gross heating value in Btu/lbmol for a combustible gas mixture of 75 mol% methane, 10 mol% propane, and 15 mol% n-butane.

8. Calculate the theoretical adiabatic flame temperature of benzene.

9. Refer to Example 5.7.13. Calculate the equilibrium partial pressure of chlorine at a temperature of 2225°F.

10. Repeat Problem **9** if the temperature is 2770°F.

11. The equilibrium constant (K_P) for the reaction (at 1-atm pressure)

$$SO_2 + \tfrac{1}{2}O_2 = SO_3$$

may be expressed in the form [5]

$$Y = A e^{BX}$$

where $A = 0.148 \times 10^{-4}$ $Y = K_P$
$\quad\quad\ B = 11,700$ $\quad\quad X = 1/T;\ T = $ kelvin (K)

Develop a more rigorous equation for K_P as a function of the absolute temperature (K).

12. Refer to Problem 11. The chemical reaction equilibrium constant based on partial pressures (K_P) was obtained as a function of temperature for the reaction

$$SO_2 + \tfrac{1}{2}O_2 = SO_3$$

The final result took the form:

$$\ln K = \frac{11,996}{T} - 0.362 \ln T + 9.36 \times 10^{-4} T - 2.969 \times 10^5 T^{-2} - 9.88$$

If the initial partial pressures of SO_2, O_2, and SO_3 are 0.000257, 0.102, and 0.0 atm, respectively, calculate the equilibrium partial pressure of the SO_3. The operating conditions are 1.0 atm and 1250 K.

13. Resolve Problem **12** if the temperature is 2770°F.

REFERENCES

1. R.C. Weast, ed., *CRC Handbook of Chemistry and Physics*, 51st ed., The Chemical Rubber Company, Cleveland, OH, 1971.
2. J.M. Smith and H.C. Van Ness, *Introduction to Chemical Engineering Thermodynamics*, 3rd ed., McGraw-Hill, New York, 1975.
3. T. McGowan, private communication, 1986.
4. W.A. Rubey, "Design Considerations for a Thermal Decomposition Analytical System," EPA-600/2-80-098, Cincinnati OH, 1980.
5. L. Theodore, personal notes, 1985.

6

Thermochemical Applications

6.1 INTRODUCTION

In Chapter 5, a number of scientific principles, mostly from the fields of chemistry and chemical engineering, were presented. The presentation was general and the principles studied pertain to almost any type of chemical reaction. In this chapter, these same concepts are applied to the incineration process. *Mass conservation* and *stoichiometry* (Sections 5.2 and 5.3) are employed in the *Stoichiometric Calculations* of Section 6.3, in which the prediction of the flue gas composition from the incineration of waste–fuel mixtures of known composition is discussed. *Thermochemistry* (Section 5.4) and *chemical reaction equilibrium* (Section 5.5) principles are applied in Section 6.4, *Thermochemical Calculations*, to predict the incinerator operating temperature as a function of the waste–fuel mixture, net heating value, and excess air ratio. *Chemical kinetics* (Section 5.6) are applied in Section 6.5, *Applications of Chemical Reaction Principles*, to predict the all-important *residence time* requirement for the destruction of a particular waste.

This chapter begins with a discussion of *Fuel Options* (Section 6.2). In practice, the wastes to be incinerated may have too low a heating value to sustain combustion. In these cases, auxiliary fuel must be burned with the waste. The types of fuel available and a discussion of selection criteria are the subjects of this section. This chapter concludes with sections on *Illustrative Examples* (Section 6.6) and *Problems*.

6.2 FUEL OPTIONS

The properties of the fuels most often used in a hazardous waste incineration facility are reviewed here. Fuels burned in an incinerator may be gaseous, liquid, or solid. Some of the more common fuels are the following:

178

- *Gaseous fuels* are principally natural gas (80–95% methane, the balance ethane, propane, and small quantities of other gases). Light hydrocarbons obtained from petroleum or coal treatment may also be used.
- *Liquid fuels* are mainly hydrocarbons obtained by distilling crude oil (petroleum). The various grades of fuel oil, gasoline, shale oil, and various petroleum cuts and residues are considered in this category.
- *Solid fuels* consist principally of coal (a mixture of carbon, water, noncombustible ash, hydrocarbons, and sulfur). Coke, wood, and solid waste (garbage) may also be employed.

These classifications are not mutually exclusive and necessarily overlap in some areas. The stoichiometric and thermochemical analyses of problems involving the different classes of fuels are similar. Consequently, the actual physical form of the fuel is not important in any calculational study.

In the following sections, fuels are discussed primarily in terms of their usefulness and economic utilization in assisting the incineration process. Properties of fuels considered to be of prime importance are composition and heating value. The heating value refers to the quantity of heat released during combustion of a unit amount of fuel gas. Thus, the heating value of a fuel is one of its most important properties. As described in Chapter 5, the major products of the complete combustion of a fuel are CO_2 and H_2O. Because of the presence of varying amounts of water in different fuels, two methods for expressing heating value are in common use. Once again, the *gross* or *higher heating value* is defined as the amount of heat evolved in the complete combustion of a fuel with any water present in the liquid state; the *net* or *lower heating value* is defined as the amount of heat evolved in the combustion of a fuel when all the products, including the water, are in the gaseous state. Thus, the gross heating value is greater than the net heating value by the latent heat of vaporization of the total amount of water originally present and that formed by oxidation of the hydrogen in the fuel.

In dealing with the incineration of some fuels, it is occasionally convenient to express the composition in terms of a single hydrocarbon, even though the fuel is a mixture of many hydrocarbons. For example, gasoline is usually considered to be octane (C_8H_{18}) and diesel fuel is considered to be dodecane ($C_{12}H_{26}$). However, the composition of most fuels employed in waste incineration is given in terms of its componental (or ultimate) analysis.

Approximate componental analyses of the three types of fuels most often employed at hazardous waste incineration facilities are given in Table 6.2.1. Estimated values for the net heating value are also included.

Natural Gas

Natural gas is perhaps the closest approach to an ideal fuel, because it is practically free from noncombustible gas or particulate residue. Of the many gaseous fuels, natural gas is the most important for use at hazardous waste incineration facilities. No fuel preparation is necessary at the site because gases are easily mixed with air, and the combustion reaction proceeds rapidly once the ignition temperature is reached.

Natural gas is found compressed in porous rock or cavities, which are sealed

TABLE 6.2.1. Componential Analysis for Common Fuels

Element	Weight Fraction (lb element/lb fuel)		
	Residual Fuel Oil (No. 6)	Distillate Fuel Oil (No. 2)	Natural Gas
C	0.866	0.872	0.693
H	0.102	0.123	0.227
N	—	—	0.08
S	0.03	0.005	—
Net Heating Value	19,500 Btu/lb	18,600 Btu/lb	1,0uu Btu/scf (60°F)

between strata under the earth's surface. When these gas-bearing pools are tapped by drilling wells, the gas is found to be under pressure, which may be as high as 2000 psig. As gas is withdrawn, this pressure gradually decreases until it eventually becomes so low that the field must be abandoned. Thus, the natural gas from a well will usually not remain constant in its elemental composition and/or heating value.

Natural gas consists mainly of methane, with smaller quantities of other hydrocarbons, particularly ethane, and trace amounts of propane. Carbon dioxide and nitrogen are usually present in small amounts and sometimes there can be appreciable amounts of hydrogen sulfide present. (This is generally removed at the field before transmission.)

The physical and chemical characteristics of natural gas are influenced by the underground conditions existing in the localities where it is found. The gas may contain heavy saturated hydrocarbons, which are liquid at ordinary pressures and temperatures. Dry natural gas contains < 0.1 gal of gasoline vapor/1000 ft^3; > 0.1 gal/1000 ft^3 is termed *wet*. Note that the terms *wet* and *dry*, when dealing with natural gas, refer to its gasoline content, not to its moisture content. Natural gases are also classified as either *sweet* or *sour*. A *sour* gas is one that contains some mercaptans and a high percentage of hydrogen sulfide, while the *sweet* gas is one in which these compounds are present in trace quantities. Most natural gases employed at a hazardous waste incineration facility are *dry* and *sweet*.

Environmental concerns with natural gas are minimal. Most gaseous fuels, with the possible exception of some waste gases, are considered to be clean fuels. Pipeline-grade natural gas is virtually free of sulfur and particulates. Its flue products do not pollute water. In addition, natural gas transportation and distribution facilities have a minimal adverse ecological impact. However, leakage of natural gas can pose a very serious explosion hazard.

The principal air contaminants from gaseous fuels, which are affected by the combustion system design and operation, are the oxidizable materials—carbon monoxide, carbon, and unburned hydrocarbons. Burner design operating temperature and excess air also affect the production of the oxides of nitrogen. In some instances, the impurity in a gaseous fuel may be hydrogen sulfide. Another important sulfur impurity is carbon disulfide. In most gases, however, the total organic sulfur content is relatively small. Other sulfur compounds that may be present in trace amounts are the thiophenes, carbon oxysulfide, mercaptans, thioesters, and organic sulfide.

Proper operation of a waste facility employing natural gas requires that the fuel rate be controlled in relation to the demand. The air supply must be appropriate to the fuel supply. This is usually accomplished by automatic control. The incoming gas supply is regulated at a constant pressure upstream of the control valve. This valve can be used to control the gas flow. Combustion air regulation is achieved through manipulating dampers or by a special draft controller. Large installations are likely to use more elaborate systems where the fuel and air flows are metered with automatic adjustment to compensate for any changes or disturbances. Natural gas passes through one or more fixed orifices before entering the incineration chamber. Since flow through an orifice is proportional to the square root of the pressure drop across it, small fluctuations of the upstream pressure will not have a very significant effect on the gas flow rate. However, should it be necessary to reduce the firing rate to 50% of its normal value (2-to-1 turndown), for example, a fourfold decrease in gas pressure would be required, with the air flow rate adjusted accordingly. This factor can present a control problem. Before initiating firing with natural gas, the gas injection orifices should be inspected to verify that all passages are unobstructed. Filters and moisture traps should be clean, in place, and operating effectively to prevent any plugging of gas orifices. Proper location and orientation of diffusers should also be confirmed.

Liquid Fuels

Fuel oil may be defined as petroleum or any of its liquid residues remaining after the more volatile constituents have been removed. Thus, the term *fuel oil* may conveniently cover a wide range of petroleum products. It may be applied to crude petroleum, to a light petroleum fraction similiar to kerosene or gas oil, or to a heavy residue left after distillation. The principal industrial liquid fuels are therefore the by-products of natural petroleum. These fuel oils are marketed in two principal classes: distillates and residuals. The principal industrial boiler fuel is residual oil, known as No. 6 or *Bunker C*. The residual oil, as the name implies, is left over after the more valuable products are distilled off. As an industrial and utility fuel, it competes with coal, although its price has fluctuated wildly over the last 15 yr. It is specified mainly by viscosity. Grades No. 1 and 2 are sometimes designated as *light* and *medium* domestic fuel oils and are specified mainly by the temperature of the distillation range and specific gravity.

Despite the multiplicity of chemical compounds found in fuel oils, there are some important physical and chemical properties that deserve comment.

1. *Specific Gravity*. For fuel oils, specific gravity is usually taken as the ratio of the density of the oil at 60°F to the density of water at 60°F. Gravity determinations are readily made by immersing a specific gravity meter into the oil and reading the scale at the point to which the instrument sinks. The specific gravity is then read directly. The American Petroleum Institute (API), the U.S. Bureau of Mines, and the U.S. Bureau of Standards agreed to recommend that only one scale be used in the petroleum industry, and that it be known as the API scale. This relationship is given by Eq. (4.3.7).

2. *Viscosity*. The Saybolt Universal viscosity is expressed as the amount of time (in seconds) that it takes to run 60 cm^3 of the oil through a standard size orifice at any desired temperature. Viscosity is commonly measured at 100, 150, and 210°F. Fuel oil (particularly No. 6) is very viscous, and it takes a long time to make a determination with the Saybolt Universal viscometer. For this reason, the viscosity of fuel oil may be measured with a Saybolt Furol viscometer, which is the same as the Saybolt Universal viscometer except that the orifice is larger. Interestingly, the viscosity of fuel oil decreases as the temperature rises and becomes nearly constant above about 250°F. Therefore, when fuel oil is heated to reduce the viscosity for good atomization, there is little gain in heating the oil beyond 250°F.

3. *Heating Value*. This may be expressed in either British thermal units per gallon or British thermal units per pound (at 60°F). The heating value per gallon increases with specific gravity because there is more weight for a given volume; values range from ~ 140,000 to 150,000 Btu/gal. The heating value per pound of fuel oil ranges in value from 18,600 to 19,500 Btu/lb (see Table 6.2.1).

4. *Flash and Fire Point*. The *flash point* of fuel oil is the lowest temperature at which sufficient vapor is given off to form a momentary flash when flame is brought near the surface of the oil in a small container called a Cleveland Cup. The *fire point* is the lowest temperature at which the oil gives off enough vapor to burn continuously. Stored fuel oils (as well as liquid wastes) should be maintained significantly below their flash points.

The rate of combustion of fuel oil is limited by vaporization. Light distillate oils readily vaporize; other (heavier) fuel oils, because of their heavier composition, require additional equipment to assure vaporization and complete combustion. In order to achieve complete combustion, oils are atomized into small droplets for rapid vaporization. The rate of evaporation is dependent on surface area, which increases as the atomized droplet size becomes smaller. The desired shape and droplet size are influenced adversely if the fuel viscosity is too high. At ambient temperature, No. 2 fuel oil may be atomized with little difficulty, but typically No. 6 fuel oil must be heated to around 210°F to assure proper atomization. Dirt and foreign matter suspended in the oil may cause wear in the oil pump and blockage of the atomizing nozzles. Strainers or replaceable filters are required in both the oil suction line and the discharge line. During combustion of a distillate fuel oil, the droplet becomes uniformly smaller as it vaporizes. By contrast, a residual oil droplet undergoes thermal and catalytic cracking, and its composition and size undergo various changes with time. Vapor bubbles may form, grow, and burst within a droplet in such a way as to shatter the droplet as it is heated in the combustion zone.

The physical and chemical properties of the oil and the characteristics of the incineration equipment influence the air pollution emissions from a waste facility. Some of these properties are

- *Sulfur Content*. Sulfur is a very undesirable element in fuel oil because its products of combustion are acidic and cause corrosion in the ducting, valves, downstream equipment, and so on.
- *Solid Impurities*. Fuel oil usually contains all the solid impurities originally present in the crude oil. If these solids contain a large proportion of salt, they are very fusible and can stick to the ducting, valves, waste heat boiler, and so on.

• *Vanadium and Sodium Contents.* The vanadium content in fuel oil may be deposited in the ash on surfaces. These deposits act catalytically in converting SO_2 to SO_3, thereby creating acid dew-point problems. Both sodium and vanadium from fuel oil may form sticky ash compounds having low melting temperatures. These compounds increase the deposition of ash and are corrosive.

Some of the advantages of using fuel oil as the fuel additive at a hazardous waste incineration facility are (1) the fuel oil requires less storage space than coal, (2) is not subject to spontaneous combustion, (3) is the most difficult to ignite of the liquid fuels, (4) is not subject to deterioration, and (5) is free from expensive manual handling.

Coal

Although coal is not usually utilized as an auxiliary fuel, it may be used in future fluidized-bed HWI applications.

There is no satisfactory definition of coal. It is a mixture of organic, chemical, and mineral materials produced by a natural process of growth and decay, accumulation of both vegetal and mineral debris—accomplished by chemical, biological, bacteriological, and metamorphic action. The characteristics of coal vary considerably with location, and even within a given mine some variation in composition is usually encountered. Coals are classified according to rank, which refers to the degree of conversion from one form of coal to another (e.g., lignite to anthracite). The following ranking of coals is employed today: anthracite, bituminous, subbituminous, and lignite.

Solid fuels, including coal, consist of free carbon, moisture, hydrocarbons, oxygen (mostly in the form of oxygenated hydrocarbons), small amounts of sulfur and nitrogen, and nonvolatile noncombustible materials designated as *ash*. Two types of analyses are in common use for expressing the composition of these fuels: the *ultimate* analysis and the *proximate* analysis.

In an *ultimate* analysis, determination is made of the carbon, moisture (water), ash, nitrogen, sulfur, net hydrogen, and perhaps combined water in the fuel. Ash content is usually determined as a whole (one quantity), although a separate analysis can be made on the ash. Since the hydrogen content is always in excess of the amount needed to form water with all the oxygen present, it is assumed that the oxygen is completely united with the available hydrogen, and this combination may be reported as *combined water*. Note that these two elements do react in the combustion process to form water. The hydrogen in excess of that necessary to combine with the oxygen in the fuel is often termed *net hydrogen*, which usually reacts with the oxygen in the combustion air, halogens (to form acids), and so on. Another common approximate method for expressing the composition of a solid fuel is to report it as moisture, volatile combustible matter, fixed carbon, and ash. This is known as a *proximate analysis*. The *moisture* in the proximate analysis is the same as that in the ultimate analysis. The *volatile combustible matter* consists of a large amount of the carbon in the fuel, which is lost as volatile hydrocarbons, plus the hydrogen and combined water, which are given off when combusted. The *fixed carbon* is the carbon left in the fuel after the volatile combustible matter has been removed. *Ash* is the noncombustible residue after complete combustion of the coal. The weight of ash is usually slightly less than that of the mineral matter originally present before burning. However, for high calcium content coals, the ash can be higher than the mineral matter due to retention of some of the sulfur present in the coal. When coal is heated, it becomes soft and sticky, and as the temperature in an incinerator continues to rise, it becomes fluid. Coals with ash that softens and fuses at comparatively low temperatures are likely to deposit the ash within the system and create problems. The sulfur is separately measured, and its amount is useful in judging the air pollution potential of the combustion of the coal. Combustion of sulfur forms oxides, which usually combine with water to form acids that may condense when the flue gas is cooled below its dew point temperature.

Use of coal as a fuel in a hazardous waste incinerator results in particulate (ash) carry-over and sulfur oxides (from sulfur) emissions. Fly ash is not considered a major problem; equipment employed for its control is presented in Chapter 10. Sulfur emissions can constitute a major and additional problem since most of the so-called hazardous wastes do not contain sulfur. The sulfur in coal is found in both organic and inorganic iron pyrites and/or marcasite. There appears to be no evidence that sulfur occurs in coal in its elemental state. The amount of sulfur found in coals varies significantly. During incineration, about 80% of the sulfur in the coal appears in the flue gas in the form of the oxides of sulfur; most of this is SO_2; < 5% is SO_3. Coal cleaning at the mine reduces the ash content and simultaneously reduces the sulfur content by removing some of the iron pyrites. Today, cleaning is accomplished by gravimetric separation, which is a successful method because pyrite is nearly five times denser than coal. Unfortunately, methods to reduce organic sulfur are not economically viable at this time.

In selecting coal as the fuel at an incineration site, its storage must be considered. Coal slowly deteriorates when exposed to weathering. Attention must be given to the manner in which the coal is stockpiled; large piles loosely formed can ignite spontaneously. This problem is most severe with smaller sizes of coal and high sulfur content. Where large amounts must be stored, such as at power stations, stockpiles are created by using heavy equipment to form piles several hundred feet wide, several thousand feet long, and ~ 20 ft high. When smaller quantities are employed and turnover is fairly rapid, conical piles with a 12-ft depth or less are used. Where open piles are not permitted, silos can be used for coal storage; these should be equipped with fugitive dust control for use during loading.

Selection

The two major classes of hazardous waste incinerators are liquid injection and rotary kiln; fluidized-bed incinerators, among other types, are used to a lesser extent (see Chapter 7). No set rule applies to the selection of a given fuel for a particular type of incinerator. However, some general comments and guidelines can be offered. Natural gas and No. 2 distillate oil are the usual options available for a liquid injection incinerator. Number 6 Bunker oil is often not used with smaller units because of its *gunky* properties. Rotary kilns can use natural gas, No. 2 or No. 6 oil, or finely pulverized coal. Number 2 oil appears to be the preferred choice for most existing small liquid injection and rotary kiln units. Due to today's economics, No. 6 oil is most often used in large units. Coal appears to be ideally suited for the fluidized-bed incinerator, but fuel oils have been widely used.

If a significant amount of fuel is to be employed at a hazardous waste incineration facility, the design engineer may require a thorough survey of the various fuels available in the location of the plant. The results of this study will assist in the fuel selection process. The factors in choosing a fuel for a proposed facility are:

Cost of fuel.

Delivered cost of fuel as fired. Transportation should be included in this cost.

Market-price history and trends. These are needed to serve as a guide for future price relationships.

Resources or reserves. These should be extensive to assure future supplies.

Relative availability. This should be considered under both normal and unusual circumstances or demands.

Long-range government conservation. This may be applied to limited reserves, especially when these reserves are in demand for specific essential uses.

6.3 STOICHIOMETRIC CALCULATIONS

In hazardous waste incineration applications, one of the key engineering calculations is the prediction of the flue gas flow rate and composition following incineration of a waste–fuel mixture. Unfortunately, no simple generalized procedure is available to accomplish this task. The resulting flue gas flow rate and composition are obviously strong functions of the amount of air and the composition of both the waste and fuel. However, this flow rate and composition also depend, to a lesser extent, on the operating temperature in the incinerator because of chemical reaction equilibrium effects.

One of the difficulties in developing a comprehensive, overall calculational procedure to determine flue gas flows and compositions is coping with the different methods that are in use to characterize wastes. As described in Chapter 1, the waste characterization often takes the form of an ultimate analysis, but there are other ways in which the waste may be described. The four categories of characterization include:

1. Ultimate or elemental analysis—mass basis, fraction or percentage.
2. Ultimate analysis—mole basis.
3. Componential or compound analysis—mass basis.
4. Componential analysis—mole basis.

If an ultimate analysis on a mass basis is provided for the waste–fuel mixture, stoichiometric (or theoretical) air mass requirements per unit mass of mixture (m_{st}) may be calculated from

$$m_{st} = 11.5m_C + 34.5m_H - 4.29m_O - 0.97m_{Cl} + 4.29m_S \qquad (6.3.1)$$

where
m_{st} = stoichiometric air requirement per unit mass of waste–fuel mixture (lb air/lb mixture)

m_C = mass fraction of carbon (lb C/lb mixture)

m_H, m_O, m_{Cl}, m_S = mass fractions of hydrogen, oxygen, chlorine, and sulfur, respectively

The composition of the flue gas produced on a mass basis (for stoichiometric combustion only) is then

$$m_{CO_2} = 3.67m_C$$

$$m_{H_2O} = 9.0m_H - 0.25m_{Cl} + \{m_{H_2O, w}\}$$

$$m_{SO_2} = 2.00m_S$$

$$m_{N_2} = 8.78m_C + 26.3m_H - 3.29m_O - 0.74m_{Cl} + 3.29m_S + m_{N, w}$$

$$m_{HCl} = 1.03m_{Cl} \tag{6.3.2}$$

where
$$m_{CO_2} = \text{mass } CO_2 \text{ in the flue gas (lb } CO_2/\text{lb mixture)}$$
$$m_{H_2O}, m_{SO_2}, m_{N_2}, m_{HCl} = \text{masses of each product compound per unit mass of the waste–fuel mixture}$$
$$m_{N, w} = \text{mass fraction of nitrogen in the waste–fuel mixture (lb N/lb mixture)}$$
$$\{m_{H_2O, w}\} = \text{mass fraction of water in the waste–fuel mixture (lb } H_2O/\text{lb mixture); used only when part of the weight fraction of the waste–fuel mixture is expressed as water or moisture.}$$

This set of equations assumes that there is enough hydrogen available in the mixture to completely convert the chlorine to hydrogen chloride. The bracketed water term accounts for any water present in the waste–fuel mixture if a mass fraction of water or moisture is included in the analysis. The $m_{N, w}$ term accounts for any nitrogen that may be present in either the fuel or waste. If the ultimate analysis is on a *mole* basis, a balanced stoichiometric equation of the following form may be used.

$$C_z H_y O_x Cl_w S_v N_u + [z + \phi + v - \tfrac{1}{2}x]O_2 + \tfrac{79}{21}(z + \phi + v - \tfrac{1}{2}x)N_2 \longrightarrow$$
$$z\,CO_2 + 2\phi\,H_2O + w\,HCl + v\,SO_2 + [\tfrac{1}{2}u + \tfrac{79}{21}(z + \phi + v - \tfrac{1}{2}x)]N_2 \tag{6.3.3}$$

where z, y, x, w, v, u = number of moles (or mole fractions) of C, H, O, Cl, S, N present in the waste–fuel mixture, respectively

$$\phi = \tfrac{1}{4}(y - w) \quad \text{when } y > w$$
$$= 0 \qquad\qquad \text{when } y \leqslant w$$

Note that the parameter ϕ is included if its value is positive, and ignored if it is zero or negative. For this approach, the mole fraction for carbon in the waste–fuel mixture (x_C) is given by

$$x_C = \frac{z}{z + y + x + w + v + u} \tag{6.3.4}$$

with similar equations applicable to hydrogen, oxygen, and so on. This assumes that the waste–fuel mixture consists only of C, H, O, Cl, S, and N. Finally, Eq. (6.3.1) through (6.3.4) are valid provided:

- Cl_2, SO_3, and NO_x formation is neglected.
- Oxygen in the waste–fuel mixture is available for combustion and is therefore treated as a *credit* in calculating stoichiometric air requirements.
- The air essentially consists of 79% nitrogen and 21% oxygen (by mole or volume).
- Combustion is complete.

Whether the calculations are performed on a mass or a mole basis, the *excess* air usage requirements for a particular application may easily be included in this analysis. For example, for 100% excess air, twice the stoichiometric air requirement would be employed; half of this air would be used in stoichiometrically combusting the waste–fuel mixture and the remaining half would be carried through the process in the flue gas as the *extra* or excess air.

The preferred choice of the waste characterization method in performing these calculations is often dictated by the contents of the waste mixture. If the waste contains an agglomeration of organics, ultimate analysis would make the calculations easier; if the waste consists of one hydrocarbon (e.g., benzene) in water, the componential analysis is preferred. The choice as to whether the analysis is on a mass or mole basis may also be dictated by the calculational procedure employed.

The EPA has provided worksheets to simplify the calculation of the mass fraction and standard volumetric flow rate of the flue gas. The worksheets in their original form are given in Tables 6.3.1 and 6.3.2 for liquid injection and rotary kiln incinerators, respectively.[1] This calculational procedure is based on the availability of an ultimate analysis for both the waste and any fuel employed. If the analysis data are based on moles or on one individual species, the user must convert the data to an ultimate analysis based on mass. In addition, this procedure does not take into account chemical reaction equilibrium effects, for example, how much of the HCl is converted to Cl_2. (See Sections 5.5 and 5.7.) Additional calculations, not provided on the worksheet, must be performed if the componential analysis is required on a mole fraction basis or if the actual volumetric flow rate is needed.

As the reader may ascertain, there are some very definite limitations to the EPAs procedure as presented in Tables 6.3.1 and 6.3.2. The preferred method is to perform the calculations on a mole basis throughout. Although a simple all-encompassing calculational scheme cannot be provided for the flue composition and flow rates, a suggested step-by-step outline for this engineering calculation is given here. It should assist the reader in obtaining these results for most engineering applications.

1. Select a basis for the waste–fuel mixture. This is a very critical step that the engineer often does not give enough thought to. A good choice of basis can often greatly simplify the calculations.
    (a) If an ultimate analysis is given on a mole basis, select 1 mol (or 100 mol) as a basis.
    (b) If a componential analysis is given on a mole basis, select 1 mol as a basis.
    (c) If an ultimate analysis is given on a mass basis, select 100 units of mass.
    (d) If a componential balance is given on mass basis, select 100 units of mass. For (c) and (d), the data should be converted to an elemental or componential mole basis.

TABLE 6.3.1. EPA Worksheet: Procedure to Calculate Stoichiometric Air Requirements, Combustion Gas Flow, and Composition (Liquid Injection Incineration)[a]

1. Identify the elemental composition and moisture content of the waste or waste mixture.

Carbon, C_w	_____	lb/lb waste
Hydrogen, H_w	_____	lb/lb waste
Moisture, H_2O_w	_____	lb/lb waste
Oxygen, O_w	_____	lb/lb waste
Nitrogen, N_w	_____	lb/lb waste
Chlorine, Cl_w	_____	lb/lb waste
Fluorine, F_w	_____	lb/lb waste
Bromine, Br_w	_____	lb/lb waste
Iodine, I_w	_____	lb/lb waste
Sulfur, S_w	_____	lb/lb waste
Phosphorus, P_w	_____	lb/lb waste

2. If auxiliary fuel is to be burned in conjunction with the waste, identify the fuel type and approximate proposed fuel-to-waste ratio from the permit application. (If auxiliary fuel is to be used only for startup, proceed to Step No. 5.)

Fuel type _____

Fuel–waste ratio, n_f _____ lb fuel/lb waste

3. Determine the approximate elemental composition of the fuel from the following table.

	lb component/lb fuel		
Component	Residual Fuel Oil (e.g., No. 6)	Distillate Fuel Oil (e.g., No. 2)	Natural Gas
C_f	0.866	0.872	0.693
H_f	0.102	0.123	0.227
N_f	—	—	0.08
S_f	0.03	0.005	—

4. Calculate the composition of the combined waste/auxiliary fuel feed.

$\text{C} \quad \dfrac{C_w + n_f C_f}{1 + n_f} =$ _____ lb/lb feed

$\text{H} \quad \dfrac{H_w + n_f H_f}{1 + n_f} =$ _____ lb/lb feed

$\text{H}_2\text{O} \quad \dfrac{H_2O_w}{1 + n_f} =$ _____ lb/lb feed

$\text{N} \quad \dfrac{N_w + n_f N_f}{1 + n_f} =$ _____ lb/lb feed

$\text{O} \quad \dfrac{O_w}{1 + n_f} =$ _____ lb/lb feed

$\text{Cl} \quad \dfrac{Cl_w}{1 + n_f} =$ _____ lb/lb feed

$\text{F} \quad \dfrac{F_w}{1 + n_f} =$ _____ lb/lb feed

TABLE 6.3.1. (*Continued*)

Br $\dfrac{Br_w}{1 + n_f}$ = _____ lb/lb feed

I $\dfrac{I_w}{1 + n_f}$ = _____ lb/lb feed

S $\dfrac{S_w + n_f S_f}{1 + n_f}$ = _____ lb/lb feed

P $\dfrac{P_w}{1 + n_f}$ = _____ lb/lb feed

5. Calculate the stoichiometric oxygen requirement.

$C \times 2.67\dfrac{lb\ O_2}{lb\ C}$ = _____ lb O_2/lb feed

$\left(H - \dfrac{Cl}{35.5} - \dfrac{F}{19}\right) \times 8.0\dfrac{lb\ O_2}{lb\ H}$ = _____ lb O_2/lb feed

$S \times 1.0\dfrac{lb\ O_2}{lb\ S}$ = _____ lb O_2/lb feed

$P \times 1.29\dfrac{lb\ O_2}{lb\ P}$ = _____ lb O_2/lb feed

-0(in feed)   = $-$ _____ lb O_2/lb feed[b]

$(O_2)_{stoich} = \Sigma =$ _____ lb O_2/lb feed

6. Calculate the combustion gas mass flows, based on the stoichiometric oxygen requirement.

CO_2 $C \times 3.67\dfrac{lb\ CO_2}{lb\ C}$ = _____ lb CO_2/lb feed

H_2O $\left[\left(H - \dfrac{Cl}{35.5} - \dfrac{F}{19}\right) \times 9.0\dfrac{lb\ H_2O}{lb\ H}\right] + H_2O$ (in feed) = _____ lb H_2O/lb feed

N_2 $\left[(O_2)_{stoich} \times 3.31\dfrac{lb\ N_2}{lb\ O_2}\ (in\ air)\right] + N$ (in feed)[c]   = _____ lb N_2/lb feed

HCl $Cl \times 1.03\dfrac{lb\ HCl}{lb\ Cl}$ = _____ lb HCl/lb feed

HF $F \times 1.05\dfrac{lb\ HF}{lb\ F}$ = _____ lb HF/lb feed

Br_2 Br = _____ lb Br_2/lb feed[d]

I_2 I = _____ lb I_2/lb feed[e]

SO_2 $S \times 2.0\dfrac{lb\ SO_2}{lb\ S}$ = _____ lb SO_2/lb feed

P_2O_5 $P \times 2.29\dfrac{lb\ P_2O_5}{lb\ P}$ = _____ lb P_2O_5/lb feed

Combustion products = CP = Σ = _____ lb/lb feed

TABLE 6.3.1. (*Continued*)

7. Identify the total excess air rate.

    EA = _____ %/100

8. Calculate the additional nitrogen and oxygen present in the combustion gases due to excess air feed.

    $(O_2)_{EA} = EA \times (O_2)_{stoich}$ = _____ lb O_2/lb waste

    $(N_2)_{EA} = 3.31 \dfrac{lb\ N_2}{lb\ O_2} (in\ air) \times (O_2)_{EA}$ = _____ lb N_2/lb waste

9. Calculate the total combustion gas flow.

    Combustion gases = CG = CP + $(O_2)_{EA}$ + $(N_2)_{EA}$ = _____ lb/lb waste

10. Calculate the mass fraction of each combustion gas component.

    CO_2 $\dfrac{CO_2}{CG}$ = _____ lb/lb gas

    H_2O $\dfrac{H_2O}{CG}$ = _____ lb/lb gas

    N_2 $\dfrac{N_2(from\ No.\ 6) + (N_2)_{EA}}{CG}$ = _____ lb/lb gas

    O_2 $\dfrac{(O_2)_{EA}}{CG}$ = _____ lb/lb gas

    HCl $\dfrac{HCl}{CG}$ = _____ lb/lb gas

    HF $\dfrac{HF}{CG}$ = _____ lb/lb gas

    Br_2 $\dfrac{Br_2}{CG}$ = _____ lb/lb gas

    I_2 $\dfrac{I_2}{CG}$ = _____ lb/lb gas

    SO_2 $\dfrac{SO_2}{CG}$ = _____ lb/lb gas

    P_2O_5 $\dfrac{P_2O_5}{CG}$ = _____ lb/lb gas

11. Identify those components that constitute < 1 to 2% of the combustion gas. These components can be eliminated from further consideration in heat and material balance calculations. In most cases, CO_2, H_2O, N_2, and O_2 will be the only combustion gas components that need to be considered.

12. Calculate the volumetric flow of the major combustion products at standard conditions of 68°F and 1 atm.[f]

    CO_2 $\dfrac{CO_2}{CG} \times CG \div 0.114 \dfrac{lb}{scf}$ = _____ scf/lb feed

TABLE 6.3.1. (*Continued*)

H_2O $\quad \dfrac{H_2O}{CG} \times CG \div 0.0467 \dfrac{lb}{scf}$ $\quad =$ _____ scf/lb feed

N_2 $\quad \dfrac{N_2}{CG} \times CG \div 0.0727 \dfrac{lb}{scf}$ $\quad =$ _____ scf/lb feed

O_2 $\quad \dfrac{O_2}{CG} \times CG \div 0.083 \dfrac{lb}{scf}$ $\quad =$ _____ scf/lb feed

Other $\quad \dfrac{other}{CG} \times CG \div (0.00259\ M) \dfrac{lb}{scf} =$ _____ scf/lb feed

where M = molecular weight[g]

Total flow, $q = \Sigma =$ _____ $\dfrac{scf}{lb\ feed}$

$q \times m_{feed}\ (lb/h) \div 60 =$ _____ scfm

The following footnotes are comments prepared by L. Theodore and J. Reynolds.

[a] The notation of the original EPA worksheet has been retained, and often does not correspond to the notation used in this text. Any illustrative examples or problems based on the use of this table will also retain the EPA notation.
[b] The quantity O in the feed need not be multiplied by 0.5 since there is 1.0 lb $O_2/1.0$ lb O. This term is subtracted in calculating stoichiometric oxygen requirements since this is a credit, that is, there is oxygen already present in the feed that is used for combustion.
[c] The N (in feed) term need not be multiplied by 0.5 since there is 1.0 lb $N_2/1.0$ lb N.
[d] The Br term need not be multiplied by 0.5.
[e] The I term need not be multiplied by 0.5.
[f] The density terms—0.114, 0.0467, and so on, lb/ft^3—can be obtained from tables or calculated directly from the ideal gas law for each component.
[g] M is the average molecular weight (abbreviated MW elsewhere in text) of the other components.

TABLE 6.3.2. EPA Worksheet: Procedure to Calculate Stoichiometric Air Requirements, Approximate Combustion Gas Flows, and Approximate Gas Compositions (Rotary Kilns)[a]

1. Identify the elemental composition and moisture content of the waste fed to the kiln.

	1. Solids (kiln)	*2. Liquids (kiln)*	
Carbon, C_w	_____	_____	lb/lb waste
Fuel hydrogen, H_w	_____	_____	lb/lb waste
Moisture, H_2O_w	_____	_____	lb/lb waste
Oxygen, O_w	_____	_____	lb/lb waste
Nitrogen, N_w	_____	_____	lb/lb waste
Chlorine, Cl_w	_____	_____	lb/lb waste
Fluorine, F_w	_____	_____	lb/lb waste
Bromine, Br_w	_____	_____	lb/lb waste
Iodine, I_w	_____	_____	lb/lb waste
Sulfur, S_w	_____	_____	lb/lb waste
Phosphorus, P_w	_____	_____	lb/lb waste

TABLE 6.3.2. (*Continued*)

2. Identify the approximate liquid and solid waste feed rates to the kiln, and calculate the liquid/solid feed fractions.

Liquid feed rate, m_1 = _____ lb/h
Solid feed rate, m_2 = _____ lb/h
Total feed, $m_{12} = m_1 + m_2$ = _____ lb/h
Liquid fraction, $n_1 = m_1/m_{12}$ = _____ lb liquid/lb waste
Solid fraction, $n_2 = 1 - n_1$ = _____ lb solid/lb waste

3. If auxiliary fuel is to be burned in conjunction with the wastes, identify the fuel type and approximate proposed fuel-to-waste ratio.

Fuel type: _____
Fuel–waste ratio in kiln; n_{fk} = _____ lb fuel/lb waste

4. Determine the approximate elemental composition of the fuel from the following table.

	lb component/lb fuel		
Component	Residual Fuel Oil (e.g., No. 6)	Distillate Fuel Oil (e.g., No. 2)	Natural Gas
C_f	0.866	0.872	0.693
H_f	0.102	0.123	0.227
N_f	—	—	0.08
S_f	0.03	0.005	—

5. Calculate the composition of the combined waste/auxiliary fuel feed to the kiln.

C_k $\dfrac{n_1 C_1 + n_2 C_2 + n_f C_f}{1 + n_f}$ = _____ lb/lb feed

H_k $\dfrac{n_1 H_1 + n_2 H_2 + n_f H_f}{1 + n_f}$ = _____ lb/lb feed

$H_2 O_k$ $\dfrac{n_1 H_2 O_1 + n_2 H_2 O_2}{1 + n_f}$ = _____ lb/lb feed

N_k $\dfrac{n_1 N_1 + n_2 N_2 + n_f N_f}{1 + n_f}$ = _____ lb/lb feed

O_k $\dfrac{n_1 O_1 + n_2 O_2}{1 + n_f}$ = _____ lb/lb feed

Cl_k $\dfrac{n_1 Cl_1 + n_2 Cl_2}{1 + n_f}$ = _____ lb/lb feed

F_k $\dfrac{n_1 F_1 + n_2 F_2}{1 + n_f}$ = _____ lb/lb feed

Br_k $\dfrac{n_1 Br_1 + n_2 Br_2}{1 + n_f}$ = _____ lb/lb feed

I_k $\dfrac{n_1 I_1 + n_2 I_2}{1 + n_f}$ = _____ lb/lb feed

TABLE 6.3.2. (*Continued*)

$$S_k \quad \frac{n_1 S_1 + n_2 S_2 + n_f S_f}{1 + n_f} \quad = \underline{\hspace{4cm}} \text{ lb/lb feed}$$

$$P_k \quad \frac{n_1 P_1 + n_2 P_2}{1 + n_f} \quad = \underline{\hspace{4cm}} \text{ lb/lb feed}$$

6. Calculate the stoichiometric oxygen requirement for the kiln.

$$C_k \times 2.67 \frac{\text{lb } O_2}{\text{lb C}} \quad = \underline{\hspace{4cm}} \text{ lb } O_2/\text{lb feed}$$

$$H_k - \frac{Cl_k}{35.5} - \frac{F_k}{19} \times 8.0 \frac{\text{lb } O_2}{\text{lb H}} = \underline{\hspace{4cm}} \text{ lb } O_2/\text{lb feed}^b$$

$$S_k \times 1.0 \frac{\text{lb } O_2}{\text{lb S}} \quad = \underline{\hspace{4cm}} \text{ lb } O_2/\text{lb feed}$$

$$P_k \times 1.29 \frac{\text{lb } O_2}{\text{lb P}} \quad = \underline{\hspace{4cm}} \text{ lb } O_2/\text{lb feed}$$

$$\frac{-O_k(\text{in feed})}{(O_2)_{\text{stoich(k)}}} \quad = \underline{\hspace{4cm}} \text{ lb } O_2/\text{lb feed}^c$$

7. Calculate the combustion gas mass flows, based on the stoichiometric oxygen requirement (assume complete combustion is achieved for purposes of gas flow estimation).

$$CO_{2_k} \quad C_k \times 3.67 \frac{\text{lb } CO_2}{\text{lb C}} \quad = \underline{\hspace{3cm}} \text{ lb } CO_2/\text{lb feed}$$

$$H_2O_k \quad H_k - \frac{Cl_k}{35.5} - \frac{F_k}{19} \times 9.0 \frac{\text{lb } H_2O}{\text{lb H}} + H_2O_k \text{ (in feed)} = \underline{\hspace{2cm}} \text{ lb } H_2O/\text{lb feed}^d$$

$$N_{2_k} \quad (O_2)_{\text{stoich(k)}} \times 3.31 \frac{\text{lb } N_2}{\text{lb } O_2} \text{ (in air)} + N_k \text{ (in feed)} = \underline{\hspace{2cm}} \text{ lb } N_2/\text{lb feed}$$

$$HCl_k \quad Cl_k \times 1.03 \frac{\text{lb HCl}}{\text{lb Cl}} \quad = \underline{\hspace{3cm}} \text{ lb HCl/lb feed}$$

$$HF_k \quad F_k \times 1.05 \frac{\text{lb HF}}{\text{lb F}} \quad = \underline{\hspace{3cm}} \text{ lb HF/lb feed}$$

$$Br_{2_k} \quad Br_k \quad = \underline{\hspace{3cm}} \text{ lb } Br_2/\text{lb feed}^e$$

$$I_{2_k} \quad I_k \quad = \underline{\hspace{3cm}} \text{ lb } I_2/\text{lb feed}^f$$

$$SO_{2_k} \quad S_k \times 2.0 \frac{\text{lb } SO_2}{\text{lb S}} \quad = \underline{\hspace{3cm}} \text{ lb } SO_2/\text{lb feed}$$

$$P_2O_{5_k} \quad P_k \times 2.29 \frac{\text{lb } P_2O_5}{\text{lb P}} \quad = \underline{\hspace{3cm}} \text{ lb } P_2O_5/\text{lb feed}$$

Kiln combustion products $= CP_k \quad = \underline{\hspace{3cm}} \text{ lb/lb feed}$

8. Identify the elemental composition and moisture content of the liquid wastes to be burned in the afterburner (if any).

$$C_3 \quad \underline{\hspace{6cm}} \text{ lb/lb waste}$$
$$H_3 \quad \underline{\hspace{6cm}} \text{ lb/lb waste}$$
$$H_2O_3 \quad \underline{\hspace{6cm}} \text{ lb/lb waste}$$

TABLE 6.3.2. (*Continued*)

O_3	_____	lb/lb waste
N_3	_____	lb/lb waste
Cl_3	_____	lb/lb waste
F_3	_____	lb/lb waste
Br_3	_____	lb/lb waste
I_3	_____	lb/lb waste
S_3	_____	lb/lb waste
P_3	_____	lb/lb waste

9. Identify the fuel type and approximate, proposed fuel-to-waste ratio for the afterburner.

Fuel type: _____

Fuel–waste rateio, n_{fa} = _____ lb fuel/lb waste

10. Determine the approximate elemental composition of the fuel from the table shown in Step 4.

C_{fa} = _____ lb/lb fuel

H_{fa} = _____ lb/lb fuel

N_{fa} = _____ lb/lb fuel

S_{fa} = _____ lb/lb fuel

11. Calculate the composition of the combined waste/auxiliary fuel feed to the afterburner.

C_a $\dfrac{C_3 + n_{fa}C_{fa}}{1 + n_{fa}}$ = _____ lb/lb feed

H_a $\dfrac{H_3 + n_{fa}H_{fa}}{1 - n_{fa}}$ = _____ lb/lb feed

H_2O_a $\dfrac{H_2O_3}{1 + n_{fa}}$ = _____ lb/lb feed

N_a $\dfrac{N_3 + n_{fa}N_{fa}}{1 + n_{fa}}$ = _____ lb/lb feed

O_a $\dfrac{O_3}{1 + n_{fa}}$ = _____ lb/lb feed

Cl_a $\dfrac{Cl_3}{1 + n_{fa}}$ = _____ lb/lb feed

F_a $\dfrac{F_3}{1 + n_{fa}}$ = _____ lb/lb feed

Br_a $\dfrac{Br_3}{1 + n_{fa}}$ = _____ lb/lb feed

I_a $\dfrac{I_3}{1 + n_{fa}}$ = _____ lb/lb feed

S_a $\dfrac{S_3 + n_{fa}S_{fa}}{1 + n_{fa}}$ = _____ lb/lb feed

P_a $\dfrac{P_3}{1 + n_{fa}}$ = _____ lb/lb feed

12. Calculate the stoichiometric oxygen requirement for the afterburner feed.

$C_a \times 2.67\dfrac{\text{lb } O_2}{\text{lb } C}$ = _____ lb O_2/lb feed

TABLE 6.3.2. (*Continued*)

$$\left(H_a - \frac{Cl_a}{35.5} - \frac{F_a}{19}\right) \times 8.0 \frac{lb\ O_2}{lb\ H} = \underline{\hspace{3cm}} \quad lb\ O_2/lb\ feed$$

$$S_a \times 1.0 \frac{lb\ O_2}{lb\ S} \qquad = \underline{\hspace{3cm}} \quad lb\ O_2/lb\ feed$$

$$P_a \times 1.29 \frac{lb\ O_2}{lb\ P} \qquad = \underline{\hspace{3cm}} \quad lb\ O_2/lb\ feed$$

$$-O_a\ (in\ feed) \qquad = \underline{\hspace{3cm}} \quad lb\ O_2/lb\ feed^g$$

$$(O_2)_{stoich(a)} = \Sigma = \underline{\hspace{3cm}} \quad lb\ O_2/lb\ feed$$

13. Calculate the combustion gas mass flows, based on the stoichiometric oxygen requirement.

$$CO_{2_a} \quad C_a \times 3.67 \frac{lb\ CO_2}{lb\ C} \qquad = \underline{\hspace{2cm}} \quad lb\ CO_2/lb\ feed$$

$$H_2O_a \quad \left[\left(H_a - \frac{Cl_a}{35.5} - \frac{F_a}{19}\right)\left(9.0 \frac{lb\ H_2O}{lb\ H}\right)\right] + H_2O_a\ (in\ feed) = \underline{\hspace{1.5cm}} \quad lb\ H_2O/lb\ feed$$

$$N_{2_a} \quad \left[(O_2)_{stoich(a)} \times 3.31 \frac{lb\ N_2}{lb\ O_2}\ (in\ air)\right] + N_a\ (in\ feed) \quad = \underline{\hspace{1.5cm}} \quad lb\ N_2/lb\ feed^h$$

$$HCl_a \quad Cl_a \times 1.03 \frac{lb\ HCl}{lb\ Cl} \qquad = \underline{\hspace{2cm}} \quad lb\ HCl/lb\ feed$$

$$HF_a \quad F_a \times 1.05 \frac{lb\ HF}{lb\ F} \qquad = \underline{\hspace{2cm}} \quad lb\ HF/lb\ feed$$

$$Br_{2_a} \quad Br_a \qquad = \underline{\hspace{2cm}} \quad lb\ Br_2/lb\ feed^i$$

$$I_{2_a} \quad I_a \qquad = \underline{\hspace{2cm}} \quad lb\ I_2/lb\ feed^j$$

$$SO_{2_a} \quad S_a \times 2.0 \frac{lb\ SO_2}{lb\ S} \qquad = \underline{\hspace{2cm}} \quad lb\ SO_2/lb\ feed$$

$$P_2O_{5_a} \quad P_a \times 2.29 \frac{lb\ P_2O_5}{lb\ P} \qquad = \underline{\hspace{2cm}} \quad lb\ P_2O_5/lb\ feed$$

Afterburner combustion products $= CP_a = \Sigma = \underline{\hspace{3cm}}$ lb/lb feed

14. Calculate the ratio of total afterburner feed to total kiln feed.

Liquid waste to kiln: m_1 $\quad = \underline{\hspace{5cm}}$ lb/h
Solid waste to kiln: m_2 $\quad = \underline{\hspace{5cm}}$ lb/h
Auxiliary fuel to kiln: $(m_1 + m_2)n_{fk} = \underline{\hspace{5cm}}$ lb/h
Liquid waste to afterburner: m_3 $\quad = \underline{\hspace{5cm}}$ lb/h
Auxiliary fuel to afterburner: $m_3 n_{fa} = \underline{\hspace{5cm}}$ lb/h

$$n_{ak} = \frac{m_3(1 + n_{fa})}{(m_1 + m_2)(1 + n_{fk})} = lb\ afterburner\ feed/lb\ kiln\ feed$$

15. Calculate the total combustion gas mass flows, based on stoichiometric oxygen requirements.

$$CO_2 \quad \frac{CO_{2_k} + CO_{2_a}n_{ak}}{1 + n_{ak}} = \underline{\hspace{4cm}} \quad lb/lb\ feed$$

TABLE 6.3.2. (*Continued*)

H_2O $\dfrac{H_2O_k + n_{ak}\, H_2O_a}{1 + n_{ak}} = $ _____ lb/lb feed

N_2 $\dfrac{N_{2_k} + n_{ak}\, N_{2_a}}{1 + n_{ak}} = $ _____ lb/lb feed

HCl $\dfrac{HCl_k + n_{ak}\, HCl_a}{1 + n_{ak}} = $ _____ lb/lb feed

HF $\dfrac{HF_k + n_{ak}\, HF_a}{1 + n_{ak}} = $ _____ lb/lb feed

Br_2 $\dfrac{Br_{2_k} + n_{ak}\, Br_a}{1 + n_{ak}} = $ _____ lb/lb feed

I_2 $\dfrac{I_{2_k} + n_{ak}\, I_{2_a}}{1 + n_{ak}} = $ _____ lb/lb feed

SO_2 $\dfrac{SO_{2_k} + n_{ak}\, SO_{2_a}}{1 + n_{ak}} = $ _____ lb/lb feed

P_2O_5 $\dfrac{P_2O_{5_k} + n_{ak}\, P_2O_{5_a}}{1 + n_{ak}} = $ _____ lb/lb feed

Combustion products = CP = Σ = _____ lb/lb feed

16. Identify the total excess air rate for the system (i.e., to be maintained in the afterburner).

    EA = _____ percent

17. Calculate the additional nitrogen and oxygen present in the combustion gases due to excess air feed.

    $(O_2)_{EA} = EA \times \dfrac{(O_2)_{stoich(k)} + n_{ak}\, (O_2)_{stoich(a)}}{1 + n_{ak}} = $ _____ lb O_2/lb waste

    $(N_2)_{EA} = 3.31 \dfrac{\text{lb } N_2}{\text{lb } O_2} \text{ (in air)} \times (O_2)_{EA} = $ _____ lb N_2/lb waste

18. Calculate the total combustion gas flow.

    Combustion gas flow = CG = CP + $(O_2)_{EA}$ + $(N_2)_{EA}$ = _____ lb/lb feed

19. Calculate the mass fraction of each combustion gas component.

    CO_2 $\dfrac{CO_2}{CG}$ = _____ lb/lb gas

    H_2O $\dfrac{H_2O}{CG}$ = _____ lb/lb gas

    N_2 $\dfrac{N_2 \text{ (from 15)} + (N_2)_{EA}}{CG}$ = _____ lb/lb gas

    O_2 $\dfrac{(O_2)_{EA}}{CG}$ = _____ lb/lb gas

    HCl $\dfrac{HCl}{CG}$ = _____ lb/lb gas

    HF $\dfrac{HF}{CG}$ = _____ lb/lb gas

TABLE 6.3.2. (*Continued*)

Br_2 $\dfrac{Br_2}{CG}$ = _____ lb/lb gas

I_2 $\dfrac{I_2}{CG}$ = _____ lb/lb gas

SO_2 $\dfrac{SO_2}{CG}$ = _____ lb/lb gas

P_2O_5 $\dfrac{P_2O_5}{CG}$ = _____ lb/lb gas

20. Identify those components that constitute < 1 to 2 % of the combustion gas. These components can be eliminated from further consideration in heat and material balance calculations. In most cases, CO_2, H_2O, N_2, and O_2 will be the only combustion gas components that need to be considered.

21. Calculate the volumetric flow of the major combustion products from the kiln at standard conditions of 68°F and 1 atm.[k]

CO_2 $\left(\dfrac{CO_2}{CG}\right) \times CG \div 0.114\,\text{lb/scf}$ = _____ scf/lb

H_2O $\left(\dfrac{H_2O}{CG}\right) \times CG \div 0.0467\,\text{lb/scf}$ = _____ scf/lb

N_2 $\left(\dfrac{N_2}{CG}\right) \times CG \div 0.0727\,\text{lb/scf}$ = _____ scf/lb

O_2 $\left(\dfrac{O_2}{CG}\right) \times CG \div 0.083\,\text{lb/scf}$ = _____ scf/lb

Other $\left(\dfrac{\text{other}}{CG}\right) \times CG \div (0.00259\,M)\,\text{lb/scf}$ = _____ scf/lb

     where M = molecular weight[l]

Total flow $q = \Sigma$ = _____ scf/lb feed

$q \times (m_1 + m_2)(1 + n_{fk})(1 + n_{ak}) \div 60$ = _____ scfm

The following footnotes are comments prepared by L. Theodore and J. Reynolds.

[a] The notation of the original EPA worksheet has been retained, and often does not correspond to the notation used in this text. Any illustrative examples or problems based on the use of this table will also retain the EPA notation.

[b] This term should read:

$[H_k - (Cl_k/35.5) - (F_k/19)]\,8.0\,(\text{lb } O_2/\text{lb H})$

[c] This term is misleading and the alignment is awkward. It should read:

$-O_k$ (in feed)   = _____ lb O_2 (in feed)/lb feed

stoich $(O_2)_k$ = Σ _____ lb O_2/lb feed

[d] The first three terms should be bracketed.

[e] These terms need not be multiplied by 0.5.

[h] The N_a term need not be multiplied by 0.5.

[i] Both terms need not be multiplied by 0.5.

[k] The lb/scf terms are densities at standard conditions.

[l] M should read the average molecular weight (abbreviated MW elsewhere in text) of the other components.

2. Convert the individual elements to their end products of stoichiometric combustion. As described earlier,

$$C \longrightarrow CO_2$$
$$H \longrightarrow H_2O$$
$$Cl \longrightarrow HCl$$
$$O \longrightarrow O_2$$
$$N \longrightarrow N_2$$
$$S \longrightarrow SO_2$$

and so on.

If the analysis is on a component basis, it may be preferable to write the balanced stoichiometric equation for the individual species participating in the combustion reaction [see Eq. (6.3.3)]. Also note that the quantity of water formed must reflect the hydrogen combined with the chlorine in HCl.

3. Based on the results of Step 2, calculate the stoichiometric oxygen or air requirement on a mole basis, being careful to correct for any oxygen present in the waste–fuel mixture. This will appear as a credit in the oxygen balance. This calculation may be performed on an elemental basis or directly from a balanced stoichiometric equation. The reader may again choose to utilize Eq. (6.3.3) for assistance.

4. Calculate and then add the excess air required for operation of the incinerator to the flue products of stoichiometric combustion. Once again, care should be exercised to adjust the quantity of nitrogen to reflect that contained in the waste–fuel mixture. This step will complete the *initial* or first determination for the flue gas at the outlet of the incinerator.

5. Based on the initial calculations in Step 4 for HCl, H_2O, O_2, SO_2, and so on, determine the amount of HCl and SO_2 converted to Cl_2 and SO_3, respectively. This will require a chemical reaction equilibrium calculation as described in Chapter 5. The incinerator operating temperature must be specified since the reaction equilibrium constant used to calculate each of these conversions is a function of the temperature.

6. Calculate the *final* componential or compound balance of the flue gas at the outlet of the incinerator. The results will be on a mole basis. Since this product is a gas, a volume basis is equivalent to a mole basis. These can then be converted to other units, including a mass basis (ppm, etc.) depending on the choice of the user.

7. The total molar flow rate can also be calculated by referring to the basis used for Steps 1 through 6. The total (or componential) volumetric flow rate at actual or standard conditions can also be calculated via either the ideal gas law or use of the appropriate flue gas density.

Several illustrative examples detailing this calculational approach are presented in Section 6.6.

6.4 THERMOCHEMICAL CALCULATIONS

The operating temperature in an incinerator is a function of many variables. For most hazardous waste incinerators, the operating temperature is calculated by determining the flame temperature under adiabatic or near-adiabatic conditions. From a calculational point of view, the flame temperature is a function of several variables, with a particularly strong dependence on the excess air requirement and the heating value of the combined waste–fuel mixture. Both a rigorous and an approximate approach to this calculation are provided in this section.

Rigorous Approach

A rigorous flame temperature calculation is schematically depicted in Fig. 6.4.1. The overall enthalpy approach introduced to the reader in Chapter 5 is now applied to this incinerator calculation. It should be kept in mind that any convenient calculation path can be selected in determining the enthalpy change between reactants (initial state) and products (final state). Since heat (enthalpy) of reaction and/or net (or gross) heating values are available at a reference temperature (T_0) only, the calculations proceed along a chosen path that includes the reaction at this reference temperature. Referring to Fig. 6.4.1, the enthalpy change (gain) associated with the cooling of the reactants (feed) is first obtained. This is followed by the enthalpy changes associated with the combustion reaction, the heating of the flue products, and the heat loss from the incinerator. These thermochemical calculations have been developed on a mass basis as is customary in incineration applications. Heating values—both higher (gross) and lower (net)—are expressed as British thermal units per pound (Btu/lb).

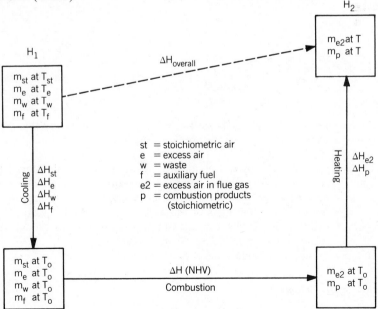

Figure 6.4.1. Flame temperature calculation schematic.

Cooling Step

$$\Delta H_{st} = m_{st} c_{Pst} (T_0 - T_{st})$$

$$\Delta H_e = m_e c_{Pe} (T_0 - T_e)$$

$$\Delta H_w = m_w c_{Pw} (T_0 - T_w)$$

$$\Delta H_f = m_f c_{Pf} (T_0 - T_f) \tag{6.4.1}$$

where ΔH_{st} = sensible enthalpy change for cooling stoichiometric air (st) from its initial temperature T_{st} to reference temperature T_0

m_{st} = mass of st per unit mass of waste–fuel mixture (lb/lb mixture)

c_{Pst} = average specific heat capacity of st over the temperature range T_{st}–T_0 (Btu/lb-°F)

T_{st} = initial temperature of st (°F)

T_0 = reference temperature (°F)

The subscripts st, e, w, f refer to the stoichiometric air, excess air, waste, and fuel, respectively.

Combustion Step

$$NHV = m_w (NHV_w) + m_f (NHV_f) \tag{6.4.2}$$

where NHV = net heating value of waste–fuel mixture at T_0 (Btu/lb mixture)

NHV_w = net heating value of waste at T_0 (Btu/lb waste)

NHV_f = net heating value of fuel at T_0 (Btu/lb fuel)

Heating Step

$$\Delta H_{e2} = m_{e2} c_{Pe2} (T - T_0)$$

$$\Delta H_p = \sum_{prod} m_i c_{Pi} (T - T_0) \tag{6.4.3}$$

where ΔH_{e2} = sensible enthalpy change for heating excess air in the flue gas (e2) from reference temperature T_0 to the incinerator operating temperature T

ΔH_p = sensible enthalpy change for heating products of combustion (p) from reference temperature T_0 to the incinerator operating temperature T

m_{e2} = mass of e2 per unit mass of waste–fuel mixture (lb/lb mixture)

m_i = mass of i per unit mass of waste–fuel mixture (lb/lb mixture)

c_{Pi} = average specific heat capacity of i over the temperature range T_i to T

e2 = subscript indicating excess air in the flue gas

p = products of combustion (CO_2, H_2O, HCl, SO_2, N_2, etc.)

Overall Enthalpy Balance

$$H_2 - H_1 = \Delta H_{st} + \Delta H_e + \Delta H_w + \Delta H_f + (1 - R)NHV + \Delta H_{e2} + \Delta H_p$$

$$= (R)NHV \tag{6.4.4}$$

The term R refers to the *fraction* (not percentage) of the net heating value associated with the combustion step that is lost by heat across the walls of the incinerator. If the system is assumed to be adiabatic, R is set equal to zero.

Equating these cooling and combustion enthalpy changes with the heating and overall enthalpy changes provides a rigorous equation relating the incinerator operating temperature (T) with the system's variables. Application of this approach is given in the Illustrative Examples Section (6.6) in this chapter.

Approximate Calculations

Some reasonable assumptions can be made to simplify the rigorous approach provided previously since the calculations are somewhat cumbersome. These are detailed in this section. When compared to the rigorous approach, a simpler (and in many instances, a more informative) set of equations results that are valid for purposes of engineering calculation.

Assumptions

1. The sensible enthalpy change associated with the cooling step can in many instances be neglected compared to the net heating value (combustion step) of the combined waste–fuel mixture. For this condition,

$$\Delta H_{st} = \Delta H_e = \Delta H_w = \Delta H_f = 0 \tag{6.4.5}$$

2. Although the products of combustion consist of many components, the major or primary components are nitrogen, carbon dioxide, and water (vapor). The average heat capacities of these components over the temperature range 60 to 2000°F (the latter being a typical incinerator operating temperature) are 0.27, 0.27, and 0.52 Btu/lb-°F, respectively. The arithmetic average of these three components is 0.35. However, since this product stream consists primarily of nitrogen, the average heat capacity of the combined mixture (not including the excess air) may be assigned a value of approximately 0.3 Btu/lb-°F. For this condition,

$$\Delta H_p = m_p(0.3)(T - T_0) \tag{6.4.6}$$

where m_p = mass of stoichiometric air, fuel, and waste entering the incinerator per unit mass of waste–fuel mixture, or equivalently, mass of products less that of excess air per unit mass of waste-fuel mixture (lb/lb mixture).

3. The average heat capacity of the (excess) air is ~ 0.27 Btu/lb-°F over the temperature range 60 to 2000°F. If this value is rounded to 0.3,

$$\Delta H_e = m_e(0.3)(T - T_0) \tag{6.4.7}$$

where m_e = mass of excess air per unit mass of waste–fuel mixture (lb/lb mixture).

4. Under truly adiabatic conditions the term R in Eq. (6.4.4) is zero. For this condition,

$$H_2 - H_1 = 0 \tag{6.4.8}$$

Under actual operating conditions, there will be some heat loss across the walls of the incinerator. However, if there is some preheat of the feed (usually the air), the effect of this assumption, to some extent, may balance the heat gained in the preheat (see Assumption 1).

5. Since most of the heating value data are available at 60°F, the reference or standard temperature should be arbitrarily set to this condition, that is,

$$T_0 = 60°F \tag{6.4.9}$$

and NHVs should be obtained at approximately this temperature. It should be noted that in incinerating liquids it is reasonable to assume 100% release of the heating value of the incoming liquid. However, it is not always valid to make the same assumption for incoming solids or for an incoming mixture of solids and liquids.

6. Perhaps the key assumption in this development is that associated with the stoichiometric air requirements for the combined waste–fuel mixture. Referring to Table 5.3.1, the stoichiometric air requirement, v_{st} (ft^3 air/lb mixture), divided by NHV for any of the listed hydrocarbons is approximately 0.01 ft^3 air/Btu (or 100 Btu/ft^3 air). For example, the ratios for methane (M), benzene (B), and toluene (T) are

$$M \quad v_{st}/NHV = 9.53/913 = 0.0104 \, \text{ft}^3 \, \text{air/Btu}$$

$$B \quad v_{st}/NHV = 35.73/3601 = 0.0099 \, \text{ft}^3 \, \text{air/Btu}$$

$$T \quad v_{st}/NHV = 42.88/4284 = 0.0100 \, \text{ft}^3 \, \text{air/Btu}$$

Using the density of air at 60°F, this ratio can be converted to approximately 750 lb air/10^6 Btu or 7.5×10^{-4} lb air/Btu. Thus, for this condition the stoichiometric air requirement (m_{st}) is given by

$$m_{st} = 7.5 \times 10^{-4} \, NHV \tag{6.4.10}$$

It should be noted that the validity and applicability of this assumption is limited to waste–fuel mixtures consisting of pure hydrocarbons, that is, organics with only hydrogen and carbon atoms, and to chlorinated organic mixtures where the mass fraction of the chlorine is low. However, for purposes of engineering calculations, this value for stoichiometric air will apply for almost all incinerator applications.

To further simplify the development to follow, the describing equations are again based on 1.0 lb of the combined waste–fuel mixture, that is,

$$m_w + m_f = m_b = 1.0 \tag{6.4.11}$$

where m_w = mass of w (waste) per unit mass of waste–fuel mixture (lb/lb mixture)

w, f, b = subscripts indicating waste, fuel, and burnable mixture, respectively

Thus, m_w and m_f now become the mass fractions of the waste and fuel, respectively, in the burnable mixture (b). Applying the six assumptions listed here with this basis results in the following equation:

$$\text{NHV} = (1.0 + 7.5 \times 10^{-4}\,\text{NHV})(0.3)(T - 60) + m_e\,(0.3)(T - 60) \quad (6.4.12)$$

or

$$\text{NHV} = (1 + m_e + 7.5 \times 10^{-4}\,\text{NHV})(0.3)(T - 60) \quad (6.4.13)$$

It is more convenient to work with the mass of excess air in terms of a *fraction* or *percentage* rather than on the basis of a unit mass of the waste–fuel mixture (m_e). The fraction of excess air (EA) is defined by

$$\text{EA} = m_e/m_{st} \quad (6.4.14)$$

Since m_{st} may be approximated by

$$m_{st} = 7.5 \times 10^{-4}\,\text{NHV} \quad (6.4.15)$$

Eq. (6.4.15) for m_e becomes

$$m_e = (\text{EA})(7.5 \times 10^{-4})(\text{NHV}) \quad (6.4.16)$$

The approximate enthalpy balance is then

$$\text{NHV} = [1 + (\text{EA})(7.5 \times 10^{-4})(\text{NHV}) + (7.5 \times 10^{-4})(\text{NHV})](0.3)(T - 60) \quad (6.4.17)$$

This may be further simplified to

$$\text{NHV} = [1 + (1 + \text{EA})(7.5 \times 10^{-4})(\text{NHV})](0.3)(T - 60) \quad (6.4.18)$$

Three parameters or variables appear in Eq. (6.4.18). Specifying any two of these constitutes a complete set for this equation. Thus, three equations may be generated expressing one variable (dependent) in terms of the other two (independent), that is,

$$T = f_1(\text{EA, NHV})$$
$$\text{EA} = f_2(T, \text{NHV})$$
$$\text{NHV} = f_3(T, \text{EA}) \quad (6.4.19)$$

The three resulting equations are

$$T = 60 + \frac{\text{NHV}}{(0.3)[1 + (1 + \text{EA})(7.5 \times 10^{-4})(\text{NHV})]} \quad (6.4.20)$$

$$\text{EA} = \frac{[\text{NHV}/0.3(T - 60)] - 1}{7.5 \times 10^{-4}\,\text{NHV}} - 1 \quad (6.4.21)$$

$$\text{NHV} = \frac{0.3(T - 60)}{1 - (1 + \text{EA})(7.5 \times 10^{-4})(0.3)(T - 60)} \quad (6.4.22)$$

Note: The units of T and NHV are °F and Btu/lb, respectively, EA is a dimension-less fraction.

The reader should also note that these equations become sensitive to the value assigned to c_P (0.3 in this case). This is a function of both the temperature (T) and excess air fraction (EA), and also depends on the flue products since the heat capacities of air and CO_2 are about half that of H_2O. In addition, the 7.5×10^{-4} term "derived" earlier may vary slightly with the composition of the waste–fuel mixture incinerated. The overall relationship between operating temperature and composition is therefore rather complex, and its prediction not necessarily as straightforward as shown here. These variabilities can be compensated for by treating the c_P term as a statistical parameter in Eqs. (6.4.20) through (6.4.22). Theodore and Reynolds[2] have generated values for c_P for excess air percentage (EA) and operating temperature (T) ranging from 0 to 100% and 1000 to 3500 K, respectively, for benzene, *n*-octane, and dichloromethane. These results are presented in Tables 6.4.1–6.4.3. In addition, regression coefficients α, β, and γ were obtained for the equation

$$c_P = \alpha + \beta T + \gamma(\text{EA}) \tag{6.4.23}$$

These may be found in Table 6.4.4.

Solutions to Eqs. (6.4.20)–(6.4.22) are graphically provided in Fig. 6.4.2. The equations and graph clearly demonstrate the interdependence of temperature, excess air, and heating value. Application of these equations to an incinerator system can also be found in the Illustrative Examples (Section 6.6).

EPA Worksheets

The EPA has also provided a procedure to estimate the excess air requirement for an incinerator. This procedure is shown in Tables 6.4.5 and 6.4.6 (with the original worksheet numbers and notation retained).[1] Note that the stoichiometric oxygen requirement $(O_2)_{stoich}$ as well as the mass composition of the flue gas must be provided. In addition to the assumed heat capacities for the CO_2, N_2, and so on, the final form of the equation (see Step No. 7 in Table 6.4.5) is based on an assumption of 5% heat loss across the walls of the incinerator. Use of this table is also demonstrated later in this chapter in Section 6.6.

Finally, the reader should note that the stoichiometric calculations presented in the previous section and the thermochemical calculations presented in this section were developed independently of one another. Strictly speaking, both of these calculations are interrelated since the products of combustion are a function of the operating temperature that impacts on the quantity of Cl_2, SO_3, and so on. The operating temperature, on the other hand, can affect the excess air requirement which, in turn, significantly affects the products of combustion. Thus, an actual industrial application may require the simultaneous solution of both the stoichio-metric and thermochemical equations since they are interdependent. A number of operating conditions are often specified to eliminate this interdependence, enabling one to perform the calculations separately.

TABLE 6.4.1. Specific Heat Capacity of Benzene (C_6H_6) as a Function of Temperature and Excess Air [Results of a Regression Analysis Using Eq. (6.4.23)][a]

Excess Air (EA) (%)	Temperature (K)										
	1000	1250	1500	1750	2000	2250	2500	2750	3000	3250	3500
0	0.34	0.35	0.37	0.38	0.38	0.39	0.40	0.40	0.41	0.41	0.41
10	0.32	0.33	0.35	0.35	0.36	0.37	0.37	0.38	0.38	0.38	0.39
20	0.31	0.32	0.33	0.34	0.35	0.36	0.36	0.36	0.37	0.37	0.37
30	0.30	0.32	0.33	0.33	0.34	0.35	0.35	0.35	0.36	0.36	0.36
46	0.30	0.31	0.32	0.33	0.33	0.34	0.34	0.35	0.35	0.35	0.35
50	0.29	0.31	0.31	0.32	0.33	0.34	0.34	0.34	0.35	0.35	0.35
60	0.29	0.30	0.31	0.32	0.33	0.33	0.34	0.34	0.34	0.34	0.34
70	0.29	0.30	0.31	0.32	0.32	0.33	0.34	0.34	0.34	0.34	0.34
80	0.29	0.30	0.31	0.31	0.32	0.33	0.33	0.33	0.34	0.34	0.34
90	0.29	0.30	0.31	0.31	0.32	0.32	0.33	0.33	0.33	0.34	0.34
100	0.28	0.29	0.30	0.31	0.32	0.32	0.33	0.33	0.33	0.33	0.34

[a] c_P = cal/g-°C or Btu/lb-°F.

TABLE 6.4.2. Specific Heat Capacity of Octane (C_8H_{18}) as a Function of Temperature and Excess Air [Results of a Regression Analysis Using Eq. (6.4.23)][a]

Excess Air (EA) (%)	Temperature (K)										
	1000	1250	1500	1750	2000	2250	2500	2750	3000	3250	3500
0	0.38	0.39	0.41	0.42	0.43	0.44	0.45	0.46	0.46	0.47	0.47
10	0.35	0.36	0.38	0.39	0.40	0.41	0.41	0.42	0.42	0.42	0.43
20	0.33	0.35	0.36	0.37	0.38	0.38	0.39	0.39	0.40	0.40	0.40
30	0.32	0.33	0.34	0.35	0.36	0.37	0.38	0.38	0.38	0.39	0.39
40	0.31	0.33	0.34	0.35	0.35	0.36	0.37	0.37	0.37	0.37	0.38
50	0.31	0.32	0.33	0.34	0.35	0.35	0.36	0.36	0.36	0.37	0.37
60	0.30	0.31	0.32	0.33	0.34	0.35	0.35	0.36	0.36	0.36	0.36
70	0.30	0.31	0.32	0.33	0.34	0.34	0.35	0.35	0.35	0.36	0.36
80	0.30	0.31	0.32	0.33	0.33	0.34	0.34	0.35	0.35	0.35	0.35
90	0.29	0.31	0.31	0.32	0.33	0.34	0.34	0.34	0.35	0.35	0.35
100	0.29	0.30	0.31	0.32	0.33	0.33	0.34	0.34	0.34	0.35	0.35

[a] c_P = cal/g-°C or Btu/lb-°F.

TABLE 6.4.3. Specific Heat Capacity of Dicloromethane (CH_2Cl_2) as a Function of Temperature and Excess Air [Results of a Regression Analysis Using Eq. (6.4.23)][a]

Excess Air (EA) (%)	Temperature (K)										
	1000	1250	1500	1750	2000	2250	2500	2750	3000	3250	3500
0	0.24	0.25	0.26	0.27	0.27	0.28	0.29	0.29	0.30	0.31	0.31
10	0.24	0.25	0.26	0.27	0.28	0.28	0.29	0.30	0.30	0.31	0.31
20	0.25	0.26	0.26	0.27	0.28	0.28	0.29	0.30	0.30	0.31	0.31
30	0.25	0.26	0.27	0.27	0.28	0.29	0.29	0.30	0.30	0.31	0.31
40	0.25	0.26	0.27	0.28	0.28	0.29	0.30	0.30	0.31	0.31	0.31
50	0.25	0.26	0.27	0.28	0.28	0.29	0.29	0.30	0.30	0.31	0.31
60	0.25	0.26	0.27	0.28	0.28	0.29	0.30	0.30	0.31	0.31	0.31
70	0.25	0.26	0.27	0.28	0.29	0.29	0.30	0.30	0.31	0.31	0.31
80	0.26	0.27	0.27	0.28	0.29	0.29	0.30	0.30	0.31	0.31	0.31
90	0.26	0.27	0.27	0.28	0.29	0.29	0.30	0.30	0.31	0.31	0.31
100	0.26	0.27	0.27	0.28	0.29	0.29	0.30	0.30	0.31	0.31	0.31

[a] c_P = cal/g-°C or Btu/lb-°F.

TABLE 6.4.4. Regression Coefficients for Use With Eq. (6.4.23)[a]

Compound	Formula	α	$\beta \times 10^5$	$\gamma \times 10^3$
Benzene	C_6H_6	0.336	2.230	−0.942
n-Octane	C_8H_{18}	0.372	2.434	−1.311
Dichloromethane	CH_2Cl_2	0.233	2.645	−0.102
Benzene and n-octane		0.354	2.332	−1.126
Overall		0.314	2.436	−0.785

[a] Temperature range: 1000–2750 K; Excess air range: 0–100%.

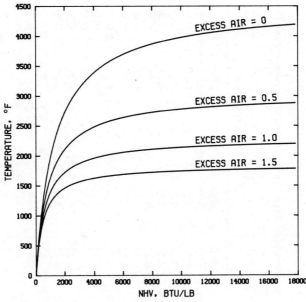

Figure 6.4.2. Operating temperature as a function of excess air and net heating value.

TABLE 6.4.5. EPA Worksheet: Procedure to Calculate Excess Air Rate for a Specified Temperature and Feed Composition[a]

1. Identify the following input variables:
   From Table 6.3.1, Step No. 5
   $(O_2)_{stoich}$ = _____ lb/lb feed
   From Table 6.3.1, Step No. 6
   CO_2 = _____ lb/lb feed
   H_2O = _____ lb/lb feed
   N_2 = _____ lb/lb feed
   Other major component(s) = _____ lb/lb feed

   From thermochemical data on the waste
   NHV_{waste} = _____ Btu/lb waste

   From proposed operating conditions
   Operating temperature, T = _____ °F
   Air preheat temperature, T_{air} = _____ °F
   (if applicable)

   If auxiliary fuel is to be burned in conjunction with the waste, also identify the following from Table 6.3.1
   n_f = _____ lb fuel/lb waste
   HV_f = _____ Btu/lb fuel

2. If air preheating is employed, calculate the corresponding enthalpy input to the incinerator. If the combustion air is not to be preheated, proceed to Step No. 3.
   $\Delta H_1' = 1.12(T_{air} - 77)(O_2)_{stoich}$ = _____ Btu/lb feed[b]
   $\Delta H_1 = \Delta H_1'(1 + EA)$

3. Calculate the heat generated by combustion of the waste or waste/auxiliary fuel mix.
   $\Delta H_2 = \dfrac{NHV_{waste} + n_f HV_f}{1 + n_f}$ = _____ Btu/lb feed

4. Calculate the heat loss through the walls of the incinerator, assuming 5% loss.
   $Q = 0.05\Delta H_2$ = _____ Btu/lb feed

5. Calculate the enthalpy of the combustion products leaving the incinerator.[c]
   $0.26(CO_2 + N_2)(T - 77)$ = _____ Btu/lb feed[d]
   $0.49(H_2O)(T - 77)$ = _____ Btu/lb feed
   other $\times \bar{C}_{P_{other}} (T - 77)$ = _____ Btu/lb feed

   $\Delta H_3 = \Sigma$ = _____ Btu/lb feed

6. Calculate the enthalpy of excess air leaving the incinerator.
   $\Delta H_4' = 1.1(T - 77)(O_2)_{stoich}$ = _____ Btu/lb feed
   $\Delta H_4 = \Delta H_4' \times EA$

7. Calculate the excess air percentage as follows:

   $EA = 100\left(\dfrac{\Delta H_1' + \Delta H_2 - Q - \Delta H_3}{\Delta H_4' - \Delta H_1'}\right)$ = _____ %[e]

The following footnotes are comments prepared by L. Theodore and J. Reynolds.
[a] The notation of the original EPA worksheet has been retained, and often does not correspond

to the notation used in this text. Any illustrative examples or problems based on the use of this table will also retain the EPA notation.

[b] The term 1.12 has been derived in the following manner. The enthalpy change for air is

$$\Delta H'_1 = (air)_{stoich} \, (C_{P, \, air})(T_{air} - 77)$$

Every lbmol of air contains approximately 0.21 lbmol of O_2. Every lb of air therefore contains approximately $(0.21)(32/29)$ lb O_2 where the 32 and 29 are the molecular weights of the oxygen and air, respectively. Thus,

$$(air)_{stoich} = (O_2)_{stoich}/(0.21)(32/29)$$

If the heat capacity of air is roughly 0.25 Btu/lb-°F, this equation reduces to that given in Step No. 2.

[c] Here 0.26 and 0.49 represent the heat capacity, in Btu/lb-°F, of CO_2 and H_2O, respectively.

[d] Care should be exercised with this term. The N_2 represents the nitrogen contribution from the *stoichiometric* air and the fuel–waste. The nitrogen in the *excess air* is treated separately in Step No. 6. Another approach is to treat this stoichiometric nitrogen through Step No. 6. There is approximately 0.77 lb N_2/lb air. In order to account for the N_2 in the flue gas associated with the stoichiometric air, this term would then read:

$$\Delta H_4 = \Delta H'_4(0.77 + EA)$$

The contribution of fuel and waste nitrogen can usually be safely neglected.

[e] This equation has been derived by an enthalpy balance across the incinerator.
Energy in:
    Enthalpy of preheat air in $= \Delta H'_1(1 + EA)$
    Enthalpy of fuel/waste combustion $= \Delta H_2$
Energy out:
    Heat loss across walls $= Q$
    Enthalpy of flue gas $= \Delta H_3$
    Enthalpy of excess air in flue gas $= (\Delta H'_4)(EA)$
Equating the "in" with the "out" terms leads to the final equation given in Step No. 7. Care should be exercised with the ΔH_2 term. Combustion reactions are exothermic, that is, energy is liberated or given off. It is standard thermodynamic (engineering) practice to record these enthalpy of combustion terms as *negative* values. The approach here uses this term as a *positive* quantity.

TABLE 6.4.6. Procedure to Calculate the Maximum Achievable Excess Air Rate for a Rotary Kiln Operating at a Specified Temperature with a Specified Feed Composition[a]

1. Identify the following input variables:
   From Table 6.3.2, Step No. 6
   $(O_2)_{stoich(k)} = $ _____ lb O_2/lb kiln feed
   From Table 6.3.2, Step No. 7
   $CO_{2(k)}$ = _____ lb/lb feed
   $H_2O_{(k)}$ = _____ lb/lb feed
   $N_{2(k)}$ = _____ lb/lb feed
   Other major combustion product(s) = _____ lb/lb feed
   From Table 6.3.2, Step No. 2

TABLE 6.4.6. (*Continued*)

Liquid waste feed fraction, n_1 = _____ lb liquid/lb waste
Solid waste feed fraction, n_2 = _____ lb solid/lb waste

From thermochemical data on the waste
Liquid waste heating value, NHV_1 = _____ Btu/lb
Solid waste heating value, NHV_2 = _____ Btu/lb

From proposed operating conditions
Kiln operating temperature, T_k = _____ °F
Air preheat temperature, T_{air} = _____ °F
  (if applicable)

If auxiliary fuel is to be burned in the kiln along with the wastes during normal operation, identify the following from Table 6.3.2:

n_{fk} = _____ lb fuel/lb waste
HV_{fk} = _____ Btu/lb fuel

2. If air preheating is used, calculate the corresponding enthalpy input to the kiln. If the combustion air is not preheated, proceed to Step No. 3:

$\Delta H'_{1_{(k)}} = 1.12(T_{air} - 77)(O_2)_{stoich(k)}$ = _____ Btu/lb feed
$\Delta H_{1_{(k)}} = \Delta H'_{1_{(k)}}(1 + EA_k)$

3. Calculate the maximum heat generated in the kiln by combustion of the wastes or waste/auxiliary fuel mix:

$$\Delta H_2 = \frac{n_1 NHV_1 + n_2 NHV_2 + n_{fk} HV_{fk}}{1 + n_{fk}} = \text{_____}$$ Btu/lb feed

4. Estimate the heat loss through the walls of the kiln, assuming 5% loss:

$Q_{(k)} = 0.05 \Delta H_{2_{(k)}}$ = _____ Btu/lb feed

5. Calculate the enthalpy of the combustion products leaving the kiln:

$0.26[CO_{2_{(k)}} + N_{2_{(k)}}](T_k - 77)$ = _____ Btu/lb feed
$0.49(H_2O_{(k)})(T_k - 77)$ = _____ Btu/lb feed
$\text{other}_k \left(\dfrac{lb}{lb\ feed}\right) \times \overline{C}_{P\text{other}}(T_k - 77)$ = _____ Btu/lb feed

$\Delta H_{3_{(k)}} = \Sigma$ = _____ Btu/lb feed

6. Calculate the enthalpy of excess air leaving the kiln:

$\Delta H'_{4_{(k)}} = 1.1(T_k - 77)(O_2)_{stoich(k)}$ = _____ Btu/lb feed
$\Delta H_{4_{(k)}} = \Delta H'_{4_{(k)}} EA_k$

7. Calculate the excess air percentage as follows:

$$EA_k = 100 \left(\frac{\Delta H'_{1_{(k)}} + \Delta H_{2_{(k)}} - Q_k - \Delta H_{3_{(k)}}}{\Delta H'_{4_{(k)}} - \Delta H'_{1_{(k)}}}\right) = \text{_____}$$ %[b]

---

[a] The notation of the original EPA worksheet has been retained, and often does not correspond to the notation used in this text. Any illustrative examples or problems based on the use of this table will also retain the EPA notation.

[b] The reader should carefully review the note comments in the previous worksheet, that is, Table 5.4.5.

6.5   APPLICATION OF CHEMICAL REACTION PRINCIPLES

There have been a host of companion terms used to describe incinerators; among these are *oxidizer*, *afterburner*, *combustor*, *combustion device*, *burner*, and *organic reactor*. Whichever term is used, the overall process of incineration is best characterized by those phenomena occurring in a *chemical reactor*. In effect, an incinerator is one of a number of units that fits into the class of what industry describes as a *chemical reactor*. With this in mind, one may apply chemical reactor principles to either design and/or predict the performance of (hazardous waste) incinerators.

Kinetic and design calculations for reactors require information on the rate of reaction and equations to describe the concentration and temperature within the reactor. Chapter 5 developed equations to describe the rate of chemical reaction, and discussed energy (temperature) effects and chemical equilibrium. The contents of those sections can be combined and the results applied to the three major classifications of reactors: batch, tank flow, and tubular flow. These three classes of reactors are discussed here.

There are three basic modes of operation for any type of reactor. Reactors may be operated: isothermally, nonisothermally, or adiabatically. In *isothermal operation*, the temperature is maintained constant during the course of the reaction. This condition can be approached in practice by providing sufficient heat exchanger facilities to account for enthalpy effects arising during reaction. This mode of operation finds its major application in laboratory kinetic studies. Most industrial reactors are described by *nonisothermal operation*. Some energy in the form of heat is added to or removed from the reactor, but isothermal conditions are not satisfied. For *adiabatic operation*, the reactor is insulated to minimize heat transfer between the reactor contents and the surroundings. Many industrial reactors, including hazardous waste incinerators, attempt to operate in this manner.

Batch Reactors

A *batch reactor* is a vessel or container that may be open or closed. Reactants are usually added to the reactor simultaneously. The contents are then mixed (if necessary) to insure no variations in the concentrations of the species present. The reaction then proceeds. During this period, there is no transfer of mass into or out of the reactor. Because the concentrations of reactants and products change with time, this is a *transient* or *unsteady-state* operation. The reaction is terminated when the desired chemical change has been achieved. The contents are then discharged and sent elsewhere for further processing. Line diagrams of three typical batch reactors are given in Fig. 6.5.1. A more detailed schematic is presented in Fig. 6.5.2.

The describing equation for this device is obtained by applying the conservation law for either mass or number of moles on a time rate basis to the contents of the reactor. It is preferable to work with moles rather than mass in this and the following subsections since the rate of reaction is most conveniently described in terms of molar concentrations. The resulting equation takes the form

$$\theta = C_{A0} \int_0^{X_A} \frac{dX_A}{-R_A} \tag{6.5.1}$$

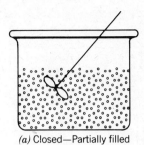

(a) Closed—Partially filled

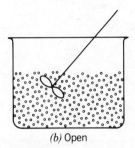

(b) Open

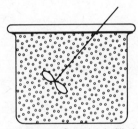

(c) Closed—Completely filled   **Figure 6.5.1.** Different batch reactors.

where R_A = rate of chemical reaction for A (lbmol/h-ft^3)

  C_{A0} = initial molar concentration of reactant A (lbmol/ft^3)

  X_A = fractional molar conversion of A in time θ (dimensionless)

  θ = residence time in the reactor (h)

If the reaction is first order and irreversible with respect to A (see Chapter 5), this equation becomes

$$\theta = -\frac{1}{k}\ln\left(\frac{C_A}{C_{A0}}\right) \quad \text{or} \quad \theta = \frac{1}{k}\ln\left(\frac{C_{A0}}{C_A}\right) \tag{6.5.2}$$

where k = reaction velocity constant (h^{-1})

  Since this equation contains the ratio of the two concentrations, the concentration may be based on either mass or number of moles. For gas phase reactions. a volume basis may also be used [e.g., ppm, ppb, or the equivalent]. The choice of basis is

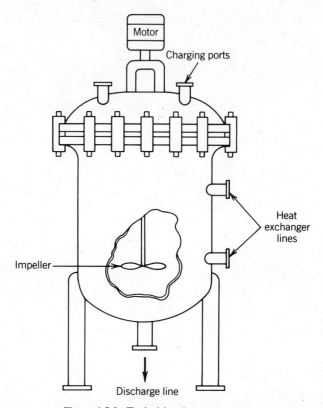

Figure 6.5.2. Typical batch reactor

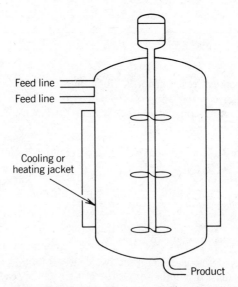

Figure 6.5.3. Typical tank flow reactor '

often a matter of convenience or preference, the form of the data, and the available expression for the rate of reaction. However, Eq. (6.5.2) applies only to first-order reactions; molar concentrations should be used for other or more complex reactions.

Continuously Stirred Tank Reactor

Another reactor where mixing is important is the *tank flow* or *continuously stirred tank reactor* (CSTR). This type of reactor also consists of a tank or kettle equipped with an agitator (see Fig. 6.5.3). It may be operated under steady-state or transient conditions. Reactants are fed continuously, and the products withdrawn continuously. The reactants and products may be in the liquid, gas, or solid state, or a combination of these. If the contents are perfectly mixed, the reactor design problem is greatly simplified for steady-state conditions because the mixing results in uniform concentration, temperature, and so on, throughout the reactor. This means that a *fluidized-bed incinerator*, operating under nearly perfect mixing conditions, may be best described by a *CSTR* model (see Chapter 7). For perfect mixing, the rate of reaction is constant; therefore the describing equations are not differential and do not require integration. High reactant concentrations can be maintained with low flow rates so that conversions approaching 100% may be achieved. A disadvantage of this practice, however, is that very large reactors must be required to obtain high conversions. The describing equation for this type of reactor takes the form

$$\theta = \frac{C_{A1} - C_{A0}}{R_{A1}} \tag{6.5.3}$$

where θ = residence time in the reactor (h)
   C_{A0} = inlet concentration of reactant A (lbmol/ft^3)
   C_{A1} = outlet concentration of reactant A (lbmol/ft^3)
   R_{A1} = rate of chemical reaction for A based on outlet conditions (lbmol/h-ft^3)

Note that the concentration of A in the reactor is equal to the outlet concentration since perfectly mixed conditions are assumed. For a first-order irreversible reaction in A, this equation becomes

$$\theta = \frac{C_{A1} - C_{A0}}{-kC_{A1}} \tag{6.5.4}$$

where k = velocity rate constant (h^{-1})

Since a ratio of concentrations appears again for this first-order reaction, the concentration may be based on either mass or number of moles.

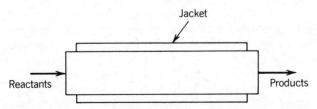

Figure 6.5.4. Tubular flow reactor, single pass.

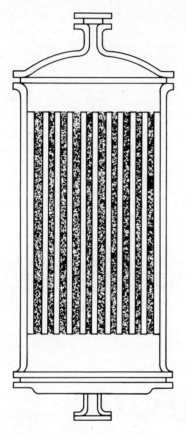

Figure 6.5.5. Multitube flow reactor.

Tubular Flow Reactors

The last and most important class of reactor to be examined is the *tubular flow reactor*. The most common type is the single-pass cylindrical tube (see Fig. 6.5.4); another important type is one that consists of a number of tubes in parallel (see Fig. 6.5.5). The reactor(s) may be vertical or horizontal. The feed is charged continuously at the inlet of the tube, and the products are continuously removed at the outlet. If heat exchange with the surroundings is required, the reactor tube is *jacketed*. If adiabatic conditions are required, the reactor is *covered* with insulation. Tubular flow reactors are usually operated under steady-state conditions so that physical and chemical properties do not vary with time. Unlike the batch and tank flow reactors, there is no mechanical mixing. Thus, the state of the reacting fluid will vary from point to point in the system, and this variation may be in both the radial r and axial z directions (see Fig. 6.5.6). The describing equations are then differential, with position as the independent variable. In terms of the final results, the reacting system is assumed to move through the reactor in plug flow (see Fig. 6.5.7a). It is further assumed that there is *no* mixing in the axial direction and *complete* mixing in the radial direction so that the concentration, temperature, and so on, do not vary through the cross section of the tube. Thus, the reacting fluid flows through the

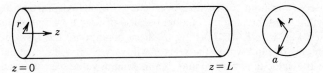

Figure 6.5.6. Reactor system, cylindrical coordinates.

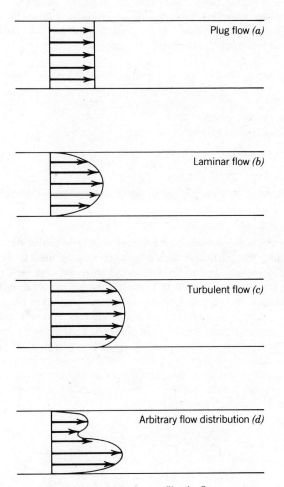

Figure 6.5.7. Velocity profiles in flow reactors.

reactor in an undisturbed plug of mass. Note that the time for this hypothetical plug to flow through this type of reactor is the same as the contact or reaction time in a batch reactor. Under these conditions, the form of the describing equations for batch reactors will also apply to tubular flow reactors. From a qualitative point of view, as the length of the reactor approaches infinity, the concentration of the limiting reactant approaches zero for irreversible reactions (except zero order) and the equilibrium concentration for reversible reactions. In actual practice, tubular flow reactors deviate from the plug flow model because of velocity variations in the radial direction (see Fig. 6.5.7b–d). For any of these conditions, the residence time for annular elements of fluid within the reactor will vary from some minimal value at a point where the velocity is a maximum to a maximum value near the wall where the velocity approaches zero. The concentration and temperature profiles, as well as the velocity profile, are therefore not constant across the reactor. Although describing equations based on the plug flow assumption are far from accurate, most practitioners still employ the plug flow model. The describing equation for this unit is obtained by applying the conservation law for mass to a differential volume element in a flow reactor. The resulting equation is

$$V = \dot{n}_{A0} \int_0^{X_A} \frac{dX_A}{-R_A} \qquad (6.5.5)$$

where V = volume of the reactor (ft^3)
$\dot{n}_{A0}$ = inlet molar feed rate of A (lbmol/h)

For first-order irreversible kinetics, Eq. (6.5.5) becomes

$$\theta = -\frac{1}{k} \ln \left(\frac{C_A}{C_{A0}} \right) \qquad (6.5.6)$$

where θ = average residence time in the reactor (h)

As mentioned previously, this equation is identical in form to that developed for a batch reactor. Finally, this particular class of reactor may be used to describe both the *liquid injection incinerator* and *rotary kiln incinerator* (see Chapter 7). These are the two most common types employed in hazardous waste incineration facilities.

As mentioned earlier, the chemical reaction sequence that takes place during the destruction of organics by thermal oxidation is a complicated process. Several intermediate products appear before the original material is completely oxidized into carbon dioxide, water, nitrogen oxides, and so on. However, most of the intermediate products exist for only a very short time, and for engineering purposes complete destruction of the organics is the principal concern. Lee et al.[3] have shown from available experimental data that the following equations fit the data better than the first-order model described previously.

$$\ln \left(\frac{C_A}{C_{A0}} \right) = -k(\theta - \theta_1), \quad \theta > \theta_1 \qquad (6.5.7)$$

$$= 0 \qquad \theta \le \theta_1$$

$$\theta_1 = x_1 - x_2 T \qquad (6.5.8)$$

where θ_1 = an *apparent delay time* (a lag prior to onset of the oxidation reaction)
x_1, x_2 = constants (h and h/°F, respectively)
T = temperature (°F)

The temperature dependence of the reaction rate constant k is correlated by the Arrhenius equation (see Chapter 5).

$$k = Ae^{-E/RT} \tag{6.5.9}$$

where E = activation energy (Btu/lbmol)
  R = universal gas constant 1.987 Btu/lbmol-°R
  T = absolute temperature (°R)

Regarding Eq. (6.5.7), the *apparent lagtime* (θ_1) is defined for each temperature as the intercept of the line representing the data with the zero reaction line, that is, $\ln(C_A/C_{A0}) = 0$, on a graph of $\ln(C_A/C_{A0})$ versus θ. The lag time (θ_1) and the reaction temperature were found to be best correlated by a linear equation (Eq. 6.5.8) using a least square analysis. The experimental data were modeled by Eq. (6.5.7), which assumes that no reaction occurs until the residence time is longer than the lag time. (The Arrhenius constants, A and E, were estimated using a nonlinear least squares estimation procedure.) The estimated parameters and some results are shown in Table 6.5.1 for a variety of organic compounds. The lower limiting temperature is indicated for each compound. Use of the parameters at temperatures lower than this value or at temperatures higher than about 1700 to 1800°F is not recommended. At temperatures > 1700 to 1800°F, the reaction rates are believed to be much faster than that predicted by the parameters in this table because different reaction mechanisms with more free radical involvement take place at higher temperatures.

Equations to predict the time–temperature kinetic rate information for untested compounds applicable in the conventional thermal oxidation temperature ranges (1200–1800°F) were developed based on the data shown in Table 6.5.1. Temperatures required for three destruction levels (99.0, 99.9, and 99.99%) at four residence times (0.5, 1, 1.5, and 2 s) were calculated for the compounds listed in Table 6.5.1. These temperatures were correlated with the characteristic compound structures, the autoignition temperature, and the residence times. The temperatures required at these three destruction levels are designated as T_{99}, $T_{99.9}$, and $T_{99.99}$. The temperature models that appear in Table 6.5.2 were developed using multiple linear regression. A *student's t* value is given for each coefficient.

The usefulness of this *reactor* approach to describing incinerators should be obvious to the reader. As described in Chapter 3, the intent of a trial burn is to determine the incinerability of a particular waste stream in a particular incinerator under a particular set of operating conditions. The purpose of selecting the POHCs for a trial burn from among the constituent compounds in the hazardous waste feed is to obtain those few compounds for analysis that are most difficult to destroy. Thus, if meaningful kinetic data are available at system and operating conditions, the destruction efficiency of these POHCs could be predicted *a priori*. This prediction would be generated without recourse to pilot or field test data.

The ability to realistically use this approach is based on the applicability of the kinetic model and reactor equation to the system in question. Note, however, that the conditions under which the data were obtained may not be practically achieved in a real incinerator operating at an industrial facility. Notwithstanding this fact, one should appreciate the assumption that the kinetics of destruction can be expressed as a psuedo-first-order reaction. This essentially assumes that the reaction is first order even though the overall destruction process may be an extremely complex function of a host of variables.

TABLE 6.5.1. Thermal Oxidation Parameters

Compound	A	E (cal/gmol)[a]	x_1	x_2	Lower Limiting Temperature (°F)	Autoignition Temperature (°F)	Calculated Destruction Temperature from Eq. (6.5.7)	
							$T_{99.99}$ at 1 s (°F)	$T_{99.99}$ at 2 s (°F)
Acrolein	3.30×10^{10}	35,900	0	0	800	453	1020	975
Acrylonitrile	2.13×10^{12}	52,100	0.375	0.000250	1250	898	1345	1297
Allyl alcohol	1.75×10^{6}	21,400	1.971	0.00146	1050	713	1176	1077
Allyl chloride	3.89×10^{7}	29,100	0.525	0.00035	1150	905	1276	1200
Benzene	7.43×10^{21}	95,900	2.59	0.00178	1275	1044	1351	1322
1-Butene	3.74×10^{14}	58,200	2.12	0.00157	1150	723	1232	1195
Chlorobenzene	1.34×10^{17}	76,600	1.22	0.0008	1350	1180	1408	1372
1,2-Dichloroethane	4.82×10^{11}	45,600	0.948	0.00073	1050	775	1216	1173
Ethane	5.65×10^{14}	63,600	7.33	0.0052	1275	959	1368	1328
Ethanol	5.37×10^{11}	48,100	2.10	0.0015	1250	793	1307	1256
Ethyl acrylate	2.19×10^{12}	46,000	0	0	1000	721	1132	1092
Ethylene	1.37×10^{12}	50,800	0	0	1200	842	1328	1281
Ethyl formate	4.39×10^{11}	44,700	0.325	0.00024	1100	851	1191	1145
Ethyl mercaptan	5.20×10^{5}	14,700	1.87	0.0022	700	570	778	704
Methane	1.68×10^{11}	52,100	1.90	0.00117	1200	999	1545	1486
Methyl chloride	7.34×10^{8}	40,900	1.518	0.00084	1500	1170	1597	1514
Methyl ethyl ketone	1.45×10^{14}	58,400	1.92	0.00136	1200	960	1290	1247
Propane	5.25×10^{19}	85,200	1.02	0.000686	1200	871	1330	1300
Propylene	4.63×10^{8}	34,200	4.54	0.00323	1200	851	1318	1247
Toluene	2.28×10^{13}	56,500	1.35	0.000922	1275	997	1340	1295
Triethylamine	8.10×10^{11}	43,200	1.20	0.0010	950	450	1101	1058
Vinyl acetate	2.54×10^{9}	35,900	1.076	0.00078	1150	800	1223	1164
Vinyl chloride	3.57×10^{14}	63,300	0	0	1250	882	1371	1332

[a] To convert cal/gmol to Btu/lbmol, divide by 1.8.

220

TABLE 6.5.2. Predictive Destruction Temperature Models[a]

$$T_{99.00} = 577 - 10.0V_1 + 110.2V_2 + 67.1V_3 + 72.6V_4 + 0.586V_5 - 23.4V_6 - 430.9V_7 + 85.2V_8 - 82.2V_9 + 65.5V_{10} - 76.1V_{11}$$

$$\quad(3.5)^a\quad(6.3)\quad(9.1)\quad(7.3)\quad(32.7)\quad(5.9)\quad(34.8)\quad(15.5)\quad(10.2)\quad(5.8)\quad(19.2)$$

$$R^2 = 0.989 \qquad S = 19.8$$

$$T_{99.90} = 594 - 12.2V_1 + 117.0V_2 + 71.6V_3 + 80.2V_4 + 0.592V_5 - 20.2V_6 - 420.3V_7 + 87.1V_8 - 66.8V_9 + 62.8V_{10} - 75.3V_{11}$$

$$\quad(4.1)\quad(6.3)\quad(9.2)\quad(7.7)\quad(31.2)\quad(4.8)\quad(32.1)\quad(15.0)\quad(7.8)\quad(5.2)\quad(18.0)$$

$$R^2 = 0.987 \qquad S = 20.9$$

$$T_{99.99} = 605 - 13.8V_1 + 122.5V_2 + 75.7V_3 + 85.6V_4 + 0.597V_5 - 17.9V_6 - 412.0V_7 + 89.0V_8 - 55.3V_9 + 60.7V_{10} - 75.2V_{11}$$

$$\quad(4.3)\quad(6.3)\quad(9.2)\quad(7.8)\quad(29.8)\quad(4.0)\quad(29.8)\quad(14.5)\quad(6.1)\quad(4.8)\quad(17.0)$$

$$R^2 = 0.985 \qquad S = 22.1$$

Variables

$T_{99.00}$ = 99.00% destruction temperature (°F)
$T_{99.90}$ = 99.90% destruction temperature (°F)
$T_{99.99}$ = 99.99% destruction temperature (°F)
R^2 = Multiple correlation coefficient
S = Standard deviation (°F)

V_1 = number of carbons
V_2 = aromatic (0 = no, 1 = yes)
V_3 = carbon–carbon double bond (0 = no, 1 = yes)
V_4 = number of nitrogens
V_5 = autoignition temperature (°F)
V_6 = number of oxygens
V_7 = number of sulfurs
V_8 = hydrogen/carbon ratio
V_9 = allyl compound (0 = no, 1 = yes)
V_{10} = carbon double bond, chlorine interaction (0 = no, 1 = yes)
V_{11} = ln (time in s)

[a] Number in parenthesis is student's t value for the coefficient.

221

6.6   ILLUSTRATIVE EXAMPLES

Example 6.6.1.   Assume perfect combustion of a contaminated fuel oil with stoichiometric air. The gravimetric weight percentage analysis of a sample of this fuel oil is

88.52% carbon	0.10% nitrogen
10.87% hydrogen	0.06% oxygen
0.40% sulfur	0.05% ash

Calculate

1. The gravimetric analysis of the flue gas.
2. Total volume of flue gas (at 500°F and 1 atm) per pound of oil burned.
3. Volume percent of CO_2 in the dry flue gas.

SOLUTION

1. Convert the gravimetric percent of each component to lbmol selecting 100 lb of fuel as a basis.

$$
\begin{array}{ll}
\text{C} & 88.52/12 = 7.38 \text{ lbmol} \\
\text{H} & 10.87/2 = 5.44 \text{ lbmol} \\
\text{S} & 0.40/32 = 0.0125 \text{ lbmol} \\
\text{O}_2 & 0.06/32 = 0.00188 \text{ lbmol} \\
\text{N}_2 & 0.10/28 = 0.00357 \text{ lbmol}
\end{array}
$$

Since the molecular weight (MW) of the ash is unknown and its content is small, it is neglected in the calculation.

Write the combustion reaction for each component in the fuel.

$$C + O_2 \longrightarrow CO_2$$
$$H_2 + \tfrac{1}{2}O_2 \longrightarrow H_2O$$
$$S + O_2 \longrightarrow SO_2$$

Determine the amount of oxygen required for each component. Use the previous results to complete the first column.

Component	lbmol in Fuel	lbmol O_2 Required
C	7.38	7.38
H_2	5.44	2.72
S	0.0125	0.0125
O_2	0.00188	−0.00188
		Total = 10.1

Note: Oxygen in the fuel reduces air requirements. Therefore, the amount of oxygen in the fuel has been subtracted from the amount of O_2 required.

Calculate the amount of nitrogen from the oxygen requirement.

$$\text{amount of } N_2 = (\text{amount of } O_2)\frac{0.79\,\text{lbmol } N_2}{0.21\,\text{lbmol } O_2}$$

Component	lbmol N_2 Required
C	27.76
H_2	10.23
S	0.047
O_2	0.0
	Total = 38.03

Determine the gravimetric composition (percentage) of the flue gas.

Component	lbmol	MW	lb	wt%
CO_2	7.38	44	324.7	21.8
H_2O	5.44	18	97.9	6.6
SO_2	0.00125	64	0.08	0.005
O_2	0.0	32	0.0	0.0
N_2	38.03	28	1064.3	71.6
	Total = 50.85		Total = 1487.0	

2. Calculate the total lbmol of flue gas produced per pound of oil burned.

$$\text{lbmol of flue gas} = \frac{\text{total lbmol from above}}{100}$$

$$= \frac{50.85}{100}$$

$$= 0.5085\,\text{lbmol gas/lb oil}$$

Remember that 100 lb of fuel was used as a basis.
    Calculate the total volume of flue gas at 500°F and 1 atm using the ideal gas law.

$$V = \frac{nRT}{P} = \frac{(\text{lbmol flue gas/lb oil})(0.7302)(500 + 460)}{1.0}$$

$$= 356.1\,\text{ft}^3/\text{lb oil}$$

3. One can now complete the following table.

Component of Dry Flue Gas	lbmol
CO_2	7.38
SO_2	0.00125
N_2	38.01
	Total = 45.39

The volume percentage of CO_2 in the dry flue gas is

$$\%CO_2 = \frac{\text{lbmol } CO_2}{\text{total lbmol dry flue gas}}(100) = \frac{7.38}{45.39}(100)$$

$$= 16.26\%$$

Example 6.6.2. A waste mixture to be incinerated in a liquid injection system has the following composition on a lb/lb basis:

$$\text{Carbon } (C_w) = 0.732$$

$$\text{Hydrogen } (H_w) = 0.109$$

$$\text{Moisture } (H_2O_w) = 0.022$$

$$\text{Oxygen } (O_w) = 0.102$$

$$\text{Chlorine } (Cl_w) = 0.035$$

The incinerator process will be assisted with 0.5 lb of natural gas per pound of waste. Thirty percent excess air is to be employed. Calculate the required quantities in the worksheet provided in Table 6.3.1.

SOLUTION
Skip Step Nos. 1 and 2 of Table 6.3.1. Identify the elemental composition of the natural gas from Step No. 3.

Components	lb component/lb fuel
C_f	0.693
H_f	0.227
N_f	0.080

Calculate the combined composition of waste and fuel feed.

$$n_C = \frac{C_w + n_f C_f}{1 + n_f}$$

$$= 0.719 \, \text{lb/lb feed}$$

$$n_H = \frac{H_w + n_f H_f}{1 + n_f}$$

$$= 0.148 \, \text{lb/lb feed}$$

$$n_{H_2O} = \frac{H_2O_w}{1 + n_f}$$

$$= 0.147 \, \text{lb/lb feed}$$

$$n_O = \frac{O_w}{1 + n_f}$$

$$= 0.0680 \, \text{lb/lb feed}$$

$$n_{Cl} = \frac{Cl_w}{1 + n_f}$$

$$= 0.0233 \, lb/lb \text{ feed}$$

$$n_N = \frac{n_f N_f}{1 + n_f}$$

$$= 0.0267 \, lb/lb \text{ feed}$$

Remember that the fuel to waste ratio (n_f) for this problem is 0.5.
Calculate the stoichiometric oxygen requirements.

$$(O_2)_{\text{stoich}} = (n_i) \frac{lb \, O_2}{lb \text{ of species } i} - n_O$$

$$= (n_C) \frac{2.67 \, lb \, O_2}{lb \, C} + \left(n_H - \frac{n_{Cl}}{35.5} \right) \frac{8.0 \, lb \, O_2}{lb \, H} - n_O$$

$$= (0.719)(2.67) + \left(0.148 - \frac{0.0233}{35.5} \right)(8.0) - 0.068$$

$$= 3.030 \, lb \, O_2/lb \text{ feed}$$

Note: Here 1 lbmol of C with 1 lbmol of O_2 forms 1 lbmol of CO_2 when combusted completely. 2.67 lb O_2/lb C is obtained by converting lbmol to lb. The same explanation applies to H. However, H also reacts to form HCl and the amount of H that forms HCl should be substracted from n_H, hence $n_H - (n_{Cl}/35.5)$.
Calculate the combustion gas mass fraction.

$$N_{CO_2} = (n_C) \frac{3.67 \, lb \, CO_2}{lb \, C}$$

$$= 2.639 \, lb \, CO_2/lb \text{ feed}$$

$$N_{H_2O} = \left(n_H - \frac{n_{Cl}}{35.5} \right) \frac{9.0 \, lb \, H_2O}{lb \, H} + n_{H_2O}$$

$$= 1.341 \, lb \, H_2O/lb \text{ feed}$$

$$N_{N_2} = (O_2)_{\text{stoich}} \left(\frac{3.31 \, lb \, N_2}{lb \, O_2} \right) + n_N$$

$$= 10.03 \, lb \, N_2/lb \text{ feed}$$

$$N_{HCl} = (n_{Cl}) \frac{1.03 \, lb \, HCl}{lb \, Cl}$$

$$= 0.0240 \, lb \, HCl/lb \text{ feed}$$

Note: Capital N is used to distinguish combustion gas mass fractions from feed mass fractions (n).

Calculate the combustion products per pound of feed.

$$CP = N_{CO_2} + N_{H_2O} + N_{N_2} + N_{HCl}$$
$$= 14.03 \, lb/lb \; feed$$

Calculate the additional nitrogen and oxygen present in the combustion gases due to the excess air feed.

$$(O_2)_{EA} = (EA)(O_2)_{stoich} \qquad (N_2)_{EA} = \frac{3.31 \, lb \, N_2}{lb \, O_2} (O_2)_{EA}$$
$$= (0.3)(O_2)_{stoich} \qquad\qquad = 3.009 \, lb \, N_2/lb \; feed$$
$$= 0.9090$$

Calculate the total combustion gas products per pound of feed.

$$CG = CP + (O_2)_{EA} + (N_2)_{EA} = 14.03 + 0.9090 + 3.009$$
$$= 17.95 \, lb/lb \; feed$$

Calculate the mass fraction of each combustion gas component.

$$CO_2 \quad \frac{N_{CO_2}}{CG} = \frac{2.639}{17.95} = 0.1470 \, lb/lb \; gas$$

$$H_2O \quad \frac{N_{H_2O}}{CG} = \frac{1.314}{17.95} = 0.0732 \, lb/lb \; gas$$

$$N_2 \quad \frac{[N_{N_2} + (N_2)_{EA}]}{CG} = \frac{13.04}{17.95} = 0.7264 \, lb/lb \; gas$$

$$O_2 \quad \frac{(O_2)_{EA}}{CG} = \frac{0.9090}{17.95} = 0.0506 \, lb/lb \; gas$$

$$HCl \quad \frac{N_{HCl}}{CG} = \frac{0.0240}{17.95} = 0.00133 \, lb/lb \; gas$$

Identify those components < 1 to 2% of the combustion gas. These components can be eliminated from heat and material balance calculations. Hydrogen chloride is the only component that constitutes < 1 to 2% of the combustion gas.

Calculate the volumetric flow of the combustion products at standard conditions of 68°F and 1 atm.

$$CO_2 \quad \frac{(N_{CO_2})}{(0.114 \, lb/scf)} = 23.15 \, scf/lb \; feed$$

$$H_2O \quad \frac{(N_{H_2O})}{(0.0467 \, lb/scf)} = 28.76 \, scf/lb \; feed$$

$$N_2 \quad \frac{[N_{N_2} + (N_2)_{EA}]}{(0.0727 \, lb/scf)} = 179.4 \, scf/lb \; feed$$

$$O_2 \quad \frac{(O_2)_{EA}}{(0.083 \, lb/scf)} = 10.96 \, scf/lb \ feed$$

$$HCl \quad \frac{N_{HCl}}{(0.0945 \, lb/scf)} = 0.25 \, scf/lb \ feed$$

$$\text{Total volumetric flow} = 242.52 \, scf/lb \ feed$$

Note: The density terms can be obtained from tables or calculated from the ideal gas law.

Example 6.6.3. A waste mixture to be incinerated by a rotary kiln has the following composition.

	Solids (1)	Liquids (2)	
Carbon (C_w)	0.74	0.30	lb/lb waste
Hydrogen (H_w)	0.10	0.12	lb/lb waste
Moisture (H_2O_w)	0.02	0.40	lb/lb waste
Oxygen (O_w)	0.11	0.10	lb/lb waste
Sulfur (S_w)	0.03	0.08	lb/lb waste

The liquid feed rate is 1500 lb/h and the solid feed rate is 2000 lb/h. The incineration process will be assisted with 0.5 lb of natural gas/lb of waste. It is to employ 50% excess air. Calculate the required quantities in Table 6.3.2.

SOLUTION. Identify the elemental composition of the natural gas. See Table 6.2.1.

Components	lb component/lb fuel
C_f	0.693
H_f	0.227
N_f	0.080

Calculate the solid waste fraction (n_1) and liquid waste fraction (n_2).

$$n_1 = (\text{solid waste rate})/(\text{liquid waste rate} + \text{solid waste rate})$$

$$= \frac{w_s}{(w_l + w_s)}$$

$$= \frac{2000}{(1500 + 2000)}$$

$$= 0.57 \, lb \ solid/lb \ waste$$

$$n_2 = \frac{w_l}{(w_l + w_s)}$$

$$= 0.43 \, lb \ liquid/lb \ waste$$

Calculate the composition of the combined waste/auxiliary fuel feed to the kiln.

$$n_C = \frac{n_1 C_1 + n_2 C_2 + n_f C_f}{1 + n_f} = 0.5982 \text{ lb/lb feed}$$

$$n_H = \frac{n_1 H_1 + n_2 H_2 + n_f H_f}{1 + n_f} = 0.1481 \text{ lb/lb feed}$$

$$n_{H_2O} = \frac{n_1(H_2O)_1 + n_2(H_2O)_2}{1 + n_f} = 0.1223 \text{ lb/lb feed}$$

$$n_N = \frac{n_f N_f}{1 + n_f} = 0.0267 \text{ lb/lb feed}$$

$$n_O = \frac{n_1 O_1 + n_2 O_2}{1 + n_f} = 0.0705 \text{ lb/lb feed}$$

$$n_S = \frac{n_1 S_1 + n_2 S_2}{1 + n_f} = 0.0343 \text{ lb/lb feed}$$

Note: Subscripts 1 and 2 designate the composition of a given component in the solid waste and liquid waste, respectively. Remember that the fuel–waste ratio (n_f) for this problem is 0.5.

Calculate the stoichiometric oxygen requirement for the kiln using the following equation.

$$(O_2)_{\text{stoich}} = \sum_{i=1}^{n} (n_i)(\text{lb } O_2/\text{lb of species } i) - n_O$$

$$= (n_C)(2.67 \text{ lb } O_2/\text{lb C}) + (n_H)(8.0 \text{ lb } O_2/\text{lb H}) + (n_S)(1.0 \text{ lb } O_2/\text{lb S}) - n_O$$

$$= 2.746 \text{ lb } O_2/\text{lb feed}$$

See the note in Example 6.6.2 explaining this equation.

Calculate the combustion gas mass fractions.

$$N_{CO_2} = (n_C)(3.67 \text{ lb } CO_2/\text{lb C}) = 2.1954 \text{ lb } CO_2/\text{lb feed}$$

$$N_{H_2O} = (n_H)(9.0 \text{ lb } H_2O/\text{lb H}) + n_{H_2O} = 1.4552 \text{ lb } H_2O/\text{lb feed}$$

$$N_{N_2} = (O_2)_{\text{stoich}}(3.31 \text{ lb } N_2/\text{lb } O_2) + n_N = 9.1160 \text{ lb } N_2/\text{lb feed}$$

$$N_{SO_2} = (n_S)(2.0 \text{ lb } SO_2/\text{lb S}) = 0.0686 \text{ lb } SO_2/\text{lb feed}$$

Calculate the amount of combustion products per lb feed.

$$CP = N_{CO_2} + N_{H_2O} + N_{N_2} + N_{SO_2}$$

$$= 12.8352 \text{ lb/lb feed}$$

Calculate the additional nitrogen and oxygen present in the combustion gases due to the excess air feed.

$$(O_2)_{EA} = (EA)(O_2)_{stoich}$$

$$= 1.373 \, lb \; O_2/lb \; feed$$

$$(N_2)_{EA} = (3.31 \, lb \; N_2/lb \; O_2)(O_2)_{EA}$$

$$= 4.545 \, lb \; N_2/lb \; feed$$

Calculate the total combustion gas products per pound feed.

$$CG = CP + (O_2)_{EA} + (N_2)_{EA} = 12.83 + 1.373 + 4.545$$

$$= 18.75 \, lb/lb \; feed$$

Calculate the mass fraction of each combustion gas component.

$$CO_2 \quad \frac{N_{CO_2}}{CG} = 0.1171 \, lb/lb \; gas$$

$$H_2O \quad \frac{N_{H_2O}}{CG} = 0.0776 \, lb/lb \; gas$$

$$N_2 \quad \frac{[N_{N_2} + (N_2)_{EA}]}{CG} = 0.7286 \, lb/lb \; gas$$

$$O_2 \quad \frac{(O_2)_{EA}}{CG} = 0.0732 \, lb/lb \; gas$$

$$SO_2 \quad \frac{N_{SO_2}}{CG} = 0.0037$$

Calculate the volumetric flow of the combustion products at standard conditions of 68°F and 1 atm.

$$CO_2 \quad \frac{(N_{CO_2})}{(0.114 \, lb/scf)} = 19.26 \, scf/lb \; feed$$

$$H_2O \quad \frac{(N_{H_2O})}{(0.0467 \, lb/scf)} = 31.16 \, scf/lb \; feed$$

$$N_2 \quad \frac{[N_{N_2} + (N_2)_{EA}]}{(0.0727 \, lb/scf)} = 187.9 \, scf/lb \; feed$$

$$O_2 \quad \frac{(O_2)_{EA}}{(0.083 \, lb/scf)} = 16.54 \, scf/ \; feed$$

$$SO_2 \quad \frac{N_{SO_2}}{(0.166 \, lb/scf)} = 0.413 \; scf/lb \; feed$$

total volumetric flow rate = 255.3 scf/lb feed

Example 6.6.4. Manhattan Waste Generators plan to incinerate pure C_6H_5Cl in a liquid injection incinerator with 100% excess air. A consulting engineer from ABC associates has assured the client that the operating temperature will exceed the permit requirement of 2100°F. Verify or dispute the consultant's claim.

SOLUTION. This problem involves an extension of the material presented in Section 5.3. However, unlike the previous development, it will require an adiabatic flame temperature calculation with 100% excess air. For this condition, the stoichiometric oxygen and nitrogen (or air) requirement is increased by a factor of two. These values and the stoichiometric equation are again given here.

$$O_2 = 14\,\text{lbmol} \quad N_2 = 52.6\,\text{lbmol}$$

$$C_6H_5Cl + 14O_2 + 52.6N_2 \longrightarrow 6CO_2 + 2H_2O + HCl + 7O_2 + 52.6N_2$$

The heat capacity term (ΔC_P) is now calculated using Eq. (5.2.9) and (5.4.11).

$$\Delta C_P = 6C_{P,CO_2} + 2C_{P,H_2O} + C_{P,HCl} + 7C_{P,O_2} + 52.6C_{P,N_2}$$

The heat capacities are obtained from Table 5.2.4.

$$C_{P,CO_2} = 10.57 + 2.10 \times 10^{-3}\,T - 2\,06 \times 10^5\,T^{-2}$$
$$C_{P,H_2O} = 7.30 + 2.46 \times 10^{-3}\,T + 0.00 \times 10^5\,T^{-2}$$
$$C_{P,HCl} = 6.27 + 1.24 \times 10^{-3}\,T + 0.30 \times 10^5\,T^{-2}$$
$$C_{P,O2} = 7.16 + 1.00 \times 10^{-3}\,T - 0.40 \times 10^5\,T^{-2}$$
$$C_{P,N_2} = 6.83 + 0.90 \times 10^{-3}\,T - 0.12 \times 10^5\,T^{-2}$$

Thus,

$$\Delta C_P = 493.67 + 0.0731\,T - 2.12 \times 10^6\,T^{-2}$$

The sensible enthalpy change for the fuel gas products is

$$\Delta H_p = \int_{298}^{T_2} \Delta C_P\, dT$$

In addition,

$$-\Delta H_c^\circ = \Delta H_p$$

ΔH_c° was calculated to be $-714{,}361$ cal/gmol from Example 5.7.9. Therefore,

$$714{,}361 = 493.67(T_2 - 298) + \frac{1}{2}(0.0731)(T_2^2 - 298^2) + 2.12 \times 10^6\left(\frac{1}{T_2} - \frac{1}{298}\right)$$

$$871{,}834 = 493.67T_2 + 0.03655T_2^2 + \frac{2.12 \times 10^6}{T_2}$$

Solving for T_2 (see Example 5.7.9 for a simplified trial-and-error procedure),

$$T_2 = 1579\,K = 2842°R = 2382°F$$

Therefore, the operating temperature does exceed the permit requirement of 2100°F.

Example 6.6.5. Calculate the operating temperature of an incinerator burning pure DDT ($C_{14}H_9Cl_5$) with 150% excess air. Assume the DDT and air enter the incinerator at ambient conditions (25°C, 1 atm).

SOLUTION. This problem is very similar to Example 5.7.10. Here, instead of stoichiometric air, 150% excess air is used. The following values can be used.

$$\Delta H_c = -1580.56\,kcal/gmol$$

$$C_{14}H_9Cl_5 + 37.5O_2 + 141.08N_2 \longrightarrow 14CO_2 + 2H_2O + 5HCl + 22.5O_2 + 141.08N_2$$

$$\Delta C_P = 14C_{P,CO_2} + 2C_{P,H_2O} + 5C_{P,HCl} + 22.5C_{P,O_2} + 141.08C_{P,N_2}$$

Substituting in the heat capacities from Table 5.2.4,

$$\Delta C_P = 1318.60 + 0.1899T - 5.327 \times 10^6\,T^{-2}$$

Since

$$-\Delta H_c = \Delta H_p = \int_{298}^{T} \Delta C_P\,dT$$

$$1580.56 \times 10^3 = 1318.60(T_2 - 298) + \frac{1}{2}(0.1899)(T_2^2 - 298^2) + 5.327 \times 10^6\left(\frac{1}{T_2} - \frac{1}{298}\right)$$

$$1{,}999{,}815.0 = 1318.60T_2 + 0.095T_2^2 + \frac{5.327 \times 10^6}{T_2}$$

Solving for T_2 by a trial-and-error method,

$$T_2 = 1377\,K$$

$$= 1104°C$$

$$= 2019°F$$

Example 6.6.6. Thirty percent by mass of benzene in water is combusted in an incinerator with stoichiometric air. Estimate the operating temperature of the incinerator using the EPA worksheet provided in Table 6.4.5. Repeat the calculation for 25, 50, 75, 100 and 150% excess air.

SOLUTION. First write down the combustion equation for 0% excess air (or stoichiometric air).

$$C_6H_6 + 7.5O_2 + 28.2N_2 \longrightarrow 6CO_2 + 3H_2O(g) + 28.2N_2$$

Converting from lbmol to pounds of flue gases (per lb of benzene incinerated),

$$\text{lb } CO_2 = \frac{(6)(\text{MW of } CO_2)}{\text{MW of } C_6H_6} = \frac{(6)(44)}{(78)}$$

$$= 3.38$$

$$\text{lb } H_2O = \frac{(3)(18)}{(78)} = 0.69$$

$$\text{lb } N_2 = \frac{(28.2)(28)}{(78)}$$

$$= 10.12$$

Since no auxiliary fuel or air preheating is employed, go to Step No. 3 of Table 6.4.5 directly.

$$\Delta H_2 = (0.3)(\text{NHV of benzene}) - (0.7)\Delta H_v$$

Note: The heat required to vaporize the water is subtracted from the NHV of benzene in the feed.

Assume the heat of vaporization of water at 77°F is 1051 Btu/lb. The NHV of benzene is 17,451 Btu/lb from Table 5.3.1.

$$\Delta H_2 = (0.3)(17,451) - (0.7)(1051)$$

$$= 4499.6 \text{ Btu/lb feed}$$

Calculate the heat loss through the walls of the incinerator, assuming 5% loss.

$$Q = (0.05)(\Delta H_2)$$

$$= 225 \text{ Btu/lb feed}$$

Calculate the enthalpy of the combustion products leaving the incinerator.

$$\Delta H_3 = (0.26)(0.3 \times 3.38 + 0.3 \times 10.12)(T - 77) + (0.49)(0.3 \times 0.69 + 0.7)(T - 77)$$

$$= 1.50(T - 77) \text{ Btu/lb feed}$$

Calculate the enthalpy of stoichiometric air.

$$\Delta H_4' = (1.1)(T - 77)(O_2)_{\text{stoich}} = (1.1)(T - 77)(0.3)(7.5)(32)/(78)$$

$$= 1.015(T - 77)$$

For 0% EA, $\Delta H_4 = 0$. From Step No. 7 of Table 6.4.5,

$$O = \Delta H_2 - Q - \Delta H_3$$
$$= 4499.6 - 225 - (1.50)(T - 77)$$

Solving for T,

$$T = 2927°F$$

For 25% excess air,

$$25 = \frac{(100)(\Delta H_2 - Q - \Delta H_3)}{\Delta H_4'}$$

$$0.25 = \frac{4499.6 - 225 - (1.5)(T - 77)}{(1.015)(T - 77)}$$

Therefore,

$$T(25\% \text{ EA}) = 2514.4°F$$

Similarly,

$$T(50\% \text{ EA}) = 2206.3°F$$
$$T(75\% \text{ EA}) = 1967.4°F$$
$$T(100\% \text{ EA}) = 1776.6°F$$
$$T(150\% \text{ EA}) = 1491.3°F$$

Example 6.6.7.   Perform the calculations required in Example 6.6.6 using the simplified approach given in Eq. (6.4.22).

SOLUTION. According to Eq. (6.4.22),

$$NHV = \frac{(0.3)(T - 60)}{1 - (1 + EA)(7.5 \times 10^{-4})(0.3)(T - 60)}$$

For 0% excess air,

$$(0.3)(17{,}451) - (0.7)(1051) = \frac{(0.3)(T - 60)}{1 - (1)(7.5 \times 10^{-4})(0.3)(T - 60)}$$

Solving for T,

$$T(0\% \text{ excess air}) = 3488.5°F$$

For 25% excess air,

$$(0.3)(17,451) - (0.7)(1051) = \frac{(0.3)(T - 60)}{1 - (1.25)(7.5 \times 10^{-4})(0.3)(T - 60)}$$

$$T(25\% \text{ excess air}) = 2934.2°F$$

Similarly,

$$T(50\% \text{ excess air}) = 2534.2°F$$

$$T(75\% \text{ excess air}) = 2231.9°F$$

$$T(100\% \text{ excess air}) = 1995.5°F$$

$$T(150\% \text{ excess air}) = 1649.4°F$$

The reader should compare the results of this illustrative example with the previous example (6.6.6). Although the reference temperatures are different—60 versus 77°F—this effect on the final calculations is negligible. The results, however, do differ significantly. Because of the high water content of the waste mixture, the average heat capacity employed in Eq. (6.4.22) is low. Using 0.35 for the heat capacity (as recommended in Table 6.4.1) in this equation leads to a temperature of 2999°F for combustion with 0% excess air. Use of this adjusted value for the other excess air calculations similarly produced (excellent) comparable results.

Example 6.6.8. A fuel–waste mixture is to be combusted in an incinerator at an operating temperature of 2000°F. Calculate the minimum net heating value of the mixture in Btu/lb if 0, 20, 40, 60, 80, and 100% excess air is employed. Use Eq. (6.4.22) to perform the calculation.

SOLUTION. Referring to Eq (6.4.22),

$$\text{NHV} = \frac{(0.3)(T - 60)}{[1 - (1 + \text{EA})(7.5 \times 10^{-4})(0.3)(T - 60)]}$$

For 0% excess air,

$$\text{NHV (0\% excess air)} = \frac{(0.3)(2000 - 60)}{[1 - (1)(7.5 \times 10^{-4})(0.3)(2000 - 60)]}$$

$$= 1032.8 \text{ Btu/lb}$$

For 20% excess air,

$$\text{NHV (20\% excess air)} = \frac{(0.3)(2000 - 60)}{[1 - (1.2)(7.5 \times 10^{-4})(0.3)(2000 - 60)]}$$

$$= 1222.2 \text{ Btu/lb}$$

Similarily,

NHV (40% excess air) = 1496.5 Btu/lb

NHV (60% excess air) = 1929.7 Btu/lb

NHV (80% excess air) = 2715.8 Btu/lb

NHV (100% excess air) = 4582.7 Btu/lb

Example 6.6.9. A plant engineer has calculated the stoichiometric oxygen requirement and combustion gas mass flow rates using the approach in Example 6.6.6.
Apply these results and calculate the excess air rate. Use Table 6.4.5 to perform the
calculation.

$$(O_2)_{\text{stoich}} = 3.195 \text{ lb } O_2/\text{lb feed}$$

Combustion gas mass fractions:

$$N_{CO_2} = 2.61 \text{ lb } CO_2/\text{lb feed}$$

$$N_{H_2O} = 1.52 \text{ lb } H_2O/\text{lb feed}$$

$$N_{N_2} = 10.60 \text{ lb } N_2/\text{lb feed}$$

ADDITIONAL DATA
NHV of the feed = 7200 Btu/lb feed

Operating temperature, $T = 1900°F$

Air preheat temperature, $T_{\text{air}} = 500°F$

No auxiliary fuel is used.

SOLUTION. Since no auxiliary fuel is used, proceed to Step No. 2 of Table 6.4.5
directly.
    Calculate the enthalpy input to the incinerator through air preheating.

$$\Delta H_1' = 1.12(T_{\text{air}} - 77)(O_2)_{\text{stoich}}$$

$$= 1513.7 \text{ Btu/lb feed}$$

$$\Delta H_1 = \Delta H_1'(1 + \text{EA})$$

Note: ΔH_1 is the enthalpy input with an excess air rate of EA. See footnote *b* in
Table 6.4.5 for the derivation of this equation.

Determine the heat generated by the combustion of the waste.

$$\Delta H_2 = 7200 \text{ Btu/lb feed}$$

Note: Since no auxiliary fuel is used, the NHV of the feed is ΔH_2.

    Calculate the heat transferred across the walls of the incinerator, assuming a 5%
loss.

$$Q = 0.05\Delta H_2$$

$$= 360 \text{ Btu/lb feed}$$

Calculate the enthalpy of the stoichiometric combustion products leaving the incinerator.

$$\Delta H_3 = 0.26(N_{CO_2} + N_{N_2})(T - 77) + 0.49(N_{H_2O})(T - 77)$$

$$= 7619 \text{ Btu/lb feed}$$

See footnote d in Table 6.4.5 for the derivation of the previous equation. Calculate the enthalpy of the excess air leaving the incinerator.

$$\Delta H_4' = 1.12(T - 77)(O_2)_{\text{stoich}}$$

$$= 6523.4 \text{ Btu/lb feed}$$

$$\Delta H_4 = (\Delta H_4')(EA)$$

$\Delta H_4'$ is the enthalpy of the air leaving the incinerator if no excess air is used. Calculate the excess air percentage by the following equation:

$$EA = 100(\Delta H_1' + \Delta H_2 - Q - \Delta H_3)/(\Delta H_4' - \Delta H_1')$$

$$= 14.7\%$$

Note: This equation is obtained by an energy balance around the incinerator.

Finally, the percentage of excess air employed in an incinerator may be estimated from the following equation:

$$EA = \frac{(95) O_2}{(21 - O_2)}$$

where O_2 is the percentage of oxygen in the discharge gas on a dry basis.

Example 6.6.10. A liquid waste consisting of 5% benzene in an aqueous solvent (with zero heating value) is to be incinerated. The combustion process will be supplemented with 0.4 lb natural gas (NHV = 19,500 Btu/lb) per lb waste. Ninety percent excess air at 300°F is to be employed. Calculate the operating temperature of the incinerator if 5.0% of the total heat released is lost across the walls of the unit. Use EPAs worksheet as presented in Table 6.4.5

SOLUTION. First refer back to Example 6.6.2. Since the waste consists of 5% benzene,

$$C_w = \frac{(0.05)(6)(\text{MW of C})}{(\text{MW of C}_6\text{H}_6)} = \frac{(0.05)(6)(12)}{(78)}$$

$$= 0.046 \text{ lb/lb waste}$$

$$H_w = \frac{(0.05)(6)(1)}{(78)} = 0.004 \text{ lb/lb waste}$$

$H_2O_w = 0.95 \text{ lb/lb waste}$

Identify the elemental composition of the natural gas (see Table 6.3.1).

$$C_f = 0.693 \text{ lb/lb fuel}$$

$$H_f = 0.227 \text{ lb/lb fuel}$$

$$N_f = 0.08 \text{ lb/lb fuel}$$

Calculate the combined composition of waste and fuel feed.

$$n_C = \frac{(C_w + n_f C_f)}{(1 + n_f)} = \frac{[0.046 + (0.4)(0.693)]}{(1 + 0.4)}$$

$$= 0.23 \text{ lb/lb feed}$$

$$n_H = \frac{(H_w + n_f H_f)}{(1 + n_f)} = \frac{[0.004 + (0.4)(0.227)]}{(1.4)}$$

$$= 0.0677 \text{ lb/lb feed}$$

$$n_N = \frac{n_f N_f}{(1 + n_f)} = \frac{(0.4)(0.08)}{(1.4)}$$

$$= 0.023 \text{ lb/lb feed}$$

$$n_{H_2O} = \frac{H_2O_w}{(1 + n_f)} = \frac{0.95}{(1.4)}$$

$$= 0.679 \text{ lb/lb feed}$$

Calculate the stoichiometric oxygen requirement.

$$(O_2)_{stoich} = (n_C)(2.67) + (n_H)(8.0) = (0.23)(2.67) + (0.0677)(8.0)$$

$$= 1.16 \text{ lb/lb feed}$$

See footnote c in Table 6.3.1 for the details of this equation.
Calculate the combustion gas mass fraction.

$$N_{CO_2} = (n_C)(3.67 \text{ lb } CO_2/\text{lb C}) = (0.23)(3.67)$$

$$= 0.844 \text{ lb } CO_2/\text{lb feed}$$

$$N_{H_2O} = (n_H)(9.0 \text{ lb } H_2O/\text{lb H}) + n_{H_2O} = (0.0677)(9) + (0.679)$$

$$= 1.289 \text{ lb } H_2O/\text{lb feed}$$

$$N_{N_2} = (O_2)_{stoich}(3.31 \text{ lb } N_2/\text{lb } O_2) + n_N = (1.16)(3.31) + 0.023$$

$$= 3.863 \text{ lb } N_2/\text{lb feed}$$

Now refer to Table 6.4.5.

Calculate the enthalpy input by the preheated air.

$$\Delta H_1' = (1.12)(T_{\text{air}} - 77)(O_2)_{\text{stoich}} = (1.12)(300 - 77)(1.16)$$
$$= 289.7 \text{ Btu/lb feed}$$

Calculate the heat generated by combustion of the waste–fuel mixture.

$$\Delta H_2 = \frac{(\text{NHV}_{\text{waste}} + n_f \text{NHV}_f)}{(1 + n_f)} - n_{\text{H}_2\text{O}}(\Delta H_v)$$
$$= \frac{(0.05 \times 17{,}451 + 0.4 \times 19{,}500)}{1.4} - (0.679)(1050)$$
$$= 5481.7 \text{ Btu/lb feed}$$

Note: The heat required to vaporize the solvent is subtracted from ΔH_2.

The heat capacity of the solvent is assumed to be 1.0 Btu/lb and the heat of vaporization is 1050 Btu/lb.

Calculate the heat loss through the walls of the incinerator.

$$Q = (0.05)(\Delta H_2)$$
$$= 274 \text{ Btu/lb feed}$$

Calculate the enthalpy of the combustion products leaving the incinerator.

$$\Delta H_3 = (0.26)(N_{\text{CO}_2} + N_{\text{N}_2})(T - 77) + (0.49)(N_{\text{H}_2\text{O}})(T - 77)$$
$$= (0.26)(0.844 + 3.863)(T - 77) + (0.49)(1.289)(T - 77)$$
$$= 1.855(T - 77) \text{ Btu/lb feed}$$

Calculate the enthalpy of the stoichiometric air.

$$\Delta H_4' = (1.1)(T - 77)(O_2)_{\text{stoich}} = (1.1)(T - 77)(1.16)$$
$$= 1.276(T - 77)$$

Use the equation in Step No. 7 of Table 6.4.5 to calculate the incinerator temperature.

$$\text{EA} = \frac{100(\Delta H_1' + \Delta H_2 - Q - \Delta H_3)}{(\Delta H_4' - \Delta H_1')}$$
$$90 = \frac{100[298.7 + 5481.7 - 274 - (1.855)(T - 77)]}{[(1.276)(T - 77) - 289.7]}$$

Solving for T,

$$T = 1994°F$$

Example 6.6.11. In order to meet recently updated pollution specifications for discharging hydrocarbons to the atmosphere, a gas stream must be reduced by 99.5% of its present hydrocarbon concentration. Due to economic considerations, it is proposed to meet this requirement by combusting the hydrocarbons in a thermal reactor operating at 1500°F. The gas and methane (fuel) are to be fed to the reactor at 80°F and 1 atm. Design the proposed reactor using kinetic principles.

DATA
Flue gas flow rate (from fuel combustion) = 2500 scfm (80°F, 1 atm)
Process gas flow rate = 7200 scfm
Hydrocarbon: Essentially toluene
Reaction rate constant, $k = 7.80\text{ s}^{-1}$ at 1500°F
Average velocity = 20 ft/s; $C_f/C_0 = 0.005/1.0$
C_f and C_0 are the final and initial concentrations, respectively.

SOLUTION. Based on the actual operating conditions of the reactor, the volumetric flow rate is

$$Q_a = Q_s\left(\frac{T_a}{T_s}\right) = (2500 + 7200)\left(\frac{(460 + 1500)}{(460 + 80)}\right)$$

$$= 35,207\text{ acfm}$$

Calculate the cross-section area of the reactor.

$$A = \frac{Q_a(\text{acfm})}{V(\text{fpm})} = \frac{(35,207)}{(20)(60)}$$

$$= 29.33\text{ ft}^2$$

Calculate the diameter of the reactor.

$$D = \left(\frac{4A}{\pi}\right)^{0.5} = \left(\frac{(4 \times 29.33)}{(3.1416)}\right)^{0.5}$$

$$= 6.11\text{ ft}$$

Calculate the residence time required assuming a plug flow reactor. Use Eq. (6.5.6).

$$\theta = -\left(\frac{1}{k}\right)\ln\left(\frac{C_f}{C_0}\right) = -\frac{1}{7.8}\ln\left(\frac{0.005}{1.0}\right)$$

$$= 0.68\text{ s}$$

Calculate the reactor volume required to achieve the residence time calculated previously.

$$V = \theta Q_a = (0.68)\frac{35,207}{60}$$

$$= 399\,\text{ft}^3$$

Calculate the length of the reactor.

$$L = \frac{V}{A} = \frac{399}{29.33}$$

$$= 13.6\,\text{ft}$$

Example 6.6.12.   A fluidized-bed incinerator is to be designed to destroy 99.99% of a unique hazardous waste. Based on laboratory and pilot plant studies, researchers have described the waste reaction by a first-order reversible mechanism. Their preliminary findings are given here.

$$A \underset{k'}{\overset{k}{\longleftrightarrow}} B$$

$$k = 1.0\exp(-10,000/T); \quad T = \text{Rankine}$$

$$k' = 9.89\exp(-35,000/T); \quad T = \text{Rankine}$$

Calculate a fluidized-bed incineration operating temperature that will minimize the volume of the incinerator and achieve the desired degree of waste conversion (destruction). The operating temperature must be in the 1400–1700°F range.

SOLUTION. A fluidized-bed incinerator is best described by a CSTR. According to Eq. (6.5.3),

$$\theta = \frac{V}{Q_a} = \frac{C_{A0} - C_{A1}}{-R_A} = \frac{C_{A0} - C_{A1}}{kC_{A1} - k'C_{B1}}$$

For 99.99% destruction of the waste A the conversion variable X_A becomes

$$X_A = 0.9999$$

So that

$$C_{A1} = 0.0001 C_{A0}; \quad C_{B1} = 0.9999 C_{A0}$$

Therefore,

$$\frac{V}{Q_a} = \frac{C_{A0} - 0.0001 C_{A0}}{[(k)(0.0001 C_{A0}) - (k')(0.9999 C_{A0})]} = \frac{0.9999}{(0.0001k - 0.9999k')} \qquad (6.6.1)$$

To minimize the incinerator volume, dV/dT is set to zero. Since Q_a is assumed to be constant, $d(V/Q_a)/dT$ can also be set to 0.

$$\frac{d(V/Q_a)}{dT} = \frac{-9999[(dk/dT) - 9999(dk'/dT)]}{(k - 9999k')^2} = 0$$

Thus,

$$\frac{dk}{dT} = \frac{10^4 \, dk'}{dT} \qquad\qquad (6.6.2)$$

Substituting in the values of the reaction velocity constants,

$$k = 1.0 \exp\left(\frac{-10,000}{T}\right)$$

$$\frac{dk}{dT} = \left(\frac{10,000}{T^2}\right) \exp\left(\frac{-10,000}{T}\right)$$

$$k' = 9.89 \exp\left(\frac{-35,000}{T}\right)$$

$$\frac{dk'}{dT} = \left(\frac{3.4615 \times 10^5}{T^2}\right) \exp\left(\frac{-35,000}{T}\right)$$

Therefore,

$$(10,000) \exp\left(\frac{-10,000}{T}\right) = (3.4165 \times 10^9) \exp\left(\frac{-35,000}{T}\right)$$

Solving for T,

$$T = 1960°R = 1500°F$$

There are simpler approaches that could have been used to solve this problem:

1. Rather the maximizing the derivative of Eq. (6.6.1), the equation could have been inverted and the derivative of the resulting equation would immediately lead to Eq. (6.6.2).
2. The rate constant, which is essentially the derivative of Eq. (6.6.1), could be maximized to obtain the minimum volume. This again would immediately lead to Eq. (6.6.2).
3. The right side of Eq. (6.6.1) could be plotted against temperature (upon which the k's depend) to determine the minimum of the function.

Example 6.6.13. It is proposed to decompose pure diethyl peroxide (A) at 225°C in a bench scale incinerator. This pollutant will be entering the incinerator at a flow rate of 12.1 L/s. It is desired to decompose 99.995% of the diethyl peroxide. The following data are available:

$$R_A = -k_A C_A \text{ gmol/L-s}; \quad k_A = 38.3 \, \text{s}^{-1} \text{ at } 225°C$$

The inside diameter of the incinerator is 8.0 cm. What should the length of the incinerator be?

SOLUTION. Assume a plug flow reactor and use Eq. (6.5.6) since the reactor is first order and irreversible.

$$\theta = -\frac{1}{k}\ln\left(\frac{C_A}{C_{A0}}\right)$$

For 99.995% destruction,

$$C_A = 5 \times 10^{-5} C_{A0}$$

$$\theta = -\left(\frac{1}{38.3}\right)\ln(5 \times 10^{-5})$$

$$= 0.259\,\text{s}$$

Calculate the incinerator volume.

$$V = (\theta)(Q_a) = (0.259)(12.1)$$

$$= 3.13\,\text{L} = 3130\,\text{cm}^3$$

Calculate the length of the reactor.

$$L = \frac{V}{(\pi D^2/4)} = \frac{3130}{[(\pi)(8^2)/(4)]}$$

$$= 62.3\,\text{cm}$$

PROBLEMS

1. Consider a gaseous fuel with the following volume fraction composition.
   $N_2 = 0.05$
   $CH_4 = 0.81$
   $C_2H_6 = 0.10$
   $C_3H_8 = 0.04$
   Calculate the standard volumetric flow rate of air required for complete combustion of 1.0 scfm of this fuel with 100% theoretical (or stoichiometric) air. Also calculate the standard volumetric flow rates of the combustion products.

2. A waste mixture to be incinerated in a liquid injection system has the following composition on a lb/lb waste basis:
   Carbon (C_w) = 0.732
   Hydrogen (H_w) = 0.109
   Moisture (H_2O_w) = 0.022
   Oxygen (O_w) = 0.102
   Chlorine (Cl_w) = 0.035
   The incineration process will be assisted with 1.0 lb of natural gas/lb of waste.

One hundred percent excess air is to be employed. Calculate the required quantities in the worksheet provided in Table 6.3.1.

3. A waste mixture to be incinerated by a rotary kiln has the following composition.

	Solids (1)	Liquids (2)	
Carbon (C_w)	0.74	0.30	lb/lb waste
Hydrogen (H_w)	0.10	0.12	lb/lb waste
Moisture (H_2O_w)	0.02	0.40	lb/lb waste
Oxygen (O_w)	0.11	0.10	lb/lb waste
Sulfur (S_w)	0.03	0.08	lb/lb waste

The liquid feed rate is 2000 lb/h and the solid feed rate is 1500 lb/h. The incineration process will be assisted with 1.0 lb of natural gas/lb of waste. One hundred and twenty percent excess air is to be employed. Use Table 6.3.2 in performing the calculations.

4. Calculate the adiabatic operating temperature of DDT ($C_{14}H_9Cl_5$) when it is incinerated with 50% excess air.

5. Estimate the incinerator temperature for the DDT in the previous problem using the simplified equation developed in Section 6.4 [Eq. (6.4.20)].

6. Repeat the calculations of Problems 4 and 5 if the combustion is accomplished with 100% excess air.

7. Pure chlorobenzene is to be incinerated in a fluid-bed unit with 150% excess air. Verify that the operating temperature of the incinerator will be in the 1500–2000°F range. Assume the feed (chlorobenzene–air mixture) is at 25°C.

8. It is planned to conduct a trial burn for a rotary kiln incinerator using pure DDT ($C_{14}H_9Cl_5$) with 200% excess air. Assuming neglible heat losses from the incinerator, calculate the operating temperature of the kiln. The feed (DDT–air mixture) is at ambient conditions.

9. Calculate the adiabatic operating temperature of C_6H_5Cl when it is incinerated with 50% excess air.

10. Estimate the adiabatic flame temperature of the C_6H_5Cl in Problem 9 using the simplified equation presented in Section 6.4 [Eq. (6.4.20)].

11. Forty percent by mass of xylene in an aqueous solvent is combusted in an incinerator with stoichiometric air. Estimate the operating temperature of the incinerator using the EPA worksheet provided in Table 6.4.5. Repeat tne calculation for 50 and 100% excess air.

12 Perform the calculations required in Problem 11 using the simplified approach given in Eq. (6.4.20).

13. A waste–fuel mixture is to be burned in an incinerator at an operating temperature of 1900°F. Calculate the minimum net heating value of the mixture in Btu/lb if 0, 20, 40, 60, 80, and 100% excess air is employed. Use Eq. (6.4.22) to perform the calculations.

14. A liquid mixture consisting of 3% waste (NHV = 20,650 Btu/lb) in an aqueous solvent (with zero heating value) is to be incinerated. The combustion process will be supplemented with 0.25-lb natural gas (NHV = 20,000 Btu/lb)/lb waste.

60% excess air at 450°F is to be employed. Calculate the operating temperature of the incinerator if 5% of the total heat released is lost across the walls of the unit. Use the EPAs worksheet as presented in Table 6.4.5.

15. Repeat Problem 14 using the simplified approach given in Eq. (6.4.20).

16. A fluidized-bed incinerator is to be designed to destroy 99.99% of a hazardous waste. The waste reaction is described by a first-order reversible mechanism where

$$A \underset{k_1'}{\overset{k_1}{\longleftrightarrow}} B$$

$$k_1 = Ae^{-E/T}, \quad T = \text{Rankine}$$

$$k_1' = A'e^{-E'/T}, \quad T = \text{Rankine}$$

$$A = 1.0\,\text{s}^{-1}, \quad E = 10,000$$

$$A' = 111\,\text{s}^{-1}, \quad E' = 40,000$$

Calculate a fluidized-bed incinerator operating temperature that will minimize the volume of the incinerator and achieve the desired degree of waste conversion.

17. Design a proposed incinerator using kinetic principles for the destruction of toluene vapor at operating conditions of 1800°F and 1 atm. The gas and methane fuel are fed to the incinerator at 60°F and 1 atm.

    DATA. Flue gas flow rate (from fuel combustion) = 2800 scfm (60°F, 1 atm)
           Process gas flow rate = 8600 scfm
           Reaction rate constant, $k = 8.2\,\text{s}^{-1}$ at 1800°F
           Average velocity = 20 ft/s
           $C_f/C_0 = 0.006/1.0$ (99.4% DRE)

REFERENCES

1. U.S. EPA, *Engineering Handbook for Hazardous Waste Incineration*, prepared by Monsanto Research Corporation, EPA Contract No. 68-03-3025, Sept., 1981.

2. L. Theodore and J. Reynolds, personal notes, 1987.

3. K-C. Lee, N. Morgan, J.L. Hanson, and G. Whipple, "Revised Predictive Model for Thermal Destruction of Dilute Organic Vapors and Some Theoretical Explanation", Paper No. 82-5.3, Air Pollution Control Association Annual Meeting, New Orleans, 1982.

III

EQUIPMENT

Part II provided the reader with some basic scientific and design principles needed to understand the incineration process. The purpose of Part III is to show how these principles are applied in practice by discussing the various pieces of equipment that constitute a hazardous waste incineration facility.

The lead chapter, Chapter 7, describes the major types of incinerators in use today and discusses the types of wastes that each one is best suited to handle; new and emerging incinerator technologies are also described. The remaining four chapters cover equipment that are always or often part of an incinerator system. The hot gases exiting the incinerator contain gaseous and particulate pollutants that cannot be permitted to escape into the atmosphere. Their removal from the gas stream requires the use of air pollution control devices (see Chapter 10). Because of materials limitations (e.g., the fabric materials used in filter baghouses), most types of air pollution control equipment cannot handle gas temperatures typical of incinerator exhaust streams. Before reaching the air pollution control equipment, therefore, the hot gases must be cooled. This is usually accomplished either through the use of a waste heat boiler (Chapter 8) in which a good portion of the high enthalpy of the incinerator exhaust gases is recovered, or by utilizing a quenching technique (Chapter 9). The choice of method to cool the gases is dictated by the economics of the particular installation. In Chapter 11, equipment ancillary to incinerator operation, for example, pumps, fans, compressors, ducts, and stacks, are covered.

7
Incinerators

Contributing author: Raymond B. Bartone

7.1 INTRODUCTION

Incineration is not a new technology and has been commonly used for treating organic hazardous waste for many years in Europe and the United States. The major benefit of incineration is that the process actually destroys most of the waste rather than disposing of or storing it. It can be used for a variety of specific wastes and is reasonably competitive in cost compared to other disposal methods. A new adaptation of this technology—ocean incineration—was introduced in the United States in 1974, although only a few test burns have been conducted in U.S. waters since.

Hazardous waste incineration involves the application of combustion processes under controlled conditions to convert wastes containing hazardous material to inert mineral residues and gases. Four conditions must be present:

1. Adequate free oxygen must always be available in the combustion zone.
2. Turbulence, the constant mixing of waste and oxygen, must exist.
3. Combustion temperatures must be maintained; exothermic reactions of combustion must provide enough heat to raise the burning mixture to a sufficient temperature to destroy all organic components.
4. Elapsed time of exposure to combustion temperatures must be adequately long in duration to ensure that even the slowest combustion reaction has gone to completion. In other words, transport of the burning mixture through the high temperature region must occur over a sufficient period of time.

Thus, four parameters influence the mechanisms of incineration: oxygen, temperature, turbulence, and residence time.

This chapter deals with the types of incinerators used to destroy hazardous waste. Emerging technologies that show promise of meeting the destruction requirements and cost considerations are also covered. This chapter includes a discussion of ocean incineration and the controversy surrounding this issue today.

7.2   ROTARY KILN INCINERATION

Rotary kiln process incinerators were originally designed for lime processing. The rotary kiln is a cylindrical refractory-lined shell that is mounted at a slight incline from the horizontal plane to facilitate mixing the waste materials with circulating air (see Fig. 7.2.1). The kiln accepts all types of solid and liquid waste materials with heating values between 1000 and 15,000 Btu/lb, and even higher. Solid wastes and drummed wastes are usually fed by a pack-and-drum feed system, which may consist of a bucket elevator for loose solids and a conveyer system for drummed wastes. Pumpable sludges and slurries are injected into the kiln through another nozzle. Temperatures for burning vary from 1500 to 3000°F. The kiln may be equipped with a lime or other caustic injection system to neutralize acid gas and combustion end products.

Published literature indicate that rotary kiln incinerators usually have a length-to-diameter ratio (*L/D*) between 2 and 10. Smaller *L/D* ratios often result in less particulate carry over, although carryover is a stronger function of the throughput velocity. Rotational speeds of the kiln are on the order of 0.2 to 1 in./s measured at the kiln periphery. Both the *L/D* ratio and the rotational speed are strongly dependent on the type of waste being combusted. In general, large *L/D* ratios along with slower rotational speeds are used when the waste requires longer residence times in the kiln for complete combustion.

Residence time of the unpumpable wastes is controlled by the rotational speed of the kiln and the angle at which it is positioned. The residence times of liquids and volatilized combustibles are controlled by the gas velocity in the incineration system. Thus, the residence time of the waste material can be controlled to provide complete burning of the combustibles. This is a critical factor in limiting releases of some air pollutants from incineration devices.

A rough estimate of the solids retention time in a particular incinerator can be determined from the following equation:

$$\theta = \frac{0.19\,L}{NDS} \tag{7.2.1}$$

where θ = retention time (min)
   L = kiln length (ft)
   N = kiln rotational velocity (rpm)
   D = kiln diameter (ft)
   S = kiln slope (ft/ft)

The coefficient of 0.19 in Eq. (7.2.1) has been estimated from limited experimental data. Other values appear in the literature. Thus, the residence time calculated is only an estimate of the actual value.[1]

Although liquid wastes are frequently incinerated in rotary kilns, kilns are primarily designed for combustion of solid wastes. They are exceedingly versatile in this regard, capable of handling slurries, sludges, bulk solids of varying size, and containerized wastes. The only wastes that create problems in rotary kilns are aqueous organic sludges that become sticky on drying and form a ring around the kiln's inner periphery, and solids (e.g., drums) that tend to roll down the kiln and are not retained as long as the bulk solids. To reduce this problem, drums and other

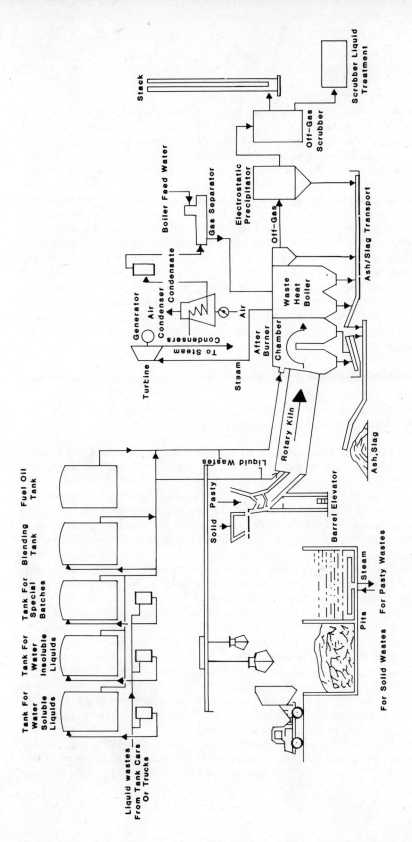

Figure 7.2.1. Rotary kiln incinerator.

Storage   Incineration   Heat Recovery   Off-Gas Cleaning. Neutralization

cylindrical containers are usually not introduced to the kiln when it is empty. Other solids in the kiln help to impede the rolling action.

Rotary kiln systems usually have a secondary combustion chamber after the kiln to ensure complete combustion of the wastes. Air-tight seals close off the high end of the kiln; the lower end is connected to the secondary combustion chamber or mixing chamber. In some cases, liquid waste is injected into the secondary combustion chamber. The kiln acts as the primary chamber to volatilize and oxidize combustibles in the wastes. Inert ash is then removed from the lower end of the kiln. The volatilized combustibles exit the kiln and enter the secondary chamber where additional oxygen is available and ignitable liquid wastes or fuel can be introduced. Complete combustion of the waste and fuel occurs in the secondary chamber. Both the secondary combustion chamber and kiln are usually equipped with an auxiliary fuel firing system to bring the units up to the desired operating temperatures. The auxiliary fuel system may consist of separate burners for auxiliary fuel, dual-liquid burners designed for combined waste–fuel firing, or single-fuel burners equipped with a premix system, whereby fuel flow is turned down and liquid waste flow is increased after the desired operating temperature is attained.

If liquid wastes are to be burned in the kiln and/or afterburner, additional considerations are

- Flame retention characteristics of the burners.
- Burner alignment to avoid flame impingement on refractory walls.
- Burner alignment to avoid interference with the operation of other burners (for multiple burner systems).

Two types of rotary kilns are currently being manufactured in the United States today: cocurrent (burner at the front end with waste feed) and countercurrent (burner at the back end). For a waste that easily sustains combustion, the positioning of the burner is arbitrary from an incineration standpoint since both types will destroy a waste. However, for a waste having low combustibility (such as a high water-volume sludge), the countercurrent design offers the advantage of controlling temperature at both ends, which all but eliminates problems such as overheating the refractory lining. The countercurrent flow technique has been reputed to carry excessive ash over into the air pollution control system due to the associated higher velocities involved. However, this condition also increases the turbulence during combustion, which is generally a desirable factor.[2]

The primary advantages of the rotary kiln for waste incineration include: a wide variety of wastes can be incinerated simultaneously, achievement of a high operating temperature, and the ability to obtain a gentle and continuous mixing of incoming wastes. Another feature of the rotary kiln is that it can operate under substoichiometric (oxygen-deficient) conditions to pyrolyze the wastes. The combustible off-gases with liquid wastes can then be incinerated in the secondary combustion chamber. This mode of operation also reduces the particulate matter carry-over in the kiln gases. The rotating action allows better volatilization of the unpumpable wastes than stationary or fixed-hearth incinerators.

Disadvantages of the rotary kiln include: high capital and operating costs; highly trained personnel required to ensure proper operation; frequent replacement of the refractory lining if very abrasive or corrosive conditions exist in the kiln; and the generation of fine particulates (which become entrained in the exhaust gases) due to the cascading action of the burning waste.

Unlike liquid injection incinerators, which have no moving parts, rotary kiln designs incorporate high temperature seals between the stationary end plates and rotating section. These seals are inherently difficult to maintain airtight, which creates the potential for release of unburned wastes. Rotary kilns burning hazardous wastes are almost always operated at a negative pressure to circumvent this problem; however, difficulties can still arise when batches of waste are fed semicontinuously. When drums containing relatively volatile wastes are fed to the kiln, for example, extremely rapid gas expansion occurs. This results in a positive pressure surge at the feed end of the kiln (even though the discharge end may still be under negative pressure), which forces unburned waste out through the end plate seals. This phenomenon is known as *puffing* and can pose a major problem if extremely toxic or otherwise hazardous materials are being burned. Fugitive emissions can also exit the kiln through the feed chute if improperly designed. Therefore, the design of the solid waste feed system is an extremely important consideration in evaluating rotary kiln incinerators.

In rotary kiln incinerators, temperature is controlled within a specific range by automatically varying the liquid waste or auxiliary fuel firing rate within the design turndown ratio and/or manually or automatically controlling the solid waste feed. Regardless of which technique is employed, provisions should be included for the following:

- *Termination of Liquid Waste Feed on Loss of Ignition in the Burner.* If more than one liquid waste burner is employed, feed need only be terminated in the burner where flameout occurs.

- *Termination of Solid Waste Feed to the Kiln when Low Temperatures Are Sensed at the Kiln Outlet.* If the feed to the kiln is automatic or semiautomatic, then the low temperature cutoff system should also be automatic. If manual feeding is employed, an alarm system is needed to warn the operator. The low temperature cutoff point should be such that solid waste burnout can be maintained, but at lower than the normal operating temperature to avoid shutdown due to routine temperature fluctuations. Engineering judgment must be used to determine acceptable minimum temperature.

The burner temperature of the afterburner can be controlled by varying the liquid or auxiliary fuel feed or by varying the secondary air flow rate. Regardless of which technique is employed, provisions should be included for at least the following:

- *Termination of Liquid Waste or Auxiliary Feed on Loss of Ignition.* This cutoff is necessary to prevent the release of unburned waste contaminants and to prevent potential explosion on release of unburned fuel. However, it also eliminates the function of the afterburner. Therefore, solid waste feed to the kiln should also be terminated on loss of ignition in the afterburner. To minimize the occurrence of flameout in the afterburner, only *clean*, homogeneous liquid wastes (or fuel) should be burned.

- *Termination of Solid Waste Feed to the Kiln if Low Temperatures Are Sensed at the Afterburner Outlet.* The afterburner feed should be maintained, however, to minimize potential release of unburned contaminants. As previously stated, the low temperature cutoff point should be such that combustion is maintained, but at lower than normal operating temperatures to avoid shutdown due to routine fluctuations.

• *Termination of Solid Waste Feed to the Kiln if High Temperatures Are Sensed at the Afterburner Outlet.* This is necessary to prevent damage to the refractory lining and the downstream air pollution control devices. The high-temperature cutoff point should be well above the normal operating temperatures, but low enough to avoid damage to the system. In the event of this cutoff, some liquid waste or fuel feed to the afterburner should be maintained to complete combustion of off-gases from solid wastes remaining in the kiln.

In addition to these criteria, all liquid waste burners in the kiln and afterburners should be equipped with manufacturer specified primary air control systems so that air–fuel stoichiometry is maintained on turndown.

Numerous hazardous wastes that previously were disposed of by potentially harmful methods (ocean dumping, landfilling, and deep-well injection) are currently being safely and economically destroyed using rotary kiln incinerators combined with proper flue gas handling. Included in this list of primarily toxic wastes are polyvinyl chloride (PVC) wastes, polylchlorinated biphenyl (PCB) wastes from capacitors, obsolete munitions, and obsolete chemical warfare agents such as GB, VX, and mustard gas. Beyond these specific wastes, the rotary kiln incinerator is generally applicable to the destruction and ultimate disposal of any form of hazardous waste material that is combustible at all. Unlikely candidates are noncombustibles such as heavy metals, high moisture content wastes, inert materials, inorganic salts, and the general group of materials having a high inorganic content.[3]

7.3   LIQUID INJECTION INCINERATION

Liquid injection incinerators are currently the most commonly used type of incinerator for hazardous waste disposal. A wide variety of units are marketed today, with the two major types being horizontally and vertically fired units; a loss common unit is the tangentially fired vortex combustor. A vertically fired unit is schematically represented in Fig. 7.3.1. As the name implies, the liquid injection incinerator is confined to hazardous liquids, slurries, and sludges with a viscosity value of 10,000 SSU (Saybolt seconds) or less. The reason for this limitation is that a liquid waste must be converted to a gas prior to combustion. This change is brought about in the combustion chamber, and is generally expedited by increasing the waste surface area through atomization. An ideal droplet size is in the 40–100-μm range, and is attainable mechanically using rotary cup or pressure atomization, or via gas-fluid nozzles and high pressure air or steam.

The key to efficient destruction of liquid hazardous wastes lies in minimizing unevaporated droplets and unreacted vapors. Just as for the rotary kiln, temperature, residence time, and turbulence may be optimized to increase destruction efficiencies. Typical combustion chamber residence time and temperature ranges are 0.5 to 2 s and 1300 to 3000°F, respectively. Liquid injection incinerators are variable dimensionally and have feed rates up to 1500 gal/h of organic wastes and 4000 gal/h of aqueous waste.

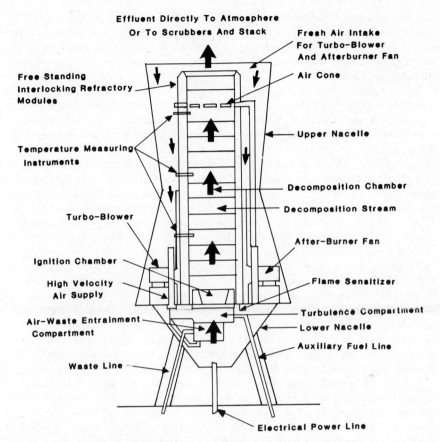

Figure 7.3.1. Vertically fired liquid injection incinerator.

Liquid waste fuel is transferred from drums into a feed tank. The tank is pressurized with nitrogen, and waste is fed to the incinerator using a remote control valve and a compatible flowmeter. The fuel line is purged with nitrogen after use. A recirculation system is used to mix the tank contents. Normally a liquid fuel or a gas (e.g., propane) preheats the incinerator system to an equilibrium temperature of approximately 1500°F before introduction of the waste liquid.

Of the three types of units discussed, the horizontal and vertical are basically similar in operating conditions. The tangentially fired unit has a much higher heat release and generally superior mixing than the other two units, making it more attractive for disposal of high water content wastes and less combustible materials. However, these conditions lend to increased deterioration of the refractory lining from thermal effects and erosion.[2]

From a combustion standpoint, liquid wastes may be classified into two types: (a) combustible liquids and (b) partially combustible liquids. Noncombustible liquids cannot be treated or disposed of by incineration. The first category contains all materials having sufficient calorific value to support combustion in a conventional combustor or burner. The second category includes materials that do not support combustion without the addition of auxiliary fuel and have a high percentage of noncombustible constituents such as water. A partially combustible waste may also include material dissolved in the liquid phase which, if inorganic in nature, will form an inorganic oxide upon combustion and require secondary collection prior to atmospheric release.

Assuming that either of these types of wastes is primarily organic in nature, even though the quantity of the organic material may be small, incineration of such materials becomes essentially a straightforward combustion problem in which air must be mixed with the waste at some temperature above its ignition temperature. When starting with a waste in liquid form, it is necessary to supply sufficient heat of vaporization in addition to raising it to its ignition temperature.

Since liquids vaporize and react more readily when finely divided in the form of a spray, atomizing nozzles are usually employed to inject waste liquids into incineration equipment whenever the viscosity of the waste permits atomization. There are many wastes that might be classified as *liquid*, which are hardly liquid in nature. Slurries, sludges, and other materials of high viscosity can be handled only in special types of incineration systems.

There are several rules that should be used in determining whether a liquid waste may be considered combustible: (a) the waste should be pumpable at ambient temperature or capable of being pumped after heating to some reasonable temperature level; and (b) the liquid must be capable of being atomized under these conditions. If it cannot be both pumped and atomized, the waste cannot be burned as a liquid but must be handled as sludge or solid. Liquid waste incineration generally involves liquids having viscosities up to approximately 10,000 SSU, although lower viscosities are desirable.

In order to be considered a combustible liquid waste, the liquid must sustain or support combustion in air without the assistance of auxiliary fuel. This means that the waste will generally have a calorific value of 8000 Btu/lb or higher. Below this value, the material would not be able to maintain a stable flame in a standard commercial combustor or burner. (However, high intensity combustors are in use that will sustain combustion down to 4500 Btu/lb.) Materials that fall into the former category (> 8000 Btu/lb) are light solvents, such as toluene, benzene, acetone, and ethyl alcohol; and heavy organic tars and still bottoms similar to residual fuel oil. The wastes may be combinations of both, which would produce a mixture having an intermediate viscosity and heating value. Such wastes come from cleaning operations in chemical plants and refineries or are the residues from distillation processes and are usually not recovered for economic reasons.

Liquid wastes having values below 8000 Btu/lb can be considered in the partially combustible category. It must again be emphasized that this is a rule of thumb and that some materials as high as 10,000 or 11,000 Btu/lb will not sustain combustion by themselves. It is also important with this type of waste that the material handling method be compatible with the equipment selected. Viscosities should be reduced to the point where the material is pumpable and atomizable at either ambient or slightly elevated temperatures. Waste material in this classification is often aqueous in nature, consisting of organic compounds miscible with water. Such waste may also contain sulfur compounds, phosphorus compounds, or combinations of organic and noncombustible inorganics. These materials may have enough organic content to exhibit visible combustion in a high temperature furnace, or they may be low in combustible material so that no visible combustion is apparent.

Normally, liquid injection incinerators consist of two stages. The primary chamber is usually a burner where combustible liquids and gaseous wastes are introduced. Noncombustible liquid and gaseous wastes usually bypass the burner and are introduced downstream of the burner into the secondary chamber. The heart of the liquid injection system is the waste atomizer or nozzle (burner), which atomizes the waste and mixes it with air into a suspension.

Before a liquid waste can be combusted, it must be converted to the gaseous state. This change from a liquid to a gas occurs inside the combustion chamber and requires heat transfer from the hot combustion gases to the injected liquid. To effect a rapid vaporization (i.e., increase heat transfer), it is necessary to increase the exposed liquid surface area. Most commonly, the amount of surface exposed to heat is increased by finely atomizing the liquid to small droplets, usually to a 40-μm diameter or smaller. Good atomization is particularly important when high aqueous wastes or other low heating value wastes are being burned. It is usually achieved in the liquid burner at the point of air–fuel mixing.

Viscosity can be reduced by heating with tank coils or in-line heaters, as mentioned previously. However, 400–500°F is normally the limit to reduce viscosity, since pumping a hot tar or similar material becomes difficult above these temperatures. Should gases be evolved in any quantity before the desired viscosity is reached, they may cause unstable fuel feed and burning. If this occurs, the gases should be trapped and then vented safely, either to the incinerator or elsewhere. Prior to heating a liquid waste stream, a check should also be made to insure that undesirable preliminary reactions such as polymerization, nitration, and oxidation will not occur. If preheating is not feasible based on these considerations, a lower viscosity and miscible liquid may be added to reduce the viscosity of the mixture, fuel oil, for example.

Solid impurities in the waste can interfere with burner operation via pluggage, erosion, and ash buildup. Both the concentration and the size of the solids relative to the diameter of the nozzle need to be considered. Filtration may be employed to remove solids from the waste prior to injection through the burner.

Liquid waste atomization can be achieved by any of the following means:

Rotary cup atomization.

Single-fluid pressure atomization.

Two-fluid, low pressure air atomization.

Two-fluid, high pressure air atomization.

Two-fluid, high pressure steam atomization.

In air or steam atomizing burners, atomization can be accomplished internally, by impinging the gas and liquid streams inside the nozzle before spraying (see Fig. 7.3.2); externally, by impinging jets of gas and liquid outside the nozzle; or by sonic means. Sonic atomizers use compressed gas to produce high frequency sound waves that are directed onto the liquid stream. The liquid nozzle diameter is relatively large, and little waste pressurization is required. Some slurries and liquids with relatively large particulates can be handled in this way without plugging problems.

The *rotary cup* consists of an open cup mounted on a hollow shaft. The cup is spun rapidly and liquid is admitted through the hollow shaft. A thin film of the liquid to be atomized is torn from the lip of the cup and surface tension reforms it into little

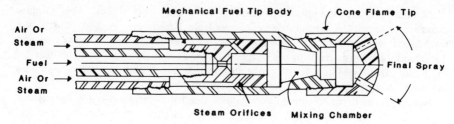

Figure 7.3.2. Internal mix nozzle.

droplets. To achieve conical shaped flames an annular high-velocity jet of air (primary air) must be directed axially around the cup. If too little primary air is admitted, the fuel will impinge on the sides of the incinerator. If too much primary air is admitted, the flame will not be stable and will be blown off the cup. For fixed-firing rates, the proper adjustment can be found and the unit operated for long periods of time without cleaning. This requires little pressurization of the liquid and is ideal for atomizing liquids with relatively high solids content. Burner turndown is about 5:1 and capacities from 1 to 250 gal/h, are available.

In *single-fluid pressure* atomizing nozzle burners, the liquid is given a swirl as it passes through an orifice with internal tangential guide slots. Moderate liquid pressures of 100 to 150 psi provide good atomization with low-to-moderate liquid viscosity. In the simplest form, the waste is fed directly to the nozzle but turndown is limited from 2.5 to 3:1 since the degree of atomization drops rapidly with decrease in pressure. In a modified form involving a return of liquid, turndown up to 10:1 can be achieved. When this type of atomization is used, secondary combustion air is generally introduced around the conical spray of droplets. The flame tends to be short, bushy, and of low velocity. Combustion tends to be slower as only secondary air is supplied and a larger combustion chamber is required. Typical burner capacities are in the range of 10 to 100 gal/h. Disadvantages of single-fluid pressure atomization are erosion of the burner orifice and a tendency toward pluggage with solid or liquid pyrolysis products, particularly in smaller sizes.

Two-fluid atomizing nozzles may be of the low pressure or high pressure variety, the latter being more common with high viscosity materials. In low pressure atomizers, air from blowers at pressures from 0.5 to 5.0 psig is used to aid atomization of the liquid. A viscous tar, heated to a viscosity of 75 to 90 SSU, requires air at a pressure of somewhat more than 1.5 psig, while a low viscosity or aqueous waste can be atomized with 0.5-psig air. The waste liquid is supplied at a pressure of 4.5 to 17.5 psig. Burner turndown ranges from 3:1 to 6:1. Atomization air required varies from 350 to 1000 ft³/gal of waste liquid. Less air is required as atomizing pressure is increased. The flame is relatively short as up to 40% of the stoichiometric air may be mixed with the liquid in atomization.

High pressure two-fluid burners require compressed air or steam at pressures from 30 to 150 psig. Air consumption is from 80 to 200 ft³/gal waste, and steam requirements may be 2 to 4 lb/gal waste with careful control of the operation. Turndown is relatively poor (3:1 or 4:1) and considerable energy is employed for atomization. The major advantage of such burners is the ability to burn barely pumpable liquids without further viscosity reduction. Steam atomization also tends to reduce soot formation with wastes that would normally burn with a smoky flame.

TABLE 7.3.1. Kinematic Viscosity and Solids Handling Limitations of Various Atomization Techniques[a]

Atomization Type	Maximum Kinematic Viscosity (SSU)	Maximum Solids	
		Mesh Size	Concentration (%)
Rotary cup	130–175	35–100	20
Single-fluid pressure	150		0
Internal low pressure air[b]	100		0
External low pressure air[b]	200–1500	200[d]	30[d]
External high pressure air[c]	150–5000	100–200	70
External high pressure steam[c]	150–5000	100–200	70

[a] From T. Bonner et al., *Hazardous Waste Incineration Engineering*, Noyes Data Corp., Park Ridge, NJ, 1981.
[b] Lower than 30 psi.
[c] Greater than 30 psi.
[d] Depends on nozzle i.d.

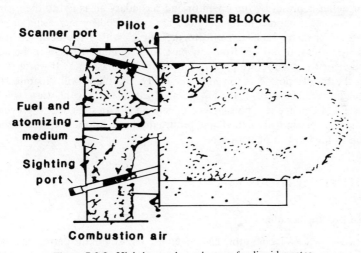

Figure 7.3.3. High heat release burner for liquid wastes.

Table 7.3.1 identifies typical kinematic viscosity and solids-handling limitations for the various atomization techniques. These data are based on a survey of 14 burner manufacturers.[3] In evaluating a specific incinerator design, however, the viscosity and solids content of the wastes should be compared with the manufacturer specifications for the particular burner employed.

Each burner, regardless of type, is usually mounted in a refractory block or ignition tile (see Fig. 7.3.3). This is necessary to confine the primary combustion air introduced through the burner, to ensure proper air–waste mixing, and to maintain ignition. Some burners and tiles are arranged to aspirate hot combustion gases back into the tile, which aids in vaporizing the liquid and increasing flame temperature more rapidly.

The dimensions of the burner block, or ignition tile, vary depending on the burner design. Manufacturers each have their own geometric specifications, which have been developed through past experience. Therefore, it is not possible to specify a single burner block geometry for design evaluation purposes. However, this aspect can be checked to eliminate systems that do not provide for any flame retention.

The location of each burner in the incinerator and its firing angle relative to the combustion chamber is another important consideration. In axial or side-fired non-swirling units, the burner is mounted either on the end, firing down the length of the chamber, or in a sidewall, firing along a radius. Such designs, while simple and easy to construct, are relatively inefficient in their use of combustion volume. Improved utilization of combustion space and higher release rates can be achieved with the utilization of swirl or vortex burners or designs involving tangential entry. Regardless of the burner location and/or gas flow pattern, the burner is placed so that the flame does not impinge on refractory walls. Impingement results in flame quenching and can lead to smoke formation or otherwise incomplete combustion. In multiple burner systems, each burner should be aligned so that its flame does not impact on other burners.

In most liquid injection incinerator designs, the desired temperature at the chamber outlet is preset by the operator and secondary air is fed to the system at a constant rate. Fluctuations in temperature are controlled by increasing or reducing the waste or auxiliary fuel feed rate to the burner within the designed turndown ratio. This turndown ratio is fixed, in part, by the limited range of liquid waste injection velocities required to prevent flame liftoff or flashback. If waste is injected through the burner nozzle at too high a velocity, the flame will separate from the burner and be extinguished. If the injection velocity is too low the waste will burn in the nozzle and damage it. The range of injection velocities needed to prevent these occurrences is determined by the flame propagation rate for the wastes and the flame retention characteristics of the burner.

7.4   OTHER METHODS

Fluidized-Bed Incineration

Fluidized-bed process incinerators have been used mostly in the petroleum and paper industries, and for processing nuclear wastes, spent cook liquor, wood chips, and sewage sludge disposal. Hazardous wastes in any physical state can be applied to a fluidized-bed process incinerator. Auxiliary equipment includes a fuel burner system, an air supply system, and feed systems for liquid and solid wastes. There are two basic bed-design modes, bubbling bed and circulating bed, distinguished by the extent to which solids are entrained from the bed into the gas stream.

The bubbling fluidized bed design, as shown in Fig. 7.4.1, takes its name from the behavior of a granular bed of nonreactive sand, stirred by passing a gaseous oxidizer (air, oxygen, or nitrous oxide) through the bed at a rate sufficiently high to cause the bed to expand and act as a fluid. Preheating of the bed to start-up temperature is accomplished by a burner located above and impinging down on the bed. The waste gas or liquid is passed directly into the sand. With the onset of ignition, a continuous

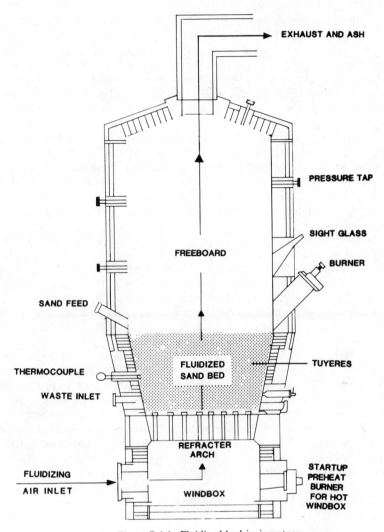

Figure 7.4.1. Fluidized-bed incinerator.

exposure of the uniformly distributed waste to the heated sand results in efficient oxidation.

Bubbling fluidized beds are designed to operate with temperatures between 840 to 1800°F in the bed. To sustain combustion temperatures of 1550°F, the waste must contain a minimum of 4500 Btu/lb of waste. In newer designs, the combustion air is preheated by an air heat exchanger to temperatures between 800 and 1200°F. This reduces the required heating value of the waste to sustain combustion, thus making the incineration more energy efficient.

Continued bed agitation by fluidizing air allows larger waste particles to remain suspended until combustion is completed. Bed depths usually range from about 1.5

to 10 ft. Variation in bed depth can affect the residence time of combustibles. It is desirable to maintain the bed depth at a minimum value consistent with complete combustion and minimum excess air. In general, a shallow fluidized-bed depth is preferred in a continuous process because this provides the lowest pressure drop and power consumption as well as maximum heat and mass transfer.

In regard to bed material, if a bed consists of small particles, the rate of growth of the air bubbles in the bed is small because of the high resistance to gas flow in the bubble. Since bed porosity is low, bubbles rise rapidly through the bed. The slow rate of the bubble growth and their short residence time in the bed decrease the tendency for slug formation. In large-particle beds, bubbles tend to be larger; gas enters at the bottom of the bubble and, as the bubble rises, is then swept around and dragged down the sides. The region around the bubble is called the *cloud*; outside of the cloud, bed particles do not tend to make good contact with the gas. As a result of this phenomenon, poor combustion is often observed in large-particle bubbling beds.[4]

The circulating bed design has not been, to date, extensively used in HWI applications. However, its advantages offer some potential for future use in this area. In this type of fluidized bed, a much higher flooding velocity (12 to 40 ft/s) is employed, with the resulting entrainment of a large portion of the solids in the incinerator. The high gas velocities, solids loadings, and back mixing produce a high degree of turbulence throughout the incinerator, which quickly and uniformly mixes the fuel-waste mixture and bed material. The associated high heat and mass transfer rates result in high combustion efficiency.

In a typical circulating bed design, fuel and limestone are fed into the lower section of the combustion chamber, and primary air is introduced through a distributor plate in the bottom. There is no fixed bed depth or bed level, as in a bubbling bed. Rather, the density of the bed varies over the entire incinerator height, with the greatest density near the bottom where the fuel and limestone are introduced. Fuel combustion occurs as it rises in the incinerator. Secondary air is introduced at higher levels in the incinerator to maintain particle entrainment throughout the height of the incinerator, to provide staged combustion for NO_x reduction, to complete combustion of fines, and to control combustion temperatures. The hot combustion gases with the entrained solids leave the top of the combustion chamber and enter a hot, refractory-lined cyclone. The cyclone collects the solids containing unburned waste and unreacted sorbent and returns them to the incinerator or, in some designs, to an external heat exchanger.

Advantages of fluidized-bed incinerators are

- The combustion design is fairly simple and its maintenance cost is low.
- It has a high combustion efficiency.
- Designs are more compact.
- Comparatively low gas temperatures and excess air requirements minimize the formation of nitric oxide.
- In some cases, the bed itself neutralizes some of the hazardous products of combustion.
- The bed mass provides a large surface area for reaction.
- Temperatures throughout the bed are relatively uniform.

- It can process aqueous waste slurries and can tolerate fluctuation in waste feed rates.
- If the waste contains sufficient calorific values, the use of auxiliary fuels is unnecessary; moreover, the excess heat may be recycled in some cases.
- The bed can function as a heat sink; start-up after weekends may require little or no preheat time.

Disadvantages of fluidized-bed incinerators are

- Bed diameters and height are limited by design technology.
- Ash removal presents a potential problem.
- Systems requiring low temperatures may have carbon build-up in the bed due to increased residence time.
- Operating costs are high.
- Waste types are limited.
- Certain organic wastes will cause the bed to agglomerate.
- Particulate emissions can be a major problem.
- It is still a relatively unproven technology.[1]

The relatively uniform temperature within a fluidized bed is due to several factors. Turbulent agitation with the fluidized mass breaks and disperses any hot or cold spots throughout the bed before they can grow to significant size; there is also a rapid movement of solids from one part of the bed to another. This does not mean that every solid particle in a fluidized bed is at the same temperature. However, departure of an individual particle from the mean temperature of a fluidized bed will be much less than in a fixed bed.

A fluidized-bed incinerator is usually high in initial cost; yet when waste quantities are high and the material cannot be handled in more conventional systems, fluidized-bed incineration is practical. The cost of the incineration system widely varies and depends on the type of waste, its caloric value, and the size of the combustor. A cost of 2 to 3 times more than a conventional incinerator equipped with a waste atomizing system would not be an unusual estimate. Potential problems with particulates requiring the addition of scrubbing systems would also increase costs; costs for analytical instrumentation to monitor residue levels are also necessary.

The type and composition of the waste is a significant design parameter that will impact on cost not only during combustion but also during storage, processing, and transport prior to incineration. If the waste is a heterogeneous mixture, operations will be more complex, and the combustor will require auxiliary fuel. Homogeneous wastes that are injected and uniformly dispersed in the bed simplify the system design and cost less to incinerate. Installation and operating costs will also vary significantly depending on the type of waste processed. Maintenance costs, with no moving mechanical parts in the reactor, would only be a small percentage of the initial capital costs. Environmental control costs would be related to the type of equipment necessary to control particulate emissions and off-gases, for example, cyclones and afterburners. Additional costs would be incurred for monitoring residues with analytical instrumentation. In some cases, depending on the character of the hazardous waste and the type of equipment available, conventional incinerators may be converted into fluidized-bed combustors. This could represent a smaller capital investment compared to the cost of erecting a fluidized-bed incinerator originally.[5]

Multiple Hearth Incineration

Single hearth incinerators, known as controlled-air or pyrolysis incinerators, are not ordinarily used for hazardous waste incineration and will not be discussed here.

The multiple hearth incinerator (commonly called the Herreshoff furnace) is a flexible unit that has been utilized to dispose of sewage sludges, tars, solids, gases, and liquid combustible wastes. This type of unit was initially designed to incinerate sewage plant sludges in 1934. In 1968, there were over 125 installations in operation with a total capacity of 17,000 tons/day (wet basis) for this application alone.[5]

A typical multiple hearth furnace includes a refractory-lined steel shell, a central shaft that rotates, a series of solid flat hearths, a series of rabble arms with teeth for each hearth, an air blower, fuel burners mounted on the walls, an ash removal system, and a waste feeding system. Side ports for tar injection, liquid waste burners, and an afterburner may also be included. An illustration of a multiple hearth incinerator and a typical flow scheme is shown in Fig. 7.4.2. Sludge and/or granulated solid combustible waste is fed through the furnace roof by a screw feeder or belt and flapgate. The rotating air-cooled central shaft with air-cooled rabble arms and teeth distributes the waste material across the top hearth to drop holes. The waste falls to the next hearth and then the next until discharged as ash at the bottom. The waste is agitated as it moves across the hearths to make sure fresh surface is exposed to hot gases.

Units range from 6 to 25 ft in diameter and from 12 to 75 ft in height. The diameter and number of hearths are dependent on the waste feed, the required processing time, and the type of thermal processing employed. Generally, the uppermost hearth is used as an afterburner. Normal incineration usually requires a minimum of six hearths, while pyrolysis applications require a greater number.

The rabble arms and teeth located on the central shaft all rotate in the same direction; additional agitation of the waste (back rabbling) is accomplished by reversing the angles of the rabble teeth. Waste retention time is controlled by the design of the rabble tooth pattern and the rotational speed of the central shaft. Liquid and/or gaseous combustible wastes may be injected into the unit through auxiliary burner nozzles; this utilization of liquid and gaseous waste represents an economic advantage because it reduces secondary fuel requirements, thus lowering operating costs.

There are three temperature zones in a multiple hearth furnace. The top hearths operate at temperatures between 570 and 1020°F to dry the waste material. Incineration takes place in the middle hearths at temperatures between 1380 and 1830°F and the incinerated wastes are cooled by incoming combustion air in the bottom hearths at temperatures between 400 and 570°F. Exit gases have good potential for heat recovery, being around 570 to 1100°F.

Advantages of multiple hearth incinerators are

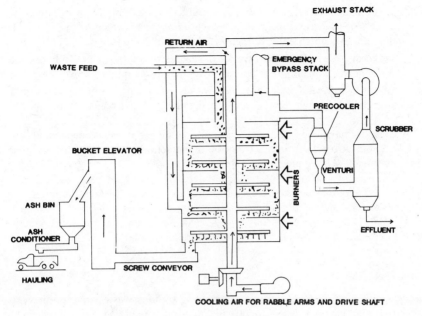

Figure 7.4.2. Multiple hearth incinerator.

- The retention or residence time in multiple hearth incinerators is usually higher for hazardous materials having low volatility than in other incinerator configurations.
- Large quantities of water can be evaporated.
- A wide variety of wastes with different chemical and physical properties can be handled.
- Multiple hearth incinerators are able to utilize many fuels including natural gas, reformer gas, propane, butane, oil, coal dust, waste oils, and solvents.
- Because of its multizone configuration, fuel efficiency is high and typically improves with the number of hearths used.
- Fuel burners can be added to any of the hearths to maintain a desired temperature profile.
- Multiple hearth incinerators are capable of a turndown ratio of 35%.
- High fuel efficiency is allowed by the multizone configuration.

Disadvantages of multiple hearth incinerators are

- Due to the longer residence times of the waste materials, temperature response throughout the incinerator when the burners are adjusted is usually very slow.
- It is difficult to control the firing of supplemental fuels as a result of this slow response.
- Maintenance costs are high because of the moving parts (rabble arms, main shaft, etc.) which are subjected to combustion conditions.

- Multiple hearth incinerators are susceptible to thermal shock resulting from frequent feed interruptions and excessive amounts of water in the feed. These conditions can lead to early refractory and hearth failures.
- If used to dispose of hazardous wastes, a secondary combustion chamber probably will be required and different operating temperatures might be necessary.
- These devices are not well suited for wastes containing fusible ash, wastes that require extremely high temperature for destruction, or irregular bulky solids.[2]

Wastes that have been incinerated in multiple hearth installations include sewage primarily, but also solid residues from the manufacture of aromatic amines, reactor bottoms from PVC manufacture, chemical sludge, oil refinery sludge, still bottoms, and pharmaceutical wastes. Potential candidates for multiple hearth incineration are halogenated organic solids or sludges and organic solids or sludges containing either sodium, silicon, sulfur, phosphorus, nitrogen, chlorine, or fluorine. Examples of wastes not compatible with this process include heavy metals, inert materials, inorganic salts, and materials with a high inorganic content.[3]

Coincineration

Coincineration refers to the combination of hazardous waste combustion with an industrial combustion process to provide additional energy or as a supplemental source of fuel. This is not a unique technology; any existing incineration process can be used for this special case of mixing waste streams to obtain better combustion of particularly intractable waste material. Coincineration most often occurs in processes such as cement kilns, utility and industrial boilers, and refuse or sludge incinerators. The advantages of coincineration include utilization of the waste's heat content to save fuel with usually little or no increase in emissions. The major disadvantage is the limited number of acceptable waste types based on the particular combustion process and the product manufactured.

Cement Kiln Incineration

Cement kiln incineration appears at present to be a satisfactory technology for treating hazardous wastes, but all operational parameters have not been as yet adequately tested. In the near future, this process may be used commercially if more accurate and reliable data from research and demonstration studies can be provided to overcome public opposition.[1]

Refer to Fig. 7.4.3 for a cement kiln process schematic. It has been demonstrated that existing cement kilns, when properly operated, can destroy most toxic organic chemical wastes, including PCBs, which can be converted to less noxious forms during normal cement kiln operation. Reports have also documented the successful burning of waste lubricating oils with toxic metal emissions (e.g., lead) being entrained in the cement in an inert form. New EPA-sponsored tests in Puerto Rico have revealed that even highly chlorinated, very toxic wastes can be completely destroyed in cement kilns without any measurable toxic air emissions.

Since many chemical wastes have significant heating values of 10,000 Btu/lb or more, cement plants can save large quantities of energy from burning such liquid chemical wastes as alternate or synthetic fuels.

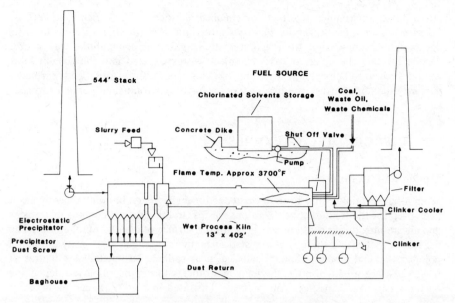

Figure 7.4.3. Cement kiln.

From 1974 to 1976, the St. Lawrence Cement Company burned waste chemicals in two separate cement kilns at their Mississauga, Ontario plant. Waste chlorinated hydrocarbons, consisting of approximately 45% PCBs, 12% aliphatics, and 33% chlorinated aromatics, were burned in a wet process cement kiln. Stack tests performed during trial burns indicated a destruction efficiency of at least 99.986% for the chlorinated hydrocarbons. About 50 ppb of volatile low molecular weight compounds (e.g., chloroform) were found in the emission samples. It appears that these low molecular weight compounds found in the kiln emissions are not materials that escaped destruction, but probably originated in the feedwater used to blend the clinker constituents. There were no detectable quantities of high molecular weight chlorinated compounds in the stack gases as determined by gas chromatograph (GC)/mass spectroscopy (MS) tests, nor were PCBs found in the cement clinkers.

An additional benefit derived from the process is that hydrochloric acid, usually generated by the pyrolysis and oxidation of chlorinated hydrocarbons, is effectively neutralized by process lime in the cement kiln. Some cement kilns have a need for a low alkali cement; in such cases, the burning of chlorinated hydrocarbons results in the lowering of the alkalinity of the cement products.

During test burns of the chlorinated waste, the average fuel replacement value was about 12%, while the heating value of these wastes averaged about 10,000 Btu/lb of waste. Fuel requirements for the kiln were reduced by about 65% of the actual energy content of the wastes burned. It was estimated from these tests that chlorine can be added to a typical wet process kiln at rates of about 0.4 to 0.7% without disrupting kiln operations.

The burning of halogenated PCB wastes at the St. Lawrence Cement Company, Mississauga was a technical success. During the 2-yr test period, PBC-containing wastes of up to 10,000 ppm of PCB were burned successfully and completely without any adverse environmental effects or chemical accidents. *The general public was, however, never informed about these practices until after the tests.* When the tests became public knowledge, the reaction of the neighboring communities was severe, despite the fact that in most cases chemical wastes injected into the cement kilns were destroyed to below analytical limits of detection. Public emotion and fear had a part in suspending further testing pending extensive environmental public hearings.[6]

7.5   EMERGING TECHNOLOGIES

Molten Salt Combustion

Molten salt technology has been in existence for many years, but only recently has it been used for the treatment of hazardous wastes. In this process, combustible wastes and air are introduced continuously beneath the surface of a bath of molten salt. The hazardous material is thereby combusted at temperatures below its normal ignition point. Individual alkali carbonate salts, such as sodium carbonate, or mixtures of these salts, are usually used as the melt, but other salts can be employed depending on the waste characteristics. This process, illustrated in Fig. 7.5.1, can be batch or continuous feed, but the capacity is small. The temperature range of operation is 1500 to 1850°F, with residence times of the order of 5 s for the gas phase and hours for the solid phase. The containers used for the molten salts are made of ceramics, alumina, stainless steel, or iron.

Ideally, during the molten salt process, organic substances are totally oxidized to CO_2 and H_2O. Acid gases resulting from the molten salt process must rise through the melt, which neutralizes the HCl, SO_2, and/or other acid gases formed. Elements such as phosphorus, sulfur, chlorine, arsenic, and silicon are reacted with the carbonate melt to form NaCl, Na_3PO_4, Na_2SO_4, Na_2SO_3, and Na_2SiO_3. Iron from metal containers forms iron oxide. Most organic substances are destroyed, leaving behind a relatively innocuous residue, while harmless levels of off-gases are emitted. Halogenated organics are converted to halogen acid gases that are controlled effectively in this system. Generally, the salt bath is stable, nonvolatile, nontoxic, and may be recycled for further use until the bath is no longer usable.

Some hydrocarbons combusted by the molten salt process are chlorinated hydrocarbons, PCBs, explosives and propellants, chemical warfare agents, rubber wastes, textile wastes, tannery wastes, various amines, contaminated ion exchange resins, tributyl phosphate, and nitroethane. As indicated earlier, halogenated organics are converted to halogen acid gases, which are controlled effectively in this system. In tests with DDT, malathion, and chlordane, 99.6 to 99.99% conversion has been achieved. Another important aspect is that in most cases the heating value of the wastes is sufficient to heat reactants to the required temperatures that maintain the salt bed in the molten state and balance heat losses.

The advantages of molten salt combustion are

- Combustion is nearly complete.
- Operating temperatures are lower than in normal incineration; thus, they are fuel efficient.
- The system is amenable to the recycling of generated heat.

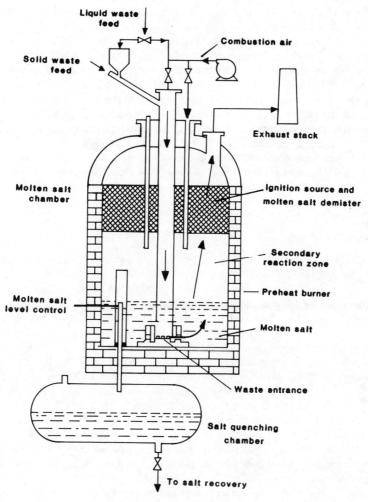

Figure 7.5.1. Molten salt destructor.

- A wide variety of wastes can be incinerated.
- Many wastes can be incinerated in compliance with the current regulations.

The problems with molten salt combustion are

- Particulate emissions for some wastes are high.
- The technology is not readily adaptable to aqueous wastes.
- Eventually waste salt and ash must be disposed of or the fluidity of the melt will be destroyed.
- A hazardous waste with > 20% ash cannot be incinerated.
- Detailed economic information for a demonstration-sized system is not currently available for many wastes.

Rockwell International in Canoga Park, California has designed and constructed a molten salt pilot plant for testing and demonstration purposes. This pilot plant combustor has a feed rate of 50 to 200 lb/h of waste. The vessel is made of Type 304 stainless steel and lined with 6-in. thick refractory blocks. The 12-ft high, 3-ft i.d. vessel contains 2000 lb of salt. Melt depth is 3 ft when air does not circulate through the bed. A natural gas fired burner is used to preheat the vessel on startup and maintain heat during standby. The heat content of the waste is usually sufficient to maintain the melt in a molten condition during combustion periods.

Because molten salt technology as applied to the destruction of hazardous wastes is in its infancy, an economic cost profile is difficult to obtain. No firm cost estimates can be made until demonstration-sized plants such as Rockwell International's have functioned for an extended period of time. It is clear, however, that the process is very efficient for combustion of certain wastes. The future will determine the economic viability of molten salt processing of hazardous waste.[4,7]

Electric Reactors

Electric reactors are designed to pyrolyze waste contaminants on particles (such as soil) through the use of an electrically heated fluid wall reactor (see Fig. 7.5.2). These units have been used successfully in other chemical processes and are just beginning to be adopted for waste destruction. The units, especially the portable versions, appear to offer a very different and potentially valuable thermal option for hazardous waste treatment.

One type of electric reactor is the high temperature fluid wall (HTFW) reactor by Thagard Research Corporation. The reactor consists of a tubular core of porous refractory material capable of emitting sufficient radiant energy to activate the reactants fed into the tubular space. The reactor has been built with cylindrical core diameters of 3, 6, and 12 in. with heated core lengths of up to 6 ft. The core material is designed to be of uniform porosity to allow the permeating of a radiation-transparent gas through the core wall into the interior. The core is completely jacketed and insulated in a fluid-tight pressure vessel. Electrodes located in the annular space between the jacket and core provide the energy required to heat the core to radiant temperatures. To achieve both high temperatures and high heat transfer rates simultaneously, (a) the reacting stream is kept out of physical contact with the reactor wall by means of a gaseous blanket formed by an inert gas flowing radially inward through the porous reactor tube (or core), and (b) the porous carbon core is heated to incandescense so that the predominant mode of heat transfer is by radiative coupling from core to stream. The reactor is heated electrically with six carbon resistance heaters to temperatures around 4000°F. Because of the extreme temperatures encountered in the operation of the device, the insulation package consists not of refractory brick but of a radiation shield made of multiple layers of graphite paper backed up with carbon felt. The short residence time associated with the reactor necessitates the pulverization of solid feeds, but also lends itself to a compact system where a reasonably high throughput for a small installation is possible and portability becomes an attainable design feature.

The advantages of the system as stated by the developer include:

• Reduced residence times (hence lower reactor costs) by virtue of rapid heat transfer.

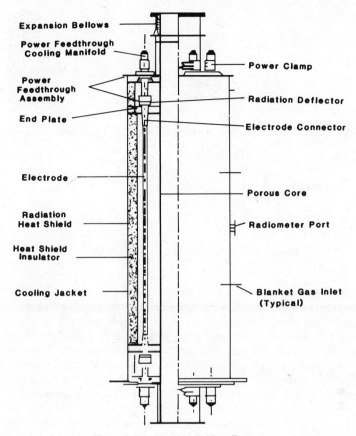

Figure 7.5.2. Electric fluid wall reactor.

- Prolonged reactor life by virtue of inert gas blanketing.
- Production of a medium-Btu combustible gas by operating in a reducing environment.

Four volatile chlorinated hydrocarbons (dichloromethane; 1,1,1-trichloro-ethane; carbon tetrachloride; and Freon-12) and one nonvolatile chlorinated hydrocarbon (hexachlorobenzene) have been decomposed in the HTFW reactor in bench-scale tests to access the applicability of the device for efficient destruction of these particular compounds. The hexachlorobenzene, loaded onto a solid radiation target, exhibited high (> 99.9999%) destruction efficiency, while the vapors, which could be heated only by secondary thermal conduction from the solid radiation target, exhibited destruction efficiencies related inversely to the compound heats of formation, indicating that vapor-phase reaction temperatures were lower than the solid reaction. Destruction efficiencies for the four vapors ranged from 99.999965% for dichloromethane to 84.99% for Freon-12. Heat transfer analysis indicated that vapor heating is dependent on the solid particle density, and that efficient heating (and destruction) of vapors can be achieved simply by increasing the particle density.

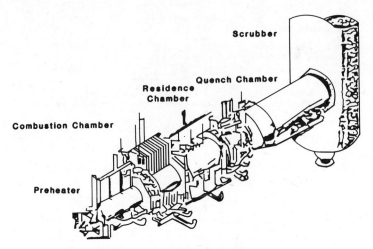

Figure 7.5.3. Plasma temperature incinerator.

The most important conclusion drawn from the tests is that chlorinated hydrocarbons introduced into the reactor in the vapor form are much more difficult to destroy than similar materials loaded onto solids, given identical residence times and reactor temperatures. While the governing factor in the case of the solids was the direct absorption of radiation by the solid surfaces and consequently extremely rapid heating, the governing factor in the case of the vapors was the conduction of the heat from the particle surface into the vapor. This is a much slower process limited in rate by the thermal conductivity of the vapor itself. Thus, as might be anticipated, the temperature levels achieved in the vapors for a given residence time will not, in general, be as high as the temperatures achieved on the solids. The experimental result of this behavior is that minimum temperatures and minimum residence times for complete destruction of vapor-phase substances will not be achieved, and that the observed destruction levels are critically dependent on the heats of formation of the substances being investigated.[7,8]

Plasma Systems

In plasma systems, the extremely high temperature of plasma is employed to destroy waste material. The plasma systems offer a very innovative approach to destroying highly toxic chemicals. A plasma temperature incinerator such as the one manufactured by Applied Energetics, Inc. (see Fig. 7.5.3), burns the toxic wastes, including PCBs, in a pressurized stream of preheated oxygen. The operation can achieve a temperature of 5000°F and maintain these temperatures at elevated levels for periods up to 1 s. The major components of a stand-alone incinerator that is designed for a feed rate of 67 ft³/h are shown schematically in Fig. 7.5.3. This incineration system consists of a preheater, combustion chamber, reaction residence chamber, quench chamber, scrubber tower, and exhaust stack. The preheater is a refractory lined, water-cooled chamber in which a clean fuel is burned to preheat oxygen to about 1800°F. The preheater consumes fuel at the rate of 265 lb/h to preheat 10,000 lb/h of oxygen. The preheated oxygen enters the combustion chamber through a disperser plate. The disperser plate contains a number of fuel nozzles and oxygen ports, and is

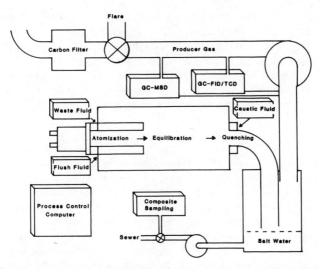

Figure 7.5.4. Pyroplasma process schematic (GC-MS, gas chromatography–mass spectroscopy; GC-FID/TCD, gas chromatography–flame ionization detection/thermal conductivity detection).

designed to achieve optimal mixing of the two reactants. The temperature of the preheated oxygen is sufficient to autoignite any fuel when the heat of combustion is at least 6000 Btu/lb. This arrangement minimizes the risk of flameout and subsequent contamination of the area with partially burned waste.

Advantages of the plasma temperature incinerator are

- In the plasma temperature operating range, many molecules are highly disassociated and have free electrons that greatly increase their reactivity.
- The combustion reaction rates have been shown to be much greater at these temperatures; in some studies a 10-fold increase has been achieved. In similar systems, solid fuels were completely consumed in 35 to 50 ms.
- The higher combustion efficiencies achieved by this design yield a more compact unit that may substantially reduce the cost of incineration. This unit can be built as a stand-alone incinerator, or can be combined with heat recovery systems to conserve the energy that is released.

Pyrolysis Systems, Inc. uses a plasma arc device in a mobile incineration system that has been demonstrated in Kingston, Ontario. In this device, a collinear electrode arrangement creates an electric arc that is stabilized by a field coil medium through which an electric current is passed. As air passes the arc, the electrical energy is converted to thermal energy by adsorption by the air molecules that are activated into ionized atomic states, losing electrons in the process. When the molecules or atoms relax from their highly activated states to lower energy levels, UV radiation is emitted. The resulting gas is in the plasma state and consists of charged and neutral particles with an overall charge near zero and with electron temperatures up to 50,000°F. As the activated components of the plasma decay, their energy is transferred to the waste material exposed to the plasma. The wastes are ultimately decayed and destroyed. A schematic of this process is shown in Fig. 7.5.4. A pilot-sized plant has been used to gather data on the destruction of chlorinated organics, primarily PCBs. Destruction efficiencies in excess of 99.999% were reported, indicating that even greater destruction can be expected among the less stable chlorinated substances. The unit is sized to destroy approximately 500 lb/h of sludge, yet it is designed to fit on a semitrailer along with the necessary auxiliary equipment. Commercial viability must be established by testing actual waste streams over an extended period on a continuous operating basis.[7]

Molten Glass

Molten glass technology uses a pool of molten glass as the heat transfer mechanism to destroy waste material. The attractiveness of molten glass is based upon the extremely good quality of residue from the process, which is essentially nonleachable glass. The combustion conditions for organics appear to be at least as good as those present in hazardous waste incinerators, and the inorganic residue and ash is incorporated into the glass. Introduction of this type of technology into the waste-management industry, especially for highly toxic organic streams containing toxic metals, could prove very attractive if it can be shown through extraction tests that the residue is nonleachable and may be delisted as a hazardous waste.

In the Electromelt pyroconverter, a molten glass process design developed by Penberthy Electromelt International in Seattle, Washington, waste (both combustible and noncombustible) is charged into a pool of molten glass in an electric furnace. The waste is burned to ashes (or dried to solids) that are immediately melted into the glass, all in the same chamber (see Fig. 7.5.5). Excess glass is drained off from time to time into canisters for ultimate disposal. The glass compositions used are very stable. Referring to Fig. 7.5.5, solid wastes, which can be in fiberboard boxes or loose, are pushed through the door opening into the glass furnace, which is heated electrically by immersed electrodes. Liquid and slurry wastes are introduced at a controlled rate through a pipe. If the liquid base is water, the water boils to steam progressively, leaving solids behind. If the liquid is oil, toluene, or other organics, the liquid burns to CO_2, H_2O, and HCl (if a chlorinated compound). Dirt and noncombustible contents drop onto the glass and melt in. The gases go up through the filter to the spray chambers where HCl is converted to common salt (NaCl).

The entire molten glass pool is kept above 2300°F, which is well above the ignition temperature for combustible materials. These materials burn with admitted air in the first half of the chamber, while the second half of the same chamber insures complete combustion. Ashes and any noncombustible materials that drop onto the surface of the glass are melted into the glass. Wet, spent, ion-exchange resins first lose the water and then burn as stated previously. Radioactive metal oxides drop and are melted into the glass. Complementary raw materials, if needed, are charged into the furnace via a screw charger.

The flue gases pass through a set of ceramic fiber filters which catch the particulates. When each filter is loaded, it is pushed into the furnace where it dissolves, capturing the dust into the glass. The flue gases then go through a gas-to-water heat exchanger and then to two water spray chambers. The first spray is kept slightly alklaine to capture acidic gases. The water in the second chamber is fairly pure to prevent loaded water from passing onward into the rest of the system. The demist chamber serves to take out practically all of the droplets. The gases are then

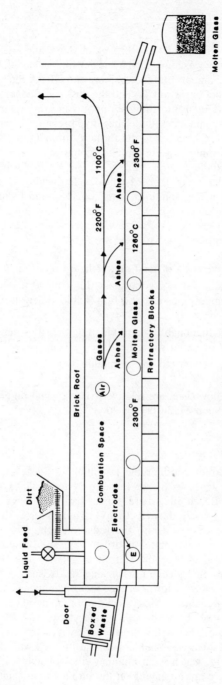

Figure 7.5.5. Penberthy pyroconverter (a molten glass reactor).

273

reheated above the dewpoint and pass through charcoal and HEPA (high efficiency particulate air) filters before going to the blower and out of the vent stack. Filter sludge from the spray chambers and the charcoal and HEPA filters are disposed of in the furnace. The charcoal burns to CO_2 and the dust and sludges are melted into the glass. The HEPA filters melt into the glass pool, capturing their contained dust.

The entire system is maintained under negative pressure by means of the exhaust blower. Combustion air is admitted to the furnace through ports that control oxygen as desired, usually 4% minimum for complete combustion.

This system can handle a large variety of wet or dry wastes. These include plastics, rubber, charcoal filters, PCBs, pesticides, herbicides, toluene, epoxy, contaminated water, electroplating solutions, dirt, concrete, and fission products (radioactive). Residence time ranges are ~ 2 s for the gas phase, and ~ 2 h for the liquid and solid phase.

Advantages of the molten glass process are

- Significant volume reduction of feed materials are realized.
- The product being shipped is a good glass, fully stable under all transportation conditions. Glass compositions are also fully stable under all reasonable and final disposal conditions.
- The equipment is small enough to be installed on the site where the waste is being generated. No loose or liquid waste needs to be transported.
- The off-gas volume is held to the minimum because no added oil or gas is required.
- The equipment and flow sheet are relatively simple, built on basic technology that has been applied in heavy industry for 30 yr.[1,7]

Infrared Technology

Infrared (IR) systems offer considerable potential for waste destruction in mobile use. An incinerator designed by Shirco Infrared Systems (Dallas) has been tested in Missouri at the Times Beach Site where dioxins had been illegally disposed of. The unit features IR lamps (powered by electricity) strung in a row that can be expanded as desired. Wastes are conveyed into the radiation zone by a steel belt, and are carried along until all volatile matter is driven off. The off-gases are burned in a secondary chamber, fired with propane at about 2400°F. The IR system has a key advantage over conventional incinerators in that solids, such as contaminated soils, are not volatilized, allowing for lower energy requirements and less particulate carry over into the secondary chamber. Most incinerators run at 50 or 100% excess air and wind up blowing a lot of the waste out of the stack. The IR system can run at atmospheric conditions or even below atmospheric with a retention time measured in minutes in the primary chamber. An element that fits well with the mobile design of the unit is its refractory, which is an alumina–silica blanket with a density of 7 to 10 lb/ft^3. Since solids are not impinging on the furnace walls, high-density refractory brick (150 lb/ft^3) is unnecessary. The entire unit can be made to fit on one trailer bed and remain well under the 80,000-lb load limit.

The cost of destroying waste by this method has been estimated to be nearly on a par with landfilling, and below the cost of rotary kiln operation. A permanently sited disposal unit could be built for an ongoing operation such as a chemical plant, and because start-up costs are amortized over a longer period of time, it is believed that lower disposal costs could make IR systems particularly attractive.[1]

7.6 OCEAN INCINERATION

Process incineration of organic chemical waste on board specially designed vessels has been practiced off European coasts since 1972, both because of a shortage of land for disposal and because ocean incineration is considered to be environmentally preferable to most land disposal options for hazardous waste. Use of ocean incineration was first proposed in the United States in 1974, when Shell Chemical Company, Houston, Texas, prohibited from continuing to dump liquid organochlorine wastes into the Gulf of Mexico, hired a Dutch company to incinerate those wastes in the Gulf.

In 1972, Congress passed the *Marine Protection, Research and Sanctuaries Act (MPRSA)* that regulates the transportation of material from the United States for the purpose of dumping into ocean waters and prohibits ocean dumping of wastes without a federal permit. The purpose of this law was to strictly limit the dumping into ocean waters of any material that would adversely affect human health, welfare, or amenities or the marine environment, ecological systems, or economic potentialities. Ocean dumping regulations were issued in 1973, and Shell was subsequently prohibited from dumping its untreated organochlorine wastes in the ocean after November, 1973. Shell then began storing its wastes in aboveground tanks at its plant in Deer Park, Texas, and in 1974 the company hired Ocean Combustion Services (OCS), operating out of the Netherlands, to incinerate their wastes using the *Vulcanus* incinerator ship.

Initially the EPA believed it did not have regulatory authority over ocean incineration based on the lack of specific directives in the *Ocean Dumping Law* (i.e., MPRSA) and its legislative history; there were doubts that Congress intended to deal with airborne pollutants in this act. However after careful consideration, the agency decided that it did have jurisdiction over incinerator ships as *indirect* ocean dumpers. This interpretation was based on the agency's concern that failure to regulate recently developed waste disposal techniques involving ocean incineration would frustrate the purposes of the MPRSA and the *London Dumping Convention.* Thus in 1974 ocean incineration without a federal permit was prohibited . This was 2 yr prior to regulation of land-based incinerators under RCRA.

The EPA issued the *Ocean Dumping Regulations* in 1973, thereby establishing criteria for evaluating permit applications. The *London Dumping Convention* (officially referred to as the *Convention on the Prevention of Marine Pollution by Dumping of Waste and Other Matter*) was established to address the growing concern of several nations that marine pollution had developed into a serious environmental problem. The United States was a leader in initiating the convention, basing its position on the newly signed MPRSA. The *London Dumping Convention* (LDC) took effect in 1975 when the minimum required number of 15 nations (including the United States) ratified it as an international treaty. The LDC requires each ratifying nation (52 countries at present time) to regulate by permit the dumping or incineration of wastes loaded from its ports.

After the LDC went into effect in 1975, the parties to the LDC adopted regulations and technical guidelines for incineration at sea. The regulations established technical standards, including combustion and destruction efficiency, operating conditions, and monitoring parameters. The United States must apply the LDCs regulations and take full account of the technical guidelines in developing permits.

The EPA has been involved in ocean incineration research for over 10 yr. Starting in 1974, a series of research burns were conducted under EPA permits for the purpose of gathering scientific information about the incineration of liquid hazardous wastes at sea and to evaluate ocean incineration as an alternative for various land-based disposal options. This technology was demonstrated in the United States off the Gulf of Mexico for the first time between October, 1974 and January, 1975. A total of 18,500 tons of organochlorine compounds from the Shell Chemical Company were burned. The burns were monitored by EPA, the *National Oceanic and Atmospheric Administration* (NOAA) and the *National Aeronautic and Space Administration* (NASA). The composition of the organochlorine waste feeds in the first two shiploads was 65% chlorine, 24% carbon, 4% hydrogen, 4% oxygen, and traces of heavy metals. Combustion chamber flame temperatures ranged from 1340 to 1610°F. The efficiency of destruction of the organochlorine compounds averaged 99.95%. The second burning operation took place during the period of March to April (1977) under an EPA special permit. A total of approximately 18,000 tons of organochlorine wastes were destroyed in four burns that were monitored by TRW, Inc., Redondo Beach, California, under contract to EPA. The average waste feed rate was 24 tons/h, and combustion efficiency was 99.95%. The third shipborne process incineration of organochlorine wastes occurred during the period of July to September (1977) in a Pacific Ocean burn zone in West Atoll where 11,500 tons of herbicide orange owned by the U.S. Air Force were burned. The burns were monitored by TRW. Because the waste contained 2,3,7,8-tetrachloro-dibenzo-*p*-dioxin (TCDD) as an impurity to the extent of an average 2 ppm, the burn required special monitoring and safety procedures. The average feed rate was 16 tons/h; the average destruction efficiency of TCDD was 99.93%.

In all three cases described, wastes were burned on board the *M/T Vulcanus*, originally a cargo ship that was converted in 1972 to a commercial tanker fitted with two liquid-injection process incinerators. Two diesel engines drive the single propeller to give cruising speeds of 10 to 30 knots. Waste is carried in 15 tanks ranging in size from 4060 to 20,300 ft³, with an overall waste capacity of 124,000 ft³. Waste is burned on board in two identical, refractory-lined furnaces that have a combined maximum feed rate of 28 tons/h. A crew of 18 man the ship, 12 to operate the ship and 6 to operate the process incinerators.

In the development of shipborne process incineration, the *Chemical Waste Management Company* bought the *M/T Vulcanus* from the Ocean Services of the Netherlands in September, 1980. In addition, *Chemical Waste Management* took delivery of its new 837,000-gal capacity ocean incineration vessel, *Vulcanus II*, in late 1982. This ship is equipped with three high-capacity incinerators that operate at temperatures between 2280 and 2730°F. The EPA has designated a burn site in the Gulf of Mexico and is in the process of designating one in the North Atlantic.[1]

Incineration vessels like the *Vulcanus* are called tanker-type ships because the liquid waste is stored in tanks below deck. The *Vulcanus I* is a converted tanker while the other ships were designed for burning wastes. These ships have double hulls and the cargo is divided among serval tanks to minimize chances of a spill in an accident. The *Vulcanus* ships have a capacity of about 3900 tons or 800,000 gal. The

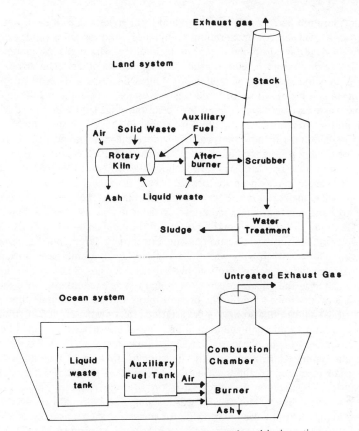

Figure 7.6.1. Land-based versus ocean-based incineration.

incinerators in the ships are first preheated with diesel fuel. Liquid waste is then injected into cylindrical combustion chambers lined with heat-resistant bricks once the desired operating temperature is reached. Combustion temperatures range from ~ 2000 to 2900°F. Gases from the vertical incinerators (three on *Vulcanus II*) flow directly to the exhaust stacks, without passing through a scrubber. The ocean's ability to buffer acidic exhaust gases eliminates the need for costly scrubbers in ocean incineration applications. Ash is reincinerated or returned to land to be placed in a hazardous waste landfill.

Ocean incineration systems currently have the capability of burning only liquid waste while land-based systems equipped with rotary kilns can burn solid and liquid waste. (Refer to Fig. 7.6.1.) Many land-based operations are fitted with secondary combustion chambers (afterburners) and are required to have air pollution control devices to remove acid gases and particulates. Most commercial land-based incinerators have about one third the annual capacity of the ocean incineration ships.

A decade ago, incineration at sea seemed the answer to destroying liquid hazardous waste. It has been over a decade since the EPA has declared the technology environmentally sound. However, ocean incineration of hazardous waste is still not a commerical reality in the United States. One reason is that the opposition to ocean incineration has been fierce. State and local government officials, environmental groups, land-based incineration companies, and business and civic organizations all have raised objections to the technology. These groups contend that the EPA is backing ocean incineration at the expense of other technologies. One argument is that the spending of millions of dollars on ships will result in aborted attempts at on-site incineration for large waste producers. In addition, the majority of burnable wastes are solids not suited for ocean burning; land-based facilities that now use liquid waste as auxiliary fuel to burn solids will have to burn valuable fuel if ocean burning siphons off the liquid wastes. Other objections involve the impact of unburned wastes and products of incomplete combustion (PICs) emitted from incinerators on marine life. Too little is known about the ocean ecosystem. The microlayer, a very thin surface skin on the ocean, is vital to oceanic life and the greatest impact from ocean incineration will involve this microlayer. The layer is hydrophobic and tends to concentrate organic materials. If damaged, it will fail to support the basic marine organisms that live on the bottom of the food chain, such as fish larvae and phytoplankton. Another question is the effect of a large spill of hazardous wastes at sea. The risk of moving wastes on land, either to a land-based incinerator or to port for transfer to a ship, are roughly balanced for any treatment technology that is not located at the generator site. An added risk is encountered once the ship is at sea. This occurs since a spill in the ocean is more difficult to contain than a spill on land. Such a spill would also have a devastating effect on the marine environment and if close to land, would damage human health and industries that depend on the sea. The probability of an ocean spill is slight, however. The EPA has estimated (although this estimate has been seriously questioned) that the probability of a spill from an ocean incineration ship such as the *Vulcanus* as only 1 in 1200 operating years. That figure is based on the historical record of spills from tank ships of similar size and class. The problem of extensive waste spillage at sea could be solved by the use of tank containers to be used worldwide for transporting chemicals by truck, rail, barge, and ship to a modular ocean incineration system. As proposed by Seaburn of Greenwich, Connecticut, 5000-gal stainless steel tanks, similar to those used to deliver gasoline to filling stations, mounted in metal frames, would be used to transport the wastes to port. There, they would be lifted by crane onto a ship or barge. The proposed Seaburn vessel, which will not be built until the final EPA regulations for ocean incineration are issued, will be an ocean-going barge that can hold up to 144 containers and four incinerators above deck. The shallow draft of the Seaburn barge will allow it to operate out of any port that has a container crane. As a result, wastes could be picked up close to where they are produced, rather than shipped to a central storage and blending location as is proposed for the tanker-type ships, a practice that increases the risk of spill.

Incineration at sea has several advantages over current land-based disposal options. It is the technical equivalent of land-based incineration since 99.99% destruction efficiencies can be achieved. In addition, an incineration ship destroys wastes away from populated areas, thereby avoiding risk to nearby communities and eliminating some community opposition to operation. Also, the acidic stack gases can be directly dispersed over the ocean surface without the elaborate scrubbing needed for land-based incinerators.

Not all types of systems are appropriate for incineration at sea. Liquid injection is the only technology well proven at sea for destruction of hazardous wastes. Rotary kilns have not been used on ships, and would require modifications to withstand the pitch and roll of shipboard operation. Fluidized-bed and molten-salt systems are both susceptible to bed shifting due to ship motion.

The EPA has explored the use of offshore platforms equipped with incinerators. Offshore platform incineration would have the following advantages associated with shipborne incineration: remoteness from populated areas and less cost than land-based incineration because the environmental controls can be based upon proximity to land and on meterological and oceanographic conditions. By using both liquid injection and rotary kiln incinerators on the offshore platform, the range of waste materials that can be handled may be broadened to include solids and slurries as well as liquids. The major disadvantage of offshore incineration is the transportation requirement from the nearest port to the platform.

7.7 ILLUSTRATIVE EXAMPLES

Example 7.7.1.   A vertical liquid injection incinerator is operating at 2200°F with a combustion gas flow rate of 6000 scfm (60°F). The incinerator is rectangular in shape and is 10 ft wide (across the front), 11 ft deep, and 28 ft high. The minimum residence time required for incineration is 2 s. Calculate (**1**) the actual combustion gas flow rate; (**2**) the volume required for minimum residence time; and (**3**) the maximum residence time.

SOLUTION

1. The actual gas flow rate is calculated by Charles' Law:

$$Q_a = Q_s \left(\frac{T_a}{T_s} \right)$$

where T_a and T_s are in absolute units.

$$Q_a = 6000 \, \text{scfm} \times \frac{(2200 + 460)}{(60 + 460)}$$

$$= 30{,}692 \, \text{acfm}$$

2. The volume required for minimum residence time is

$$V = Q_a \theta = (30{,}692 \, \text{acfm})(2 \, \text{s}) \frac{\text{min}}{60 \, \text{s}}$$

$$= 1023 \, \text{ft}^3$$

3. The maximum residence time takes into account the actual incinerator volume in the equation:

$$\theta_{\text{max}} = \frac{V}{Q_a}$$

Here

$$V = (10)(11)(28) = 3080\,\text{ft}^3$$

$$\theta_{max} = \frac{3080}{30{,}692}$$

$$= 0.1\,\text{min} = 6\,\text{s}$$

The minimum residence time requirement is satisfied. Note that the gas residence time, calculated by dividing the volume of the incinerator chamber by the combustion gas flow rate, is, however, an approximate value. This does not include flow rate variations that arise due to chemical reaction and temperature changes through the unit. Thus a residence time distribution exists with real systems.

Example 7.7.2.   An engineer involved in the analysis of a rotary kiln unit suspects that the kiln is operating below the desired residence time of 1 h. The available operating data are

$$\text{kiln length} = 25\,\text{ft}$$

$$\text{kiln diameter} = 10\,\text{ft}$$

$$\text{slope of kiln} = 0.01\,\text{ft/ft of length}$$

The present kiln rotation velocity is 0.9 rpm.

1. Is this velocity providing the necessary residence time?
2. What minimum kiln rotation velocity would meet the desired residence time?

SOLUTION

1. Equation (7.2.1) applies.

$$\theta = \frac{0.19\,L}{NDS}$$

where θ = retention time (min)
   L = kiln length (ft)
   N = kiln rotational velocity (rpm)
   D = kiln diameter (ft)
   S = kiln slope (ft/ft of length)

Calculate the residence time

$$\theta = \frac{(0.19)(25\,\text{ft})}{(0.9\,\text{rpm})(10\,\text{ft})(0.01\,\text{ft/ft length})}$$

$$= 53\,\text{min}$$

Therefore, the necessary residence time of 60 min is not being met.

2. Set the residence time of this equation to 60 min to find the minimum kiln rotation rate. Rearrange this equation and solve for N.

$$60\,\text{min} = \frac{(0.19)(25\,\text{ft})}{(N)(10\,\text{ft})(0.01\,\text{ft/ft length})}$$

$$N = \frac{(0.19)(25\,\text{ft})}{(60\,\text{min})(10\,\text{ft})(0.01\,\text{ft/ft length})}$$

$$= 0.8\,\text{rpm}$$

Therefore, a minimum rotation velocity of 0.8 rpm is required to achieve the desired residence time of 1 h in this rotary kiln.

The lead coefficient of 0.19 in this equation has been estimated from limited experimental data. Other values appear in the literature. Therefore, the residence time calculated through the use of this coefficient is at the very best, a rough estimate.

Example 7.7.3.   A 1000 scfm (60°F) effluent at 150°F needs to be treated to control a very low concentration of hazardous pollutant at the *parts-per-million* level. This is to be accomplished with a somewhat outdated thermal incinerator at 1200°F for at least 0.3 s.* The intake combustion air is at 60°F. Assuming a length-to-diameter ratio of 2.0 and a throughput velocity of 12 ft/s, calculate (1) the natural gas flow rate (NHV = 1059 Btu/scf) required for preheating the contaminated effluent to 1200°F using all fresh combustion air intake (primary air); (2) the afterburner throat diameter to give 20-ft/s throat velocity for good mixing (approximately 11.0 scf of combustion gases are produced from 1.0 scf of natural gas combusted); and (3) the length of the afterburner.

The available heat of a natural gas with a NHV of 1059 Btu/scf may be estimated from the equation[9]

$$HA_T = -0.237T + 981$$

where HA_T is in Btu/scf and T is in °F.

SOLUTION

1. First, the waste effluent mass flow rate is calculated. The gas is assumed to have the physical properties of air:

$$w = Q_{a,w}\rho = (1000\,\text{scfm})(0.0766\,\text{lb/scf})$$

$$= 76.6\,\text{lb/min} = 4596\,\text{lb/h}$$

---

* The reader should note that some outdated incinerators can operate legally at 1200°F only because they are protected by a grandfather clause. In modern fume incinerators, 1500 to 1600°F is required for CO oxidation and 1800°F for halogenated hydrocarbon flumes.

The enthalpy of air at 1200 and 150°F is obtained from Table 5.2.2:

$$\text{enthalpy at } 1200°\text{F}, \ h_1 = 294.2 \, \text{Btu/lb}$$
$$\text{enthalpy at } 150°\text{F}, \ h_2 = 28.4 \, \text{Btu/lb}$$

The heat required to increase the effluent waste stream temperature from 150 to 1200°F, allowing for a 10% loss, is calculated as follows:

$$q = (1.10) \, w \, (h_1 - h_2) = (1.10)(4596)(294.2 - 28.4)$$
$$= 1.34 \times 10^6 \, \text{Btu/h}$$

The available heat of the natural gas at 1200°F is next calculated:

$$\text{HA}_T = -0.237T + 981$$
$$= 697 \, \text{Btu/scf}$$

It should be kept in mind that HA_T is the amount of heat remaining after the combustion products from 1 scf of gas are raised to the afterburner temperature (T). This heat is then available for heating the waste effluent to the afterburner temperature.

The natural gas flow rate required is determined as follows:

$$Q_{s,\text{ng}} = \frac{q}{\text{HA}_T} = \frac{1.34 \times 10^6}{697}$$
$$= 1.929 \times 10^3 \, \text{scf/h}$$

2. The volumetric flow rate of combustion products at 1200°F is given by

$$Q_{a,c} = Q_{s,\text{ng}} \frac{11.0 \, \text{scf of combustion gas}}{\text{scf ng}} \frac{(460 + 1200)}{(460 + 60)}$$
$$= \frac{1.929 \times 10^3 \, \text{scf}}{\text{h}} \frac{11.0 \, \text{scf combustion gas}}{\text{scf ng}} (3.19)$$
$$= 6.774 \times 10^4 \, \text{ft}^3/\text{h} = 18.8 \, \text{ft}^3/\text{s}$$

The volumetric flow rate of waste effluent at 1200°F is calculated using Charles' Law.

$$Q_{a,w,1200°\text{F}} = (1000 \, \text{scfm}) \frac{(460 + 1200)}{(460 + 60)}$$
$$= 3192 \, \text{ft}^3/\text{min} = 53.2 \, \text{ft}^3/\text{s}$$

The total volumetric flow rate of gases to the afterburner is given by

$$Q_{a,t} = Q_{a,c} + Q_{a,w,1200°\text{F}} = 18.8 + 53.2$$
$$= 72.0 \, \text{ft}^3/\text{s}$$

The afterburner throat diameter is calculated as follows:

$$A = \frac{\pi D_t^2}{4}$$

In addition,

$$A = \frac{Q_{a,t}}{v_t}$$

Therefore,

$$D_t = \left(\frac{4Q_{a,t}}{v_t \pi}\right)^{0.5} = \left[\frac{(4)(72.0)}{20\pi}\right]^{0.5}$$

$$= 2.14\,\text{ft}$$

The chamber diameter is given by

$$D = \left[\frac{(4)(72)}{12\pi}\right]^{0.5}$$

$$= 2.76\,\text{ft}$$

3. The length of the afterburner chamber is

$$L = 2D = (2)(2.76)$$

$$= 5.5\,\text{ft}$$

The residence time is now calculated to determine if it is greater than the required minimum residence time of 0.3 s.

$$\theta = \frac{L}{v_{\text{throughput}}} = \frac{5.5}{12}$$

$$= 0.46\,\text{s}$$

Since the residence time of 0.46 s is greater than the minimum required residence time of 0.3 s, this design is satisfactory. Because the hazardous waste concentration in the gas stream is very low (in the ppm range), there is no contribution to the heating value associated with the combustion process.

Note: The effluent gas stream has been assumed to have the same physical and chemical properties as air. In addition, combustion of the fuel was accomplished with fresh (primary) air intake. Energy savings could be realized by employing secondary air (air from the waste effluent stream) for combustion. This would reduce fuel requirements to a considerable degree.

Example 7.7.4. An industrial organization proposes to build a hazardous waste liquid injection incinerator. The incinerator is 12 ft wide (across the front), 14 ft deep, and 28 ft high. Approximately 100×10^6 Btu/h will be liberated during the combustion process. Determine if the incinerator volume is adequate. The local regulatory agency has suggested that a heat release rate of 25,000 Btu/h-ft^3 be employed in the calculation(s).

SOLUTION. The furnace volume is calculated using the dimensions given.

$$V = (\text{length})(\text{width})(\text{height}) = (14)(12)(28)$$
$$= 4704 \, \text{ft}^3$$

The energy release rate per cubic foot of the incinerator is

$$q_H = \frac{(100 \times 10^6)}{V} = \frac{100 \times 10^6}{4704}$$
$$= 2.13 \times 10^4 \, \text{Btu/h-ft}^3$$

Since the calculated energy release rate is lower than the recommended heat release rate of 25,000 Btu/h-ft^3 for this liquid injection incinerator, the incinerator size is adequate.

Although the incinerator in this example is rectangular, the reader should note that most incinerators are cylindrical (either vertical or horizontal) in shape.

Example 7.7.5. For preliminary design purposes, determine the size of an incinerator to operate with an average energy release rate (q_H) of 25,000 Btu/h-ft^3 of incinerator volume. It is estimated that during operation, 5000 lb/h of waste with an approximate heating value of 8000 Btu/lb is to be combusted. Assume the incinerator to be a rotary kiln with a L/D ratio of 3.5.

SOLUTION. The amount of heat q released by the waste during operation is

$$q = (5000 \, \text{lb/h})(8000 \, \text{Btu/lb})$$
$$= 4 \times 10^7 \, \text{Btu/h}$$

This result is used to calculate the furnace volume required.

$$V = \frac{q}{q_H} = \frac{4 \times 10^7}{25,000}$$
$$= 1600 \, \text{ft}^3$$

The length and diameter of the kiln with a L/D ratio of 3.5 are calculated as follows:

$$V = \frac{\pi D^2}{4} L = \frac{\pi D^2}{4}(3.5D)$$

$$D = \left(\frac{V}{0.875\pi}\right)^{1/3} = \left(\frac{1600}{0.875\pi}\right)^{1/3}$$

$$= 8.35 \, \text{ft}$$

$$L = (D)(3.5) = (8.35)(3.5)$$

$$= 29.2 \, \text{ft}$$

Note: The heat release rate of 25,000 Btu/h-ft^3 in this example is close to the upper limit. Typical kiln release rates vary from 10,000 to 25,000 Btu/h-ft^3.

PROBLEMS

1. A proposed design for a liquid injection incinerator is to be evaluated. The incinerator is 11 ft wide (across the front), 14 ft deep, and 29 ft high. Approximately 1.0×10^8 Btu/h will be liberated during the combustion process. Is the furnace volume adequate for the recommended heat release rate of 25,000 Btu/h ft^3 for a liquid injection incinerator?

2. A rotary kiln with a L/D ratio of 2.45 is operating at temperatures high enough to achieve slagging. A residence time of 1 h is desired. The slope of the kiln should not exceed 0.01 ft/ft of length. (a) Determine the kiln velocity of rotation at the maximum allowable slope. (b) If the rotation rate in (a) was increased by 0.2 rpm, what would be the percentage change in the residence time?

3. A rotary kiln incinerator is operating with an average energy release rate (q_H) of 28,000 Btu/hr-ft^3 of furnace volume. During operation 4500 lb/h of waste with an approximate heating value of 8000 Btu/lb is to be combusted. Assume the L/D ratio of the rotary kiln to be 3.5. (a) Calculate the furnace volume required. (b) What are the dimensions of the kiln?

4. An exhaust process gas stream at 23,000 scfm (60°F) is to be heated from 200 to 1800°F. Assume that there are no heat losses and that the natural gas fuel has an available heat (HA) of 950 Btu/scf at 1800°F. The average heat capacity of air over the temperature range may be assumed to be equal to 7.4 Btu/lbmol-°F. (a) Determine the volumetric flow rate (in scfm) of natural gas. (b) Find the actual total gas flow rate (in acfm) at the maximum final temperature.

REFERENCES

1. T. Shen, U. Choi, and L. Theodore, *Hazardous Waste Incineration Manual*, U.S. EPA Air Pollution Training Institute, Research Triangle Park, NC, to be published.

2. U.S. EPA, *Engineering Handbook for Hazardous Waste Incineration*, prepared by Monsanto Research Corporation, Dayton, OH, EPA Contract No. 68–03–3025, September, 1981.

3. T. Bonner, B. Desai, J. Fullenkamp, T. Hughes, E. Kennedy, R. McCormick, J. Peters, and D. Zanders, *Hazardous Waste Incineration Engineering*, Noyes Data Corp., Park Ridge, NJ, 1981.

4. B.H. Edwards, *Emerging Technologies for the Control of Hazardous Wastes*, Noyes Data Corp., Park Ridge, NJ, 1983.

5. C. Brunner, *Design of Sewage Sludge Incineration Systems*, Noyes Data Corp., Park Ridge, NJ, 1980.

6. L.P. MacDonald, D.J. Skinner, F.J. Hopton and G.H. Thomas, "Burning Hazardous Waste Chlorinated Hydrocarbons in a Cement Kiln," Canadian Environmental Protection Service Report No. EAP-WP-77-22, March 1977.

7. H. Freeman, *Innovative Thermal Hazardous Organic Waste Treatment Processes*, Noyes Data Corp., Park Ridge, NJ, 1985.

8. J. Radimsky and A. Shah, *Evaluation of Emerging Technologies for the Destruction of Hazardous Wastes*, prepared by Hazardous Waste Engineering Research Lab, Cincinnati, OH, EPA/600/2-85/069 January, 1985.

9. L. Theodore, personal notes, 1987.

8

Waste Heat Boilers

8.1 INTRODUCTION

An obvious by-product of the incineration process is thermal energy—in most cases, a large amount of thermal energy. The total heat load generated by a typical hazardous waste incinerator is in the range of 10 to 150 million Btu/h. While waste heat boilers are capable of recovering 60 to 70% of this energy, the effort may or may not be justified economically. In assessing the feasibility of recovery, a number of factors must be taken into account; among these are the amount of heat wasted, the fraction of that heat that is realistically recoverable, the irregularity in availability of heat, the cost of equipment to salvage the heat, and the cost of energy. The last factor is particularly critical, and may be the most important consideration in a decision whether or not to harness the energy generated by a particular hazardous waste incinerator. Other important considerations are the incinerator capacity and nature of the waste being handled. Generally, heat recovery on incinerators < 2 to 7 million Btu/h may not be economical because of capital cost considerations. Larger capacity incinerators may also be poor candidates for heat recovery if steam is not needed at the plant site or if the combustion gases are highly corrosive; in the latter case, the maintenance cost of the heat recovery equipment may be prohibitive.

Because waste heat boilers are basically heat exchangers, the first part of this chapter covers some of the fundamentals of heat transfer. The remaining sections deal with the operation and design of waste heat boilers.

8.2 FUNDAMENTALS OF HEAT TRANSFER

A difference in temperature between two bodies in close proximity or between two parts of the same body results in a heat flow from the higher temperature to the lower temperature. There are three different mechanisms by which this heat transfer can occur: conduction, convection, and radiation. When the heat transfer is the

287

result of molecular motion (e.g., the vibrational energy of molecules in a solid being passed along from molecule to molecule), the mechanism of transfer is *conduction*. When the heat transfer results from macroscopic motion, such as currents in a fluid, the mechanism is that of *convection*. When heat is transferred by electomagnetic waves, *radiation* is the mechanism.

In most industrial processes, more than one mechanism is usually involved in the transmission of heat. However, since each mechanism is governed by its own set of physical laws, it is beneficial to discuss them independently of each other.

Conduction

As the temperature of a solid increases, the molecules that make up the solid experience an increase in vibrational kinetic energy. Since every molecule is bonded in some way to neighboring molecules, this energy can be passed through the solid. Thus, heating a wire at one end eventually results in raising the temperature at the other. This type of heat transfer is called *conduction* and is the principle mechanism by which solids transfer heat. Fluids are capable of transporting heat in a similar fashion. Conduction in a stagnant liquid, for example, occurs by the movement not only of vibrational kinetic energy, but also of translational kinetic energy as the molecules move throughout the body of the liquid. The ability of a fluid to flow, mix, and form internal currents on a macroscopic level (as opposed to the molecular mixing just described) allow fluids to carry heat energy by convection as well.

The rate of heat flow by conduction is given by Fourier's law:

$$q = -kA\frac{dT}{dx} \tag{8.2.1}$$

where q = heat flow rate (Btu/h)
$\quad\quad x$ = direction of heat flow (ft)
$\quad\quad k$ = thermal conductivity (Btu/h-ft-°F)
$\quad\quad A$ = heat transfer area, a plane perpendicular to the x direction (ft^2)
$\quad\quad T$ = temperature (°F)

The negative sign reflects the fact that heat flow is from high to low temperature and therefore the sign of the derivative is opposite that of the heat flow. If the k can be considered constant over a limited temperature range (ΔT) Eq. (8.2.1) may be integrated to give

$$q = -\frac{kA(T_h - T_c)}{(x_h - x_c)} = \frac{kA\Delta T}{L} \tag{8.2.2}$$

where T_h = higher temperature at point x_h (°F)
$\quad\quad T_c$ = lower temperature at point x_c (°F)
$\quad\quad \Delta T = T_h - T_c$
$\quad\quad L$ = distance between points x_h and x_c (ft)

It should be pointed out that the thermal conductivity (k) is a property of the material through which the heat is passing, and as such, does vary somewhat with temperature. Equation (8.2.2) should therefore be employed only for small values of

TABLE 8.2.1. Thermal Conductivities of Some Common Insulating Meterials[a]

Asbestos-cement boards	0.43
Sheets	0.096
Asbestos cement	1.202
Celotex	0.028
Corkboard, 10 lb/ft^3	0.025
Diatomaceous earth (Sil-o-cel)	0.035
Fiber, insulating board	0.028
Glass wool, 1.5 lb/ft^3	0.022
Magnesia, 85%	0.039
Rock wool, 10 lb/ft^3	0.023

[a] k = Btu/h-ft-°F.

ΔT with an average value of the conductivity used for k. Equation (8.2.1) is the basis for the definition of thermal conductivity as given in Chapter 4, that is, the amount of heat (Btu) that flows in a unit of time (1 h) through a unit area of surface (1 ft^2) of unit thickness (1 ft) by virtue of a unit difference in temperature (1°F). Values of k for some insulating solids are given in Table 8.2.1.

Equation (8.2.2) may be written in the form of the general rate equation:

$$\text{rate} = \frac{\text{driving force}}{\text{resistance}} \tag{8.2.3}$$

Since q in Eq. (8.2.2) is the heat flow rate and ΔT the driving force, the L/kA term may be considered to be the resistance to heat flow. This approach is useful when heat is flowing by conduction through different materials in sequence. Consider, for example, a flat incinerator wall made up of three different layers, an inside insulating layer, a; a steel plate, b; and an outside insulating layer, c. The total resistance to heat flow through the incinerator wall is the sum of the three individual resistances.

$$R = R_a + R_b + R_c \tag{8.2.4}$$

At steady state, the rate of heat flow through the wall is, therefore, given by

$$q = \frac{T_a - T_d}{(L_a/k_a A_a) + (L_b/k_b A_b) + (L_c/k_c A_c)} \tag{8.2.5}$$

where k_a, k_b, k_c = thermal conductivity of each section (Btu/h-ft-°F)

      A_a, A_b, A_c = area of heat transfer of each section (ft) (these are equal for constant cross section of heat conduction)

      L_a, L_b, L_c = thickness of each layer (ft)

      T_a = temperature at inside surface of insulating wall a (°F)

      T_d = temperature at outside surface of insulating wall d (°F)

In this example, the heat is flowing through a slab of constant cross section. In many cases of industrial importance, however, this is not the case. For example, in heat flow through the walls of a cylindrical vessel such as a rotary kiln incinerator, the heat transfer area increases with distance from the center of the cylinder. The heat flow in that case is given by

$$q = \frac{k A_{lm} \Delta T}{L} \tag{8.2.6}$$

A_{lm} in this equation represents the average heat transfer area, or more accurately, the log-mean average heat transfer area. The log-mean average can be calculated by

$$A_{lm} = \frac{A_2 - A_1}{\ln(A_2/A_1)} \tag{8.2.7}$$

where A_2 = outer surface area of cylinder (ft^2)
A_1 = inner surface area of cylinder (ft^2)

Convection

When a pot of water is heated on a stove, the portion of water adjacent to the bottom of the pot is the first to experience a temperature rise. Eventually, the water at the top will also become hotter. Although some of the heat transfer from bottom to top is explainable by conduction through the water, most of the heat transfer is due to the second mechanism of heat transfer, *convection*. As the water at the bottom is heated, its density becomes lower. This results in convection currents as gravity causes the low density water to move upwards to be replaced by the higher density cooler water from above. This macroscopic mixing is a far more effective mechanism for moving heat energy through fluids than conduction. This process is called *natural* convection because no external forces, other than gravity, need be applied to move the heat energy. In most industrial applications, however, it is more economical to speed up the mixing action by artifically generating a current by the use of a pump, agitator, or some other mechanical device. This is referred to as *forced* convection.

The flow of heat from a hot fluid to a cooler fluid through a solid wall is a situation often encountered in engineering equipment; examples of such equipment are heat exchangers, condensers, evaporators, boilers, and economizers. The heat absorbed by the cool fluid or given up by the hot fluid may be *sensible* heat causing a temperature change in the fluid, or it may be *latent* heat causing a phase change, such as vaporization or condensation. In a typical waste heat boiler, for example, hot flue gas gives up heat to water through thin metal tube walls separating the two fluids. As the flue gas loses heat, its temperature drops. As the water gains heat, its temperature quickly reaches the boiling point where it continues to absorb heat with no further temperature rise as it changes into steam. The rate of heat transfer between the two streams, assuming no heat loss to the surroundings, may be calculated by the enthalpy change of either fluid:

$$q = w_h(h_{h1} - h_{h2}) = w_c(h_{c2} - h_{c1}) \tag{8.2.8}$$

where q = rate of heat flow (Btu/h)
w_h = mass flow rate of hot fluid (lb/h)

w_c = mass flow rate of cold fluid (lb/h)
h_{h1} = enthalpy of entering hot fluid (Btu/lb)
h_{h2} = enthalpy of exiting hot fluid (Btu/lb)
h_{c1} = enthalpy of entering cold fluid (Btu/lb)
h_{c2} = enthalpy of exiting cold fluid (Btu/lb)

Equation (8.2.8) is applicable to the heat exchange between two fluids whether a phase change is involved or not. In the waste heat boiler example, the enthalpy change of the flue gas is calculated from its temperature drop:

$$q = w_h(h_{h1} - h_{h2}) = w_h c_{Ph}(T_{h1} - T_{h2}) \tag{8.2.9}$$

where c_{Ph} = specific heat of hot fluid (Btu/lb-°F)
T_{h1} = temperature of entering hot fluid (°F)
T_{h2} = temperature of exiting hot fluid (°F)

The enthalpy change of the water involves a small amount of sensible heat to bring the water to its boiling point plus a considerable amount of latent heat to vaporize the water. Assuming all of the water is vaporized and no superheating of the steam occurs, the enthalpy change is

$$q = w_c(h_{c2} - h_{c1}) = w_c c_{Pc}(T_{c2} - T_{c1}) + w_c \lambda_c \tag{8.2.10}$$

where c_{Pc} = specific heat of cold fluid (Btu/lb-°F)
T_{c1} = temperature of entering cold fluid (°F)
T_{c2} = boiling point of cold fluid (°F)
λ_c = heat of vaporization of cold fluid (Btu/lb)

In order to design a piece of heat transfer equipment it is not enough to know the heat transfer rate calculated by the enthalpy balances described previously. The rate at which heat can travel from the hot fluid, through the tube walls, into the cold fluid, must also be considered in the calculation of the contact area, that is, the area of the tube walls is a direct function of this rate. The slower this rate is, for given hot and cold fluid flow rates, the more contact area is required. The rate of heat transfer through a unit of contact area is referred to as the *heat flux density* and, at any point along the tube length, is given by

$$\frac{dq}{dA} = U(T_h - T_c) \tag{8.2.11}$$

where dq/dA = local heat flux density (Btu/h-ft^2)
U = local overall heat transfer coefficient (Btu/h-ft^2-°F)

The use of the overall heat transfer coefficient (U) is a simple yet powerful concept. In most applications it combines both conduction and convection effects, although heat transfer by radiation can also be included. Methods for calculating the overall heat transfer coefficient are presented later in this section. In actual practice it is not uncommon for the vendors to provide a numerical value for U. A typical value for U for estimating heat losses from an incinerator is approximately 0.1 Btu/h-ft^2-°F.

With reference to Eq. (8.2.11) the temperatures T_h and T_c in Eq. (8.2.11) are actually averages. When a fluid is being heated or cooled the temperature will vary throughout the cross section of the stream. If the fluid is being heated, its temperature will be highest at the tube wall and decrease with increasing distance from the tube wall. The average temperature across the stream cross section is, therefore, T_c, that is, the temperature that would be achieved if the fluid at this cross section was suddenly mixed to a uniform temperature. If the fluid is being cooled, on the other hand, its temperature will be lowest at the tube wall and increase with increasing distance from the wall.

In order to apply Eq. (8.2.11) to an entire heat exchanger, the equation must be integrated. This cannot be done unless the geometry of the exchanger is first defined. For simplicity, one of the simplest geometries will be assumed here—the double pipe heat exchanger. This device consists of two concentric pipes. The outer surface of the outer pipe is well insulated so that no heat exchange with the surroundings may be assumed. One of the fluids flows through the center pipe, the other flows through the annular channel between the pipes. The fluid flows may be either cocurrent, where the two fluids flow in the same direction, or countercurrent, where the flows are in opposite directions. The countercurrent arrangement is more efficient and is more commonly used. For this heat exchanger, integration of Eq. (8.2.11) along the exchanger length, applying several simplifying assumptions, yields:

$$q = UA\Delta T \qquad (8.2.12)$$

The simplifying assumptions are that U and ΔT do not vary along the length. Since this is not actually the case, both U and ΔT must be regarded as averages of some type. A more careful integration of Eq. (8.2.11), assuming that only U is constant, would show that the appropriate average for ΔT in Eq. (8.2.12) is the log-mean average, that is,

$$\Delta T_{lm} = \frac{\Delta T_2 - \Delta T_1}{\ln(\Delta T_2/\Delta T_1)} \qquad (8.2.13)$$

where ΔT_1 and ΔT_2 are the temperature differences between the two fluids at the ends of the exchanger. The area term (A) in Eq. (8.2.12) is the cylindrical contact area between the fluids. However, since a pipe of finite thickness separates the fluids, the cylindrical area (A) is not definite. Any one of the infinite number of areas between and including the inside and outside surface areas of the pipe may be arbitrarily chosen for this purpose. The usual practice is to use either the inside (A_i) or outside (A_o) surface areas; the outside area is the more commonly used of the two. Since the value of the overall heat transfer coefficient depends on the area chosen, it should be subscripted to match the area on which it is based. Equation (8.2.12) now becomes:

$$q = U_o A_o \Delta T_{lm} \qquad (8.2.14)$$

Comparing Eq. (8.2.14) to the general rate equation (8.2.3), it can be seen that $(U_o A_o)^{-1}$ may be regarded as the resistance to heat transfer between the two fluids, i.e.,

$$R_t = \frac{1}{U_o A_o} \qquad (8.2.15)$$

TABLE 8.2.2. Typical Film Coefficients[a]

	Inside Pipes	Outside Pipes[b,c]
Gases	10–50	1–3 (n), 5–20 (f)
Water (liquid)	200–2000	20–200 (n), 100–1000 (f)
boiling water	500–5000	300–9000
condensing steam		1000–10000
Nonviscous liquids	50–500	50–200 (f)
boiling liquid		200–2000
condensing vapor		200–400
Viscous liquids	10–100	20–50 (n), 10–100 (f)
condensing vapor		50–100

[a] $h = $ Btu/h-ft^2-°F
[b] (n) = natural convection.
[c] (f) = forced convection.

In practice, the flow of the fluids in a heat exchanger is turbulent, which means that the main bulk of the stream is well mixed and therefore of fairly uniform temperature. Immediately adjacent to the pipe wall, however, is a thin layer of fluid (referred to as the *boundary layer*) that is in laminar flow and is therefore not mixed. In the absence of macroscopic mixing, heat must be transferred through the film by conduction alone, and hence the laminar film presents a considerable contribution to the total resistance (R_t). The total resistance may therefore be divided into three contributions: the inside film, the tube wall, and the outside film.

$$R_t = R_i + R_w + R_o \qquad (8.2.16)$$

or

$$\frac{1}{U_o A_o} = \frac{1}{h_i A_i} + \frac{x}{k A_{lm}} + \frac{1}{h_o A_o} \qquad (8.2.17)$$

where $h_i = $ inside film coefficient (Btu/h-ft^2-°F)
$h_o = $ outside film coefficient (Btu/h-ft^2-°F)
$A_o = $ outside surface area of pipe (ft^2)
$A_i = $ inside surface area of pipe (ft^2)
$A_{lm} = $ log-mean average surface area of pipe (ft^2)
$x = $ pipe thickness (ft)
$k = $ thermal conductivity of pipe (Btu/h-ft-°F)

Film coefficients are usually determined experimentally. Many empirical correlations can be found in the literature for a wide variety of types of fluids and exchanger geometries. Typical ranges of film coefficients are given in Table 8.2.2. For fully developed turbulent flow in smooth tubes, the following relationship is recommended by Dittus and Boelter[1] for determining the inside film coefficient, h_i:

$$\text{Nu} = 0.023 \ \text{Re}^{0.8}\text{Pr}^n \qquad (8.2.18)$$

where Nu = Nusselt number = $h_i D_i/k$ (dimensionless)
      Re = Reynolds number = $D_i G/\mu$ (dimensionless)
      Pr = Prandlt number = $c_P \mu/k$ (dimensionless)

and

D_i = inside tube diameter (ft)
k = thermal conductivity of the fluid (Btu/h-ft-°F)
c_P = specific heat of fluid (Btu/lb-°F)
μ = viscosity of fluid (lb/ft-h)
$G = w/A_{cs}$ = fluid mass flux (lb/h-ft^2)
n = constant (see below)
w = fluid mass flow rate (lb/h)
A_{cs} = inside cross-sectional area for fluid flow (ft^2)

The properties in Eq. (8.2.16) are evaluated at the fluid bulk temperature, and the exponent n has the following values:

$n = 0.4$ for heating
$n = 0.3$ for cooling

Heat transfer by natural convection is another mechanism that comes into play in calculating heat losses from incinerators. If the incinerator can be considered a stationary vertical cylinder, the heat transfer coefficient may be approximated by

$$h = 0.29\left(\frac{T_w - T_a}{L}\right)^{0.25}$$

where h = natural convection film coefficient (Btu/h-ft^2-°F)
      T_w = temperature at incinerator wall (°F)
      T_a = ambient air temperature (°F)
      L = incinerator length (ft)

For stationary horizontal cylinders,

$$h = 0.27\left(\frac{T_w - T_a}{D}\right)^{0.25}$$

where D = incinerator diameter (ft)

For rotating horizontal cylinders (as with a rotary kiln), the film coefficient may be estimated by

$$h = 0.18(T_w - T_a)^{0.333}$$

After a period of service, thin films of foreign materials such as dirt, scale, or products of corrosion build up on the tube wall surfaces. As shown in Eq. (8.2.19) and (8.2.20.), these films introduce added resistances to heat flow, R_{fi} and R_{fo} and reduce the overall heat transfer coefficient.

TABLE 8.2.3. Typical Fouling Factors

	$h_d{}^a$
Steam	1500–3000
Clean water	500–2000
Dirty water	150–500
Petroleum vapors and condensates	200–2000
Petroleum residuals and crudes	100–200

$^a h_d = $ Btu/h-ft^2-°F.

$$R_t = R_i + R_{fi} + R_w + R_{fo} + R_o \tag{8.2.19}$$

For this condition,

$$\frac{1}{U_o A_o} = \frac{1}{h_i A_i} + \frac{1}{h_{di} A_i} + \frac{x}{k A_{lm}} + \frac{1}{h_{do} A_o} + \frac{1}{h_o A_o} \tag{8.2.20}$$

where $h_{di} = $ inside fouling factor (Btu/h-ft^2-°F)
$h_{do} = $ outside fouling factor (Btu/h-ft^2-°F)

Typical values of fouling factors are shown in Table 8.2.3.

Radiation

In the heat transfer mechanisms of conduction and convection, the movement of heat energy takes place through a material medium—a fluid in the case of conduction. This transfer medium is not required for the third mechanism, where the energy is carried by electromagnetic radiation. If a piece of steel plate is heated in a furnace until it is glowing red and then placed several inches away from a cold piece of steel plate, the temperature of the cold steel will rise, even if the process takes place in an evacuated container.

Radiation becomes important as a heat transfer mechanism only when the temperature of the source is very high. The driving force for *conduction* and *convection* is the temperature difference between the source and the receptor; the actual temperatures have only a minor influence. For these two mechanisms, it does not make too much difference whether the temperatures are 100 and 50°F or 500 and 450°F. Radiation on the other hand is strongly influenced by the temperature level; as the temperature level increases, the effectiveness of radiation as a heat transfer mechanism increases rapidly. It follows that, at very low temperatures, conduction and convection are the major contributors to the total heat transfer; at very high temperatures, radiation is the controlling factor. At temperatures in between, the fraction contributed by radiation depends on such factors as the convection film coefficient and the nature of the radiating surface. To cite two extreme examples,[2] for large pipes losing heat by natural convection, the temperature at which radiation accounts for roughly one half of the total heat transmission is around room temperature; for fine wires, this temperature is above a red heat. In an industrial boiler the radiation heat transfer from the burning gases to the solid surfaces inside the combustion chamber is considerable; in waste heat boilers, however, the contribution of radiation as a heat transfer mechanism is usually minor.

A perfect or ideal radiator, referred to as a *black body*, emits energy at a rate proportional to the fourth power of the absolute temperature of the body. When two bodies exchange heat by radiation, the net heat exchange is then proportional to the difference between each T^4. This is shown by Eq. (8.2.21), which is based on the Stefan–Boltzmann law of thermal radiation:

$$q = \sigma FA (T_h^4 - T_c^4) \tag{8.2.21}$$

where $\sigma = 0.1714 \times 10^{-8}$ Btu/h-ft^2-R^4 (Stefan–Boltmann constant)
   A = area of either surface (chosen arbitrarily) (ft^2)
   F = view factor (dimensionless)
   T_h = absolute temperature of the hotter body (°R)
   T_c = absolute temperature of the colder body (°R)

The view factor (F) depends on the geometry of the system, that is, the surface geometries of the two bodies plus the spacial relationship between them, and on the surface chosen for A. Values of F are available in the literature for many geometries. It is important to note that Eq. (8.2.21) applies only to black bodies and is valid only for thermal radiation, that is, that type of electromagnetic radiation resulting from a temperature difference.

Other types of surfaces besides black bodies are not as capable of radiating energy, although the T^4 law is generally obeyed. The ratio of energy radiating from one of these *gray* bodies to that radiating from the black body under the same conditions is defined as the *emissivity* (ε). For gray bodies, Eq. (8.2.21) becomes

$$q = \sigma F F_\varepsilon A (T_h^4 - T_c^4) \tag{8.2.22}$$

where F_ε = emissivity function dependent on the emissivity of each body and the geometry of the system (dimensionless)

Relationships for F_ε for various geometries are available in the literature. In many applications, the system may be approximated by two long concentric cylinders where the energy exchange is between the outside surface of the inner cylinder (A_h) and the inside surface of the outside cylinder (A_c). An example would be heat radiation from the outside surface of a hot pipe to the surrounding air. In this case, the view factor (F) (based on either area) is 1.0 and

$$F_\varepsilon = 1 \bigg/ \left[\frac{1}{\varepsilon_h} + \frac{A_h}{A_c} \left(\frac{1}{\varepsilon_c} - 1 \right) \right] \tag{8.2.23}$$

For radiation to the surrounding air, (A_h/A_c) approaches zero and Eq. (8.2.23) reduces to

$$F_\varepsilon = \varepsilon_h \tag{8.2.24}$$

Substitution for F and F_ε in Eq. (8.2.22) gives

$$q = \sigma \varepsilon_h A (T_h^4 - T_c^4) \qquad (8.2.25)$$

In the development of convection heat transfer, the film heat-transfer coefficient (h_c) was defined by

$$q_c = h_c A (T_h - T_c) \qquad (8.2.26)$$

where T_h and T_c are the temperatures of the two bodies exchanging heat by convection. In the case of the pipe surface convecting energy to the surrounding air, T_h would be the outside pipe wall temperature and T_c, the average (bulk) temperature for the air. Since radiation heat transfer often accompanies convection and the total heat transfer is the sum of both contributions, it is worthwhile to put both processes on a common basis by defining a radiation heat transfer coefficient (h_r) from

$$q_r = h_r A (T_h - T_c) \qquad (8.2.27)$$

where q_r = the heat transfer rate by radiation[3] (Btu/h)

The total heat transfer is then the sum of the convection and radiation contributions.

$$q = (h_c + h_r) A (T_h - T_c) \qquad (8.2.28)$$

For the example of the hot pipe radiating and convecting to the surrounding air, h_r can be evaluated by solving Eqs. (8.2.25) and (8.2.27) simultaneously to obtain

$$h_r = \sigma \varepsilon_h \frac{(T_h^4 - T_c^4)}{(T_h - T_c)} \qquad (8.2.29)$$

Note: T_h and T_c are absolute temperatures in degrees Rankine (°R).

In most instances, the convection heat transfer coefficient is not strongly dependent on temperature; it should be apparent from Eq. (8.2.29) that this is not the case for the radiation heat transfer coefficient. At the gas temperatures normally employed in waste heat boilers the radiation coefficient is relatively small in comparison to the convection coefficient.

8.3   OPERATION OF WASTE HEAT BOILERS

The main purpose of a boiler is to convert a liquid, usually water, into a vapor. In most industrial boilers, the energy required to vaporize the liquid is provided by the direct firing of a fuel in the combustion chamber. The energy is transferred from the burning fuel by convection and radiation to the metal wall separating the liquid from the combustion chamber. Then, conduction through the metal wall and conduction–convection into the body of the vaporizing liquid takes place. In a waste heat boiler, no combustion occurs in the boiler itself; the energy for vaporizing the liquid is provided by the sensible heat of hot gases which are usually product (flue) gases generated by a combustion process occurring elsewhere in the system. The waste heat boilers found at many incinerator facilities make use of the incineration flue gases for this purpose.

In a typical waste heat boiler installation, the water enters the unit after it has passed through a water treatment plant or the equivalent. This boiler feedwater is passed through heaters/economizers and then into a steam drum. Steam is generated in the boiler by indirectly contacting the water with the hot combustion (flue) gases. These hot gases are typically around 2000°F. The steam is separated from the water in the steam drum, may pass through a superheater, and is then available for internal use or export. The required steam rate for the process or facility and the steam temperature and pressure are the key design and operating variables on the water side. The inlet and outlet flue gas temperatures also play a role, but it is the chemical properties of the flue gas that can significantly impact on boiler performance. For example, acid gases arise due to the presence of chlorine in the waste. As described earlier, the principal combustion product of chlorine is hydrogen chloride, which is extremely corrosive to most metal heat transfer surfaces. This problem is particularly aggravated if the temperature of the flue gas is below the dewpoint temperature of HCl (i.e., the temperature at which the HCl will condense). This usually occurs at temperatures of about 300°F. In addition to acid gases, problems may also arise from the ash of hazardous waste incineration. Some wastes contain fairly high concentrations of alkali metal salts that have melting points below 1500°F. These lower metling point salts can slag and ultimately foul (and in some cases corrode) boiler tubes and/or heat transfer surfaces.

Boilers may be either fire tube or water tube (water-wall). Both are commonly used in hazardous waste incineration; the fire-tube variety is generally employed for smaller applications ($< 15 \times 10^6$ Btu/h). In the fire-tube waste heat boiler, the hot gases from the incinerator are passed through the boiler tubes. The bundle of tubes are immersed in the water to be vaporized; the vaporizing water and tube bundle are encased in a large insulated container called a shell. The steam generated is stored in a surge drum usually located above the shell and connected to the shell through vertical tubes called *risers*. Because of construction constraints, steam pressure in fire-tube boilers is usually limited to around 1000 psia.

In the water-tube waste heat boiler, the water is contained inside the tubes and the hot flue gases flow through the tube bundle usually in a direction perpendicular to the tubes (cross flow). Because of the increased turbulence that accompanies cross flow, the overall heat transfer coefficient for water-tube boilers is higher than that for fire-tube boilers. This advantage is somewhat offset, however, because it is more difficult to clean the outside surfaces of the tubes than the inside surfaces. As a result, heat transfer losses and maintenance problems due to flue gas fouling tend to be greater in water-tube boilers.

8.4 DESIGN OF WASTE HEAT BOILERS

The design of waste heat boilers involves calculations that are based on energy balances and estimations of rates of heat transfer. Although some units operate in an unsteady state or cyclical mode, the calculational procedures are invariably based on steady-state conditions.

In heat transfer equipment, there is no shaft work and potential and kinetic energy effects are small in comparison with the other terms in an energy balance equation. Heat flow to or from the surroundings is not usually desired in practice, and it is usually reduced to a small magnitude by suitable insulation. It is customary to consider this heat loss or gain negligible in comparison to the heat transfer through the walls of the tubes from the hot combustion gases to the water in the boilers. Thus, all the sensible heat lost by the hot gases may be assumed transferred to the steam. In this case, Eq. (8.2.8) may therefore be applied.

The heart of the heat transfer calculation is the heat flux, which is based on the area of the heating surface and is a function of the temperature-difference driving force and the overall heat transfer coefficient, as shown in Eq. (8.2.11). This relatively simple equation can be used to estimate heat transfer rates, area requirements, and temperature changes in a number of heat transfer devices. However, problems develop if more exact calculations are required. The properties of the fluid (viscosity, thermal conductivity, specific heat, and density) are important parameters in these calculations. Each of these, especially viscosity, is temperature dependent. Since a temperature profile, in which the temperature varies from point to point, exists in a flowing stream undergoing heat transfer, a problem appears in the choice of temperature at which the properties should be evaluated. When temperature changes within the stream become large, the difficulty of calculating heat transfer quantities is increased. Because of these effects, the entire subject of heat transfer to fluids with phase change (as in a waste heat boiler) is complex, and in practice is treated empirically rather than as a general theory.

The rigorous design and/or performance evaluation of a waste heat boiler is an involved procedure. Fortunately, several less rigorous methods are available in the literature. One approach that is fairly simple and yet reasonably accurate is that devised by Ganapathy.[4] This method provides a technique for sizing waste heat boilers of the fire-tube type and involves the use of a performance evaluation chart (see Fig. 8.4.1) that is based on fundamental heat transfer equations plus some simplifying assumptions.

Details regarding Ganapathy's method are now discussed. Equations (8.2.9), (8.2.13), and (8.2.14) may be combined and applied to the boiler where the temperature outside the tubes (T_c) is constant due to the water-to-steam phase change. The result is

$$q = w_h c_{Ph}(T_{h1} - T_{h2})$$

$$= U_i A_i \frac{(T_{h1} - T_c) - (T_{h2} - T_c)}{\ln[(T_{h1} - T_c)/(T_{h2} - T_c)]} \qquad (8.4.1)$$

where U_i = overall heat transfer coefficient based on the inside area (Btu/h-ft^2-°F)
A_i = inside heat transfer area (ft^2)
T_c = boiling temperature of the water (°F)

Note that $U_i A_i$ was used in Eq. (8.4.1) instead of $U_o A_o$. The two expressions are equal. In this procedure it is slightly more convenient to use the inside heat transfer area rather than the outside as the basis for U. Because the outside film coefficient (h_o) associated with a boiling liquid is much greater than the inside coefficient for the flue gas, the inside resistance predominates and is responsible for about 95% of the total resistance. Equation (8.2.17) may be simplified to the approximate relationship

$$U_i = 0.95 h_i \qquad (8.4.2)$$

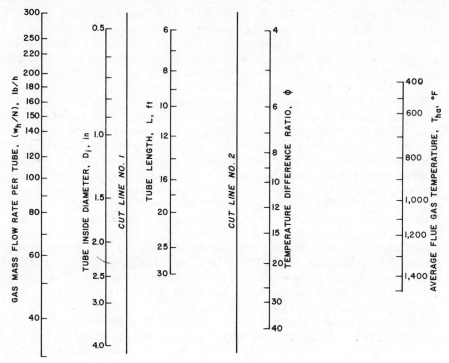

Figure 8.4.1. Waste heat boiler performance evaluation chart.

The total inside heat transfer area is given by

$$A_i = \pi D_i N L \tag{8.4.3}$$

where D_i = tube inside diameter (ft)
N = total number of tubes
L = tube length (ft)

Equation (8.2.18) may be solved explicity for h_i:

$$h_i = 0.023 \frac{k^{0.6} G^{0.8} c_P^{0.4}}{D_i^{0.2} \mu^{0.4}} \tag{8.4.4}$$

and the gas mass flux (G) may be expressed in terms of w_h

$$G = \frac{4w_h}{D_i^2 N} \tag{8.4.5}$$

Equations (8.4.1) through (8.4.5) may be combined to give

$$\ln\left(\frac{T_{h1} - T_c}{T_{h2} - T_c}\right) = \frac{U_i A_i}{w_h c_P} = \frac{CLF(T)}{D_i^{0.8}(w_h/N)^{0.2}} \tag{8.4.6}$$

where $F(T) = k^{0.6}/c_P^{0.6}\mu^{0.4}$

C = a constant equal to 0.0833 when D_i is in feet, or 0.608 when D_i is in inches

Keeping in mind that this procedure is to be used for engineering design purposes and is not intended for rigorous analytical calculations, the value of $F(T)$ does not, in practice, vary over a very large range for most incinerator flue gas streams. The performance evaluation chart shown in Fig. 8.4.1 is based on Eq. (8.4.6) for a typical value of $F(T)$.

The following describes a procedure for the use of the chart as a design tool. It is assumed in this procedure that the tube inside diameter and the number of tubes have been chosen and the tube length is to be determined. Illustrative problems in Section 8.5 further demonstrate the use of Fig. 8.4.1.

1. From the inlet and desired outlet gas temperatures, and the water saturation (boiling) temperature, calculate the arithmetic average gas temperature (T_{ha}) and the value of the temperature difference ratio (ϕ) where

$$\phi = \frac{T_{h1} - T_c}{T_{h2} - T_c} \qquad (8.4.7)$$

   Mark those points on the appropriate axes.
2. Draw a straight line connecting these points, and extend the line to the left to *cut line No. 2*. Mark this point B.
3. Mark the value of the flue gas mass flow rate per tube (w_h/N) and the value of the inside diameter on the appropriate axes. [For design purposes, a good starting value of (w_h/N) is 80–150 lb/h.]
4. Draw a straight line connecting these points and extend the line to the right to *cut line No. 1*. Mark this point A.
5. Connect points A and B by a straight line. The intersection of this line with the L axis yields the appropriate tube length.

The chart could be used in similar fashion as a performance evaluation tool. In this case, the outlet gas temperature would be unknown and hence would have to be estimated in order to determine the average gas temperature. In order to avoid a time-consuming trial-and-error procedure involving the gas outlet temperature and the average gas temperature, Eq. (8.4.8) may be used to estimate T_{ha} without too much loss of accuracy.

$$T_{ha} = 0.6(T_{h1} + T_c) \qquad (8.4.8)$$

8.5 ILLUSTRATIVE EXAMPLES

Example 8.5.1.   An incinerator is enclosed with flat walls each constructed of a 5-in. layer of a special firebrick (thermal conductivity = 0.028 Btu/h-ft-°F) backed by a 8-in. layer of brick (conductivity = 0.82 Btu/h-ft-°F). The inside and outside surface temperatures of the wall are 1900 and 190°F, respectively. (a) Calculate the heat loss per unit area (heat flux) through the wall, and (b) the temperature at the brick–fire-brick interface.

SOLUTION

a. For 1 ft^2 of wall, the resistances are

$$R_f = \frac{L_f}{k_f A} = \frac{\frac{5}{12}}{0.028(1)} = 14.88$$

$$R_b = \frac{\frac{8}{12}}{0.82(1)} = 0.813$$

$$R(\text{total}) = R_f + R_b = 15.69 \ \text{h-°F/Btu}$$

where the subscripts f and b refer to the firebrick and the brick, respectively.

The temperature drop across the wall is

$$\Delta T = 1900 - 190 = 1710°F$$

The heat loss for 1 ft^2 of area is given by

$$q = \frac{\Delta T}{\Sigma R} = \frac{1710}{15.69} = 109 \ \text{Btu/h}$$

or

$$q/A = 109 \ \text{Btu/h-ft}^2$$

b. For the brick layer,

$$\Delta T_b = q R_b = (109)(0.813) = 89°F$$

At the interface,

$$T = 190 + 89 = 279°F$$

Example 8.5.2. Estimate the heat loss in Btu/h from a hazardous waste incinerator that is rated at approximately 60×10^6 Btu/h. The incinerator is a 14-ft i.d., 40-ft long carbon steel rotary kiln. Six inches of firebrick line the inside of the incinerator followed by 4 in. of insulating brick. The 1 in. of carbon steel is surrounded by 9 in. of insulating–protection material that can be assumed to have the properties of asbestos. (Note: Some steel shell incinerators do not have external insulation.) If the flue gas velocity and incinerator operating temperature are 15 ft/s and 2235°F, respectively, calculate the heat loss from the unit. Neglect the rotational effect of the incinerator and assume the ambient temperature to be 45°F. Average thermal conductivity data (independent of temperature) are provided here by the vendor.

k (firebrick) = 1.09 W/m-°C
k (insulating brick) = 0.26 Btu/h-ft-°F
k (insulating asbestos) = 0.166 W/m-°C

SOLUTION. The describing equation is obtained from Eqs. (8.2.5) and (8.2.14)

$$q = \frac{\Delta T}{\Sigma R} = U_o A_o \Delta T$$

where ΔT = difference between incinerator operating temperature and ambient air
temperature
ΣR = sum of six resistances to heat flow

The six resistances that need to be examined are

$$R \text{ (flue gas film)} = \frac{1}{h_i A_i} \tag{8.5.1}$$

$$R \text{ (firebrick)} = \frac{L_f}{k_f A_{lm,f}} = \frac{\ln(r_{fo}/r_{fi})}{2\pi L' k_f} \tag{8.5.2}$$

$$\text{Note: } A_{lm,f} = 2\pi L' \left(\frac{r_{fo} - r_{fi}}{\ln(r_{fo}/r_{fi})} \right)$$

$$R \text{ (insulating brick)} = \frac{\ln(r_{bo}/r_{bi})}{2\pi L' k_b} \tag{8.5.3}$$

$$R \text{ (carbon steel)} = \frac{\ln(r_{co}/r_{ci})}{2\pi L' k_c} \tag{8.5.4}$$

$$R \text{ (insulating asbestos)} = \frac{\ln(r_{ao}/r_{ai})}{2\pi L' k_a} \tag{8.5.5}$$

$$R \text{ (ambient air film)} = \frac{1}{h_o A_o} \tag{8.5.6}$$

where
L_f = thickness of firebrick
k_f = thermal conductivity of firebrick
$A_{lm,f}$ = log-mean heat transfer area of the firebrick
L' = length of incinerator
r_{fi}, r_{fo} = distance from incinerator axis to inside and outside surfaces of firebrick, respectively
f, b, c, a = subscripts representing firebrick, insulating brick, carbon steel, and insulating asbestos, respectively

Based on the geometry of the incinerator and the thicknesses given in the problem statement:

$$r_{fo} = 6.667 \text{ ft} \quad r_{fi} = 6.167 \text{ ft}$$
$$r_{bo} = 7.000 \text{ ft} \quad r_{bi} = 6.667 \text{ ft}$$
$$r_{ao} = 7.833 \text{ ft} \quad r_{ai} = 7.083 \text{ ft}$$

The thermal conductivities are converted to engineering units as follows:

$$k \text{ (firebrick)} = 1.09 \text{ W/m-°C} = 0.608 \text{ Btu/h-ft-°F}$$
$$k \text{ (insulating brick)} = 0.260 \text{ Btu/h-ft-°F}$$
$$k \text{ (asbestos)} = 0.166 \text{ W/m-°C} = 0.096 \text{ Btu/h-ft-°F}$$

The resistance associated with the steel wall [Eq. (8.5.4)] is obviously close to zero. The inside (flue gas) film resistance [Eq. (8.5.1)] is difficult to quantify. A convective coefficient can be calculated from the Dittus–Boelter equation. The companion radiation coefficient will be at least an order of magnitude greater than the convective coefficient, and nearly impossible to accurately calculate. The combined coefficients, however, will produce a resistance that will be negligible relative to that of the two bricks and the asbestos. The outside resistance [Eq. (8.5.6)] can be estimated from a free convection coefficient. A trial-and-error calculation is necessary since the outside temperature of the asbestos is required. This resistance may, however, also be neglected relative to the three main resistances. It is therefore concluded that only the firebrick [Eq. (8.5.2)], insulating brick [Eq. (8.5.3)], and the asbestos [Eq. (8.5.5)] resistances need be calculated. The three (primary) resistances are

$$R \text{ [Eq. (8.5.2)]} = \frac{\ln(6.667/6.167)}{2\pi(0.608)(40)} = 5.10 \times 10^{-4} \text{ h-°F/Btu}$$

$$R \text{ [Eq. (8.5.3)]} = \frac{\ln(7.000/6.667)}{2\pi(0.26)(40)} = 7.46 \times 10^{-4} \text{ h-°F/Btu}$$

$$R \text{ [Eq. (8.5.5)]} = \frac{\ln(7.833/7.083)}{2\pi(0.096)(40)} = 4.172 \times 10^{-3} \text{ h-°F/Btu}$$

The sum of the three resistances (ΣR) is given by

$$\Sigma R = R \text{ [Eq. (8.5.2)]} + R \text{ [Eq. (8.5.3)]} + R \text{ [Eq. (8.5.5)]} = 54.28 \times 10^{-4} \text{ h-°F/Btu}$$

The heat loss is therefore

$$q = \frac{\Delta T}{\Sigma R} = \frac{2235 - 45}{5.428 \times 10^{-3}} = 403{,}500 \text{ Btu/h}$$

This indicates that nearly 1% of the heat generated in the incinerator is lost across the walls. One can conclude that, based on these calculations, the unit is well insulated since losses from most industrial incinerators have been reported in the 2 to 4% range.

Example 8.5.3.   120,000 lb/h of flue gas from a hazardous waste incinerator is to be cooled from 1800 to 500°F in a waste heat boiler. If 2-in. i.d. tubes and a flow rate of 150 lb/h through each tube is to be used, estimate the required heat transfer area, tube length, heat duty, and water mass flow rate. Water at 200°F is available for the steam generator; saturated steam at 80 psia is needed. The average specific heat of the flue gas is 0.26 Btu/lb-°F. Use Ganapathy's method.

SOLUTION. From steam tables or Table 4.5.1, the boiling point of water at 80 psia is 312°F. The temperature difference ratio (ϕ) is

$$\phi = \frac{T_{h1} - T_c}{T_{h2} - T_c} = \frac{(1800 - 312)}{(500 - 312)} = 7.9$$

The average gas temperature is

$$T_{ha} = \tfrac{1}{2}(T_{h1} + T_{h2}) = \tfrac{1}{2}(1800 + 500) = 1150°F$$

On the nomograph (Fig. 8.4.1), connect $(w_h/N) = 150$ with $D_i = 2$ and extend the line to intersect *cut line No. 1* at point A. Connect ϕ (the temperature difference ratio) = 7.9 with $T_{ha} = 1150$ and extend the line to intersect *cut line No. 2* at point B. The line connecting points A and B intersects the L line close to 21.5 ft.
 The number of tubes (N) is

$$\frac{w_h}{(w_h/N)} = \frac{120,000}{150} = 800$$

The total heat transfer area (inside) is given by

$$A_i = NL\pi D_i = (800)(21.5)(2/12) = 9005 \text{ ft}^2$$

Note: This is only one design; several alternatives are possible by changing (w_h/N) or L. If L has to be limited because of space constraints, the nomograph may be used in reverse to calculate (w_h/N). The heat duty from Eq. (8.2.9) is

$$q = w_h c_P(T_{h1} - T_{h2}) = (120,000)(0.26)(1800 - 500)$$
$$= 4.06 \times 10^7 \text{ Btu/h}$$

The required water flow rate is obtained from an enthalpy balance. The enthalpy of water at 200°F is 168.1 Btu/lb. Note that, since pressure has very little effect on the enthalpy of liquids, the enthalpy of liquid water at 200°F and ambient air pressure is essentially the same as that of liquid water at 200°F and its saturation pressure. The value of h (168.1 Btu/lb) was obtained from the saturated steam tables. The enthalpy of steam at 80 psia is 1183.1 Btu/lb. The required water flow is, therefore,

$$w_c = \frac{q}{(h_2 - h_1)} = \frac{4.06 \times 10^7}{1183.1 - 168.1} = 40,000 \text{ lb/h}$$

Example 8.5.4. If 90,000 lb/h of flue gas at 1400°F is used in the waste heat boiler designed in Example 8.5.3, calculate the exit gas temperature, duty, and steam rate. Assume that the average specific heat of the flue gas is the same.

SOLUTION. Calculate (w_h/N).

$$(w_h/N) = \frac{90,000 \text{ lb/h}}{800 \text{ tubes}} = 112.5 \text{ lb/h per tube}$$

Connect $(w_h/N) = 112.5$ with $D_i = 2$ and extend the line to point A on *cut line No. 1*. Connect point A with $L = 21.5$ and extend the line to point B on *cut line No. 2*. Since the outlet gas temperature is not known, the average gas temperature can be approximated using Eq. (8.4.8).

$$T_{ha} = 0.6(T_{h1} + T_c) = 0.6(1400 + 312) = 1027°F$$

Connect point B with $T_{ha} = 1027$. This line intersects the ϕ line at about 8.7. Calculate T_{h2} from

$$8.7 = \frac{1400 - 312}{T_{h2} - 312}$$

which yields a value of 437°F. The duty (q) and steam mass flow rate (w_c) are calculated as in Example 8.5.3.

$$q = (900,000)(0.26)(1400 - 437) = 2.25 \times 10^7 \text{ Btu/h}$$

$$w_c = \frac{2.25 \times 10^7}{(1183.1 - 168.1)} = 22,200 \text{ lb/h}$$

PROBLEMS

1. Using the resistances calculated in Illustrative Example 8.5.2, determine the maximum temperature of the 1-in. carbon steel incinerator wall.

2. A flat incinerator wall with a surface area of 480 ft^2 consists of 6 in. of firebrick with a thermal conductivity of 0.61 Btu/h-°F and an 8-in. outer layer of rock wool insulation with a density of 10 lb/ft^3 (see Table 8.2.1). If the temperature of the inside face of the firebrick and the outside surface of the rock wool insulation are 1900 and 140°F, respectively, calculate (a) the heat loss through the wall in Btu/h and (b) the temperature at the interface between the firebrick and the rock wool.

3. A rotary kiln incinerator is 30 ft long, has a 12-ft i.d. and is constructed of $\frac{3}{4}$-in. carbon steel. The inside of the steel shell is protected by 10 in. of firebrick ($k = 0.608$ Btu/h-ft-°F) and 5 in. of Sil-o-cel insulation (see Table 8.2.1) covers the outside. The ambient air temperature is 85°F and the average inside temperature is 1800°F. The present heat loss through the furnace wall is 6% of the heat generated by combustion of the waste. Calculate the thickness of Sil-o-cel insulation that must be added to cut the losses to 3%.

4. A 12-ft long waste heat boiler contains 500 1-in. tubes through which 72,500 lb/h of hot incinerator flue gas at 1700°F (inlet temperature) flows in parallel. The boiler is to generate steam at 70 psia and the available water may be assumed to be at its boiling point for that pressure. The average specific heat capacity of the flue gas is 0.30 Btu/h-ft-°F. Using Ganapathy's method, estimate (a) the outlet temperature of the flue gas, and (b) the rate of steam generation in lb/h.

5. Using Ganapathy's method, determine the required length of a waste heat boiler to be used to cool hot gases (average specific heat = 0.279 Btu/h-ft-°F) from 2000 to 550°F and generate 30,000 lb/h of steam at 330°F from water at 140°F. The boiler contains 800 1.5-in. i.d. tubes.

REFERENCES

1. F.W. Dittus and L.M.K. Boelter, *Univ. Calif.* (*Berkeley*) Pub. Eng., Vol. 2, 1930.

2. R. Perry and D. Green, *Perry's Chemical Engineering Handbook*, 6th ed., McGraw-Hill, New York, 1984.

3. J.P. Holman, *Heat Transfer*, 5th ed., McGraw-Hill, New York, 1963.

4. V. Ganapathy, "Size or Check Waste Heat Boilers Quickly," *Hydrocarbon Processing*, 169–170, September (1984).

9

Quenchers

9.1 INTRODUCTION

Hot gases from combustion operations must be cooled before entering air pollution control devices, which normally are not designed for very high temperature operation (>500°F). These gases are usually cooled either by recovering the energy in a waste heat boiler, as discussed in Chapter 8, or by quenching. Both methods may be used in tandem. For example, when scrubbing is used for air pollution control, a waste heat boiler can reduce the incinerator exit gas temperature down to about 500°F; a water quench can then be used to further reduce the gas temperature to around 200°F, as well as saturate the gas with water. This secondary cooling and saturation eliminates the problem of water evaporation in the scrubbing system and also alleviates potential problems associated with particulate generation in caustic scrubbers.

Although quenching and the use of a waste heat boiler are the most commonly used methods for combustion gas cooling in hazardous waste incinerator applications, there are several other techniques for cooling the hot gases. All methods may be divided into two categories: *direct-contact* and *indirect-contact* cooling. The direct-contact cooling methods include: (a) dilution with ambient air, (b) quenching with water, and (c) contact with high heat capacity solids. Among the indirect-contact methods are (d) natural convection and radiation from ductwork, (e) forced-draft heat exchangers, and (f) waste heat boilers. With the *dilution* method, the hot gaseous effluent from the incinerator is cooled by adding sufficient ambient air to result in a mixture of gases at the desired temperature. The *water quench* method uses the heat of vaporization of water to cool the gases. Water is sprayed into the hot gases under conditions conductive to evaporation, the heat in the gases evaporates the water, and this results in a cooling of the gases. (Submerged exhaust quenching is another technique that is sometimes employed.) In the *solids contact* method, the hot gases are cooled by giving up heat to a bed of ceramic elements. The bed in turn is cooled by incoming air to be used in the incinerator. *Natural convection and radiation* occur whenever there is a temperature difference between the gases inside a duct and the atmosphere surrounding it. Cooling hot gases by this method requires only the provision of enough heat transfer area to obtain the desired amount of cooling. In *forced-draft heat exchangers*, the hot gases are cooled by forcing cooling fluid past the barrier separating the fluid from the hot gases.

Of these five methods (excluding waste heat boilers), the most commonly used is water quenching—a method given extensive coverage in this chapter. Because the other four techniques are also occasionally used, some attention is given to them also.

9.2 DILUTION WITH AMBIENT AIR

The cooling of gases by dilution with ambient air is the simplest method that can be employed. Essentially, it involves the mixing of ambient air with a gas of known volume and temperature to produce a low temperature mixture that can be admitted to the air pollution control device. In designing such a system, the volume and temperature of gas necessary to ensure particulate capture and the amount of ambient air required to provide a gaseous mixture of the desired temperature are first determined.

Although little instrumentation is required, a gas temperature indicator with a warning device, at the very least, should be used ahead of the air pollution control device to ensure that no damage occurs owing to sudden, unexpected surges of temperature. The instrumentation may be expanded to control either the fuel input to the incinerator or the volume of ambient air to the exhaust system.

The amount of air needed is that necessary to cool the gases to approximately 500°F, which then permits the use of high temperature air pollution control devices. When the volume of the hot gases is small, this method may be used economically. When large volumes of hot gases require cooling, as is normally the case with hazardous waste incinerators, the size of the control device becomes excessively large with dilution cooling. As a result, this method is rarely used in HWI applications.

The quantity of air required for cooling may be calculated directly from an enthalpy balance, that is, the heat "lost" by the combustion (flue) gas is equal to that "gained" by the dilution air. The design of the tank to accomplish this mixing process is based on a residence time of approximately 0.8 to 1.5 s (based on the combined flow rate).

9.3 QUENCHING WITH LIQUIDS

When a large volume of hot gas is to be cooled, a method other than dilution with ambient air should be used. This is usually the case in HWI and the cooling method most often used is quenching.

Cooling by liquid quenching is essentially accomplished by introducing a liquid (usually water) directly to the hot gases. When the water evaporates, the heat of vaporizing the water is obtained at the expense of the hot combustion gas, resulting in a reduction in the gas temperature. The temperature of the combustion gases discharged from the quencher is at the adiabatic saturation temperature of the combustion gas if the operation is adiabatic and the gas leaves the quencher saturated with water vapor. (A saturated gas contains the maximum water vapor possible at that temperature; any increase in water content will result in condensation.) Simple calculational and graphical procedures are available for estimating the adiabatic saturation temperature of a gas (see Chapter 4).

There are three types of quenchers that may be employed: spray towers, packed towers (generally a poor choice), and venturi scrubbers. The venturi scrubber is actually an air pollution control device, but can also be used simultaneously for both particulate removal from, and quenching of, the hot incinerator gases. When a venturi scrubber is the primary air pollution control device on an incinerator, therefore, precooling of the gases may not be necessary. The three types of quenchers are discussed next.

Spray Towers

In one spray tower design, the quench cooling is accomplished simply by spraying water at the top of the tower as the hot gases travel upwards through the tower. When the unvaporized water reaches the bottom of the tower, it is recirculated by pumping back up to the top. Since about 10% of the water stream is vaporized during contact with the hot gases, makeup water must constantly be added. For efficient evaporation of the water, the gas velocity should be from 400 to 600 fpm (ft/min) and the entire cross section of the gas stream should be covered with a fine spray of water.

Water spray pressures generally range from 50 to 150 psig; however, to reduce the amount of moisture collected, some installations have employed pressures as high as 400 psig. Since the moisture collected in spray towers readily corrodes steel, these devices are frequently lined with materials resistant to corrosion. Since the gases discharged from the incinerator are exceptionally hot, the first portion of the duct carrying the gases from the incinerator to the quench tower are usually refractory lined or made from stainless steel. In some cases, stainless steel ducts with water sprays have been used between the incinerator and the quench tower.

For controlling the gas temperature leaving the quench chamber, a temperature controller is generally used to regulate the amount of water sprayed into the quench chamber. For emergency conditions, a second temperature controller can be used to divert excessively hot gases away from the air pollution control device.

Cooling hot gases with a water quench is relatively simple and requires very little space. Quench towers are easy to operate and, with automatic temperature controls, only that amount of water is used that is needed to maintain the desired temperature of the gases at the discharge. Their installation and operating costs are generally considered to be less than that for other cooling methods. Quench towers should not be used when the gases to be cooled contain a large amount of gases or fumes that become highly corrosive when wet. This creates additional maintenance problems, not only in the tower itself, but in the remainder of the ductwork, the control device, and the blower.

Design techniques of spray towers are available in the literature. The following procedure, due in part to Fair[1,2] and in part to Farrell,[3] is recommended.

Before the spray tower can be sized, the water flow rates necessary to bring the flue gas to the desired temperature must first be fixed. This is accomplished by means of energy and material balances around the tower, not only overall balances, but componental balances as well. Componental balances, where the mass and enthalpy of each component entering and exiting the tower is accounted for, are necessary because material will be exchanged between the water and gas streams. For example, some water vapor from the water stream will end up in the gas stream and any soluble gases in the gas stream will be dissolved by the water. In normal operations, about 10% of the water stream is vaporized. For design purposes, therefore, a 9:1 ratio between the water recirculation rate and the makeup water input rate may be assumed. It is also customary to fix the temperature differential between the incoming makeup water and the exiting cooled gases at 2 or 3°F.

Once the necessary flow rates and temperatures have been defined, the tower may be sized. The basic design parameters for the spray tower are the diameter and height. The diameter of a spray tower may be calculated from the superficial velocity (v_t), which is given by Eq. (9.3.1) in units of feet per second.

$$v_t = 0.2 \left(\frac{\rho_L - \rho_G}{\rho_G} \right)^{0.5} \tag{9.3.1}$$

where ρ_L = density of liquid (lb/ft^3)
ρ_G = density of gas (lb/ft^3)

The area of the tower is obtained by dividing the volumetric flow rate of the combustion gases by v_t.

$$A = \frac{w_G}{3600 \rho_G v_t} \tag{9.3.2}$$

where w_G = mass flow rate of the gas (lb/h).

The diameter is then given by

$$D = \left(\frac{4A}{\pi} \right)^{0.5} \tag{9.3.3}$$

In order to calculate the tower height (Z) (ft) a volumetric heat transfer coefficient (U) (Btu/h-ft^3-°F) is employed. This coefficient is then used to establish the required contact volume (V) via Eq. (9.3.4).

$$V = \frac{q}{U \Delta T_{lm}} \tag{9.3.4}$$

where V = tower volume (ft^3)
q = heat rate required to vaporize the water to the combustion gas discharge temperature (Btu/h)
ΔT_{lm} = log-mean temperature difference driving force across the quencher

If the liquid-side resistance to heat transfer is neglected, U is given by the hot gas volumetric film heat transfer coefficient (h) as provided in Eq. (9.3.5).

$$h = \frac{0.43 G^{0.8} L^{0.4}}{Z^{0.5}} \tag{9.3.5}$$

where h = gas-side volumetric heat transfer coefficient (Btu/h-ft^3-°F)
G = gas superficial mass velocity at the bottom of tower (lb/h-ft^2)
L = liquid superficial mass velocity at the bottom of tower (lb/h-ft^2)
Z = height of tower (ft)

The *superficial* mass velocity of a stream is calculated by dividing the mass flow rate (at or near the tower bottom) by the cross-sectional area. If there were no other material in the tower besides that particular stream, the *superficial* and *true* mass velocities would be equal. The mass flow rates are specified at the bottom of the tower because these quantities change as the streams move through the tower (due to the exchange of material) and reach their highest values at the bottom. The tower volume is given by Eq. (9.3.6).

$$V = AZ \qquad\qquad (9.3.6)$$

Equations (9.3.4) through (9.3.6) may be solved simultaneously for Z.

Packed Towers

Packed columns find their main HWI application in the cleaning of combustion gases and hence will be covered in detail in Chapter 10 (see Section 10.5 on *Absorbers*). However, these devices may also be used for quenching the hot flue gases and, hence, some descriptive material plus design (as a quencher) information on this topic will be presented in this chapter.

Packed columns, used in general for continuous contacting of liquid and gas streams, are usually vertical columns that have been filled with packing or other artifacts of large surface area. The liquid is distributed over, and trickles down through the packed bed, thus exposing a large surface area to the gas stream. When a packed tower is used for quenching, the hot gas stream is introduced near the bottom and flows upward through the packed bed against the downward-flowing cooling water stream. The water stream is recirculated and a makeup water stream is continuously added at the top. The makeup water is needed to replace the evaporated water that is picked up by the gas stream.

Consider a packed tower operating at a given liquid rate, as the gas rate is gradually increased. After a certain point the gas rate becomes so high that the drag on the liquid is sufficient to keep the liquid from flowing freely down the column. Liquid begins to accumulate and tends to block the entire cross section for flow (so-called *loading*). This, of course, increases both the pressure drop and prevents the packing from mixing the gas and liquid effectively, and ultimately some liquid is even carried back up the column. This undesirable condition, known as *flooding*, occurs fairly abruptly, and the superficial gas velocity at which it occurs is called the *flooding velocity*. The calculation of column diameter is based on flooding conditions, the usual operating range being taken as 50 to 75% of the flooding rate.

Before the packed tower can be sized, the flow rates and temperatures must be determined. This is accomplished by using the energy and material balances described earlier for the spray tower. Equation (9.3.7) relates the liquid and gaseous flow rates in the tower with the column cross-sectional area at the flooding point.[4]

$$Y = -3.84 - 1.06X - 0.119X^2 \qquad\qquad (9.3.7)$$

where $Y = \ln\left[w_G^2 F\psi\mu_L^{0.2}/\rho_G\rho_L g_c A_f^2\right]$

$\quad\quad X = \ln\left[(L/G)(\rho_G/\rho_L)^{0.5}\right]$

$\quad\quad L$ = liquid superficial mass velocity at bottom of packing (lb/h-ft^2)

$\quad\quad G$ = gas superficial mass velocity at bottom of packing (lb/h-ft^2)

$\quad\quad \rho_L$ = liquid density (lb/ft^3)

$\quad\quad \rho_G$ = gas density (lb/ft^3)

$\quad\quad F$ = packing factor

$\quad\quad A_f$ = cross-sectional area of column (ft^2)

$\quad\quad \mu_L$ = liquid viscosity (cP)

$\quad\quad \psi$ = ratio, density of water/liquid density

$\quad\quad g_c$ = constant = 4.173×10^8 lb-ft/lb$_f$-h^2

This equation is applicable in the range $0.01 < e^X < 10$ or $-4.6 < X < 2.3$.

In the following design procedure, Eq. (9.3.7) is used to determine the column diameter:

1. From the superficial mass velocities, calculate a value for the parameter Y.
2. From the value of Y, determine the cross-sectional area of the column that would cause flooding for the given flow rate. Equation (9.3.8), which is based on the definition of Y above, may be used for this purpose

$$A_f = \left[\frac{w_G^2 F\psi\mu^{0.2}}{\rho_G\rho_L g_c e^Y}\right]^{0.5} \tag{9.3.8}$$

The packing factor (F) is a parameter based on the *specific surface* of the packing used (ft^2 surface/ft^3 volume) and the *void fraction* of the packed region of the column. Its value is usually supplied by the vendor, although F's for most common types of packing can be found in the literature.

3. Calculate the column cross-sectional area (A) for the fraction of the flooding velocity (f) chosen for operation. Usually f is chosen as 0.60.

$$A = \frac{A_f}{f} \tag{9.3.9}$$

4. The tower diameter can then be calculated from Eq. (9.3.3).

Note: The constants in Eq. (9.3.7) are not dimensionless. The only units that may be used, therefore, are those designated in the list of symbol definitions below the equation.

A somewhat similar procedure to that employed for a spray tower is used to calculate the height of a packed tower quencher. Unfortunately, very little data are available for estimating the hot gas-side heat transfer coefficient. The recommended procedure is to calculate h (which is again assumed equal to U) using Eq. (9.3.10).

$$h = \left(\frac{Sc_G}{Pr_G}\right)^{2/3}\left(\frac{c_{PG}G}{H_G}\right) \tag{9.3.10}$$

where Sc_G = Schmidt number for the gas (dimensionless) = $\mu_G/\rho_G D'$
Pr_G = Prandtl number for the gas (dimensionless) = $c_{PG}\mu_G/k_G$
c_{PG} = combustion gas specific heat capacity (Btu/lb-°F)
G = gas superficial mass velocity (lb/h-ft^2)
H_G = height of a gas phase transfer unit (ft)
μ_G = gas viscosity (lb/ft-h)
ρ_G = gas density (lb/ft^3)
D' = diffusion coefficient (ft^2/h)
k_G = gas conductivity (Btu/h-ft-°F)

The term H_G is calculated from

$$H_G = \phi Sc^{0.5}\frac{D^n}{(Lf_1f_3f_3)^m} \tag{9.3.11}$$

where $\quad D$ = tower diameter (ft)
$f_1 = \mu_L^{0.16}$
$f_2 = (62.4/\rho_L)^{1.25}$
$f_3 = (72.8/\sigma)^{0.8}$
μ_L = liquid viscosity (lb/ft-s)
σ = surface tension (lb$_f$/ft)
ρ_L = liquid density (lb/ft^3)
ϕ, n, m = empirical coefficients that depend on the type of packing and flow conditions

Once an estimate for U has been made, V is calculated directly from Eq. (9.3.4). The packing height is given by

$$Z = V/A \tag{9.3.12}$$

A simplified approximate calculational procedure for designing either a spray or packed-tower water quencher is available.[5] The tower area (A) is given by

$$A = bw_{in}/\rho_G \tag{9.3.13}$$

where w_{in} = total mass flow rate of hot gas entering quencher (lb/s)
ρ_G = hot gas density (lb/ft^3)
b = dimensionless constant in the 0.1–0.3 range

The tower diameter is then

$$D = \left(\frac{4A}{\pi}\right)^{0.5} \tag{9.3.14}$$

The tower height may be estimated by

$$Z = 2.0D \tag{9.3.15}$$

with the tower volume (V) given by

$$V = 2D^3$$

The units of A, D, Z, and V are ft^2, ft, ft, and ft^3, respectively. This procedure provides an approximate 25 to 50% safety factor relative to the rigorous calculations.

Venturi Scrubber

Although primarily utilized as an air pollution control device (and hence described in greater detail in Chapter 10), the venturi scrubber is also sometimes used at incinerator facilities as a quencher. The device consists of three sections of ductwork: the first, a converging section, followed by the second, a short length of constant cross section called the *throat*, followed by the third, a diverging section. These scrubbers utilize the kinetic energy of a moving gas stream to atomize the scrubbing liquid into droplets. Liquid, almost always water, is injected into the high velocity gas stream either at the inlet to the converging section or at the venturi throat. In the process, the water is atomized into extremely small liquid droplets. These droplets have extremely large surface areas, which is an ideal situation for heat and mass transfer. There is no satisfactory generalized design correlation for this type of quencher. Design should therefore be based on full-scale data or at least laboratory- or pilot-scale data. Limited data[6] suggest that a water flow rate of 10 gpm/100 scfm of hot flue gas at approximately 2000°F is sufficient to cool the gases to the adiabatic saturation temperature (which is usually between 150 to 200°F). A minimum pressure drop of 5 in. H_2O is recommended. An average liquid droplet size of approximately 100 microns (μm) is produced, which has high enough surface area to accommodate the simultaneous heat and mass transfer quench process.

This method of water quenching has found wide application in hazardous waste incineration operations. Although pressure drop is a major (negative) consideration, there is particulate capture and some gaseous removal (despite the elevated temperatures). Most quench calculations do not include the "credit" associated with particulate and gaseous control.

9.4   CONTACT WITH HIGH HEAT CAPACITY SOLIDS

Inert ceramics with high heat capacities have, in a few cases, been used to cool hot incinerator effluent gases from fume incinerators to temperatures suitable for air pollution control devices. Besides cooling the gases by absorbing heat, this method also allows a certain amount of energy recovery by using the heat absorbed by the solids to subsequently preheat the air being introduced to the incinerator. In one such system, the solid elements and packed into beds called *energy recovery chambers*. These chambers operate on a *regenerative* principle—alternately absorbing, storing, and recycling heat energy. The chambers are thus used in cyclic fashion. In the first half-cycle, ambient air is passed through the previously heated ceramic elements in a particular chamber, simultaneously cooling the bed and heating the air that is then moved on to the incinerator. At any point during the cycle, the temperature at the bed varies over 1000°F from the cold side (i.e., the inlet side for the ambient air) to the hot side. In the second half-cycle, the hot incinerator effluent gases are introduced to the hot side of the bed, flow through the bed in the opposition direction of the ambient air flow in the first half-cycle, and exit the cold side; the heat lost by the gases is absorbed by the bed that is brought to its maximum average temperature for the start of the next cycle. Typical cycle times range from 2 to 10 min. Manufacturers claim over 90% thermal energy recovery with this device.

9.5   NATURAL CONVECTION AND RADIATION

When a hot gas flows through a duct, the duct becomes hot and heats the surrounding air. As the air becomes heated, natural drafts are formed carrying the heat away from the duct. This phenomenon is *natural convection*. Heat is also discharged from the hot duct to its surroundings by *radiant energy*. Both of these modes of heat transfer have already been discussed in Chapter 8.

Equation (8.2.14) shows that the rate of heat transfer is a function of the resistances to heat flow, the log-mean temperature difference between the hot gas and the air surrounding the duct, and the surface area of the duct. The rate of heat transfer (q) is determined by the amount of heat to be removed from the hot gaseous effluent entering the exhaust system. For any particular basic process, the mass flow rate of gaseous effluent and its maximum temperature are fixed. The cooling system must, therefore, be designed to dissipate sufficient heat to lower the effluent temperature to the operating temperature of the air pollution control device to be used. The rate of heat transfer can be calculated by the enthalpy difference of the gas at the inlet and outlet of the cooling system.

$$q = w_G \Delta h \tag{9.5.1}$$

where q = rate of heat transfer
w_G = mass flow rate of gas (lb/h)
Δh = enthalpy change between inlet and outlet conditions (Btu/lb)

The log-mean temperature difference in this case is the difference in temperature between the air surrounding the duct, and the inlet and outlet temperature of the gas. This term, too, is fixed for a particular process and is calculated by applying Eq. (8.2.13) as follows:

$$\Delta T_{lm} = \frac{(T_{h1} - T_c) - (T_{h2} - T_c)}{\ln\left[(T_{h1} - T_c)/(T_{h2} - T_c)\right]} \tag{9.5.2}$$

where T_{h1} = gas temperature of inlet (°F)
T_{h2} = gas temperature at outlet (°F)
T_c = air temperature (°F)

In many incinerator operations the temperature of the gaseous effluent is not constant but varies as different wastes are burned. The atmospheric temperatures also vary a great deal. In such cases, the cooling system must be designed for the worst conditions that prevail to ensure adequate cooling at all times. The inlet temperature (T_{h1}) chosen for the design calculations must be the maximum temperature of the gas entering the system; T_{h2} must be the maximum allowable temperature of the gas discharged from the cooling system; and T_c must be the maximum expected atmospheric temperature.

The overall heat transfer coefficient (U_o) is calculated from the individual film coefficients, h_i and h_o as shown in Eq. (8.2.17). Equation (8.2.18) may be employed to calculate the inside coefficient (h_i). As already described, this coefficient is a measure of the flow of heat through the inside film. An increase in h_i will, therefore, increase the rate of heat transferred from the gas to the atmosphere. It can be seen that an increase in the Reynolds number (Re) will increase the rate of heat transfer. Since the weight of gas flowing is fixed, Re can be increased only by increasing the velocity of the gas. An increase in velocity will increase the power required to move the gases through the exhaust system. Consequently, the optimum velocity for good heat transfer at reasonable blower-operating costs must be determined. It is known that a sacrifice in heat transfer rate to obtain lower blower horsepower can result in a more economical cooling system. Owing to the many variables involved, however, each system must be calculated on its own merits.

The outside film coefficient (h_o) is the sum of the coefficient due to natural convection (h_c) and the coefficient due to radiation (h_r).

$$h_o = h_c + h_r \qquad (9.5.3)$$

An empirical equation for h_c for vertical pipes more than 1-ft high and for horizontal pipes is[7]

$$h_c = 0.27 \left(\frac{\Delta T}{D_o}\right)^{0.25} \qquad (9.5.4)$$

where ΔT = the temperature difference between the outside duct wall and the air $(T_w - T_c)$
D_o = outside duct diameter (ft)

The radiation coefficient (h_r) may be calculated by Eq. (8.2.29). Since the emissivity of the surface is a function of the surface condition, and a black surface generally gives the highest emissivity, blackening of the ductwork is recommended.

To calculate h_o by Eqs. (8.2.29), (9.5.3), and (9.5.4), the temperature of the duct wall (T_w) must first be assumed and then checked. The following equation may be used by this purpose.

$$T_w = T_a - \left(\frac{h_o}{h_o + h_i}\right)(T_a - T_c) \qquad (9.5.5)$$

where T_a = average flue gas temperature (°F)
T_c = ambient air temperature (°F)

If the calculated T_w is not the same as the assumed T_w, a new T_w must be estimated and h_o recalculated. When the assumed T_w and calculated T_w are the same, the corresponding h_o is then used in Eq. (8.2.17) to calculate U_o. Typical values for the overall heat transfer coefficient are in the 0.5 to 2.0 Btu/h-ft²-°F range.

The heat transfer area (A_o) can now be found; the length of duct needed to give the necessary area is then calculated by using the outside diameter. If the length of duct needed is large, the ductwork will probably be arranged in vertical columns to conserve floor space. The columns require several 180° bends, which will offer a large resistance to the flow of gas. To minimize these losses, the gas velocity should be low, preferably less than normal dust-conveying velocities. By joining the bottoms of the columns with hoppers, any dust settling out as a result of low velocities can be collected without fouling the system. If the cooling area is such that a single loop around the plant or across a roof is sufficient, sharp bends should be avoided. When gases are cooled through a large temperature range, the volume will be considerably reduced, so that smaller diameter ductwork may be needed as the gases proceed through the cooling system. With cooling columns, the diameter of the duct joining the last column and the air pollution control device must be sized properly to provide suitable conveying velocities for the cooled effluent.

For most convection–radiation cooling systems, the only equipment used is sufficient ductwork to provide the required heat transfer area. Unless the temperature of the gases discharged from the incinerator is exceptionally high, or if there are corrosive gases or fumes present, black iron ductwork is generally satisfactory. The temperature of the duct wall can be determined for any portion of the ductwork by using the method previously described for determining T_w. If T_w proves to be greater than black iron can withstand, either a more heat resistant material should be used for that portion of the system or a portion of the cooled gas should be recirculated to lower the gas temperature at the cooling system inlet.

With this type of cooling, flexibility in controlling the gas temperature is limited. When either the gas stream temperature or air temperature, or both, are lower-than-design values, the gases discharged from the cooling device will be less than that calculated, and condensation of moisture from the effluent within the control device might result. Conversely, when design temperatures are exceeded, the temperature of the gases discharged from the cooling system could become too high.

The radiation–convection cooling system is in operation whenever hot gases are flowing through the exhaust system. Since gases being cooled are not diluted with any cooling fluid, the air pollution control device need not be sized for an extra volume of gases. Since no water is used, there is no need for pumps, and corrosion problems are nonexistent. On the other hand, these cooling systems require considerable space, and system blower horsepower requirements are increased owing to the additional resistance to gas flow. This type of cooling system is seldom used in HWI, mainly because of the excessive space requirement.

9.6   FORCED-DRAFT COOLING

A discussed in Chapter 8, heat transfer by convection is due to fluid motion. Cold fluid adjacent to a hot surface receives heat, which is imparted to the bulk of the fluid by mixing. With natural convection, the heated fluid adjacent to the hot surface rises and is replaced by colder fluid. By agitating the fluid, mixing occurs at a much higher rate than with natural currents, and heat is taken away from the hot surface at a much higher rate. In most process applications, this agitation is induced by circulating the fluid at a rapid rate past the hot surface. This method of heat transfer is called *forced convection*. Since forced convection transfers heat much faster than natural convection, most process applications use forced-convection heat exchangers. Whenever possible, heat is exchanged between hot and cold streams to reduce the heat input to the process. There are, however, many industrial applications where it is not feasible to exchange heat in this fashion, and so a cooling fluid such as water or air is used, and the heat removed from the stream by the coolant is dissipated to the atmosphere. When water is used, the heat is taken from the process stream in a shell-and-tube cooler, and the heat picked up by the water is dissipated to the atmosphere in a cooling tower. When air is used as the cooling medium in either shell-and-tube or fin tube coolers, the heated air is discharged to the atmosphere and is not recirculated through the cooler.

With forced-convection cooling, the temperature of the cooled stream can be controlled within narrow limits even with widely varying atmospheric or water temperatures. Heat transfer area is greatly reduced from that needed with natural convection. Power requirements to force the process stream through the cooler are generally less. On the other hand, either a pump or a blower is needed to circulate the cooling fluid through the cooler. With water cooling, a cooling tower may be needed and additional maintenance is required to clean scale from the tubes.

Because there are so many different types of forced convection heat exchangers, it is impossible to present a specific design procedure unless a specific type is chosen for discussion. The following describes a general approach for the calculation of the total heat transfer surface required ($A_{o,\text{total}}$) which is the most critical parameter in heat exchanger design. To use this approach, however, much information on the specific type of exchanger must be obtained either from the literature or from the vendor.

Generally, the hot gas flow rate, inlet and desired outlet gas temperatures and the inlet coolant temperature are already known. The first three quantities can be used to calculate the heat duty. The coolant flow rate is chosen on the basis of pressure drop information and heat transfer coefficient data, both of which are highly dependent on the coolant velocity and heat exchanger configuration. These data are often available from the vendor. An enthalpy balance (Eq. 8.2.8) may then be applied to determine the coolant outlet temperature. In order to calculate the overall heat transfer coefficient (U_o) both the gas-side and coolant-side film coefficients must be determined. Procedures to evaluate each of these coefficients (except when the fluid is flowing inside tubes or pipes) must be obtained for the specific flow configuration either from the literature or from the vendor. If fouling is a potential problem, the appropriate fouling factors should also be obtained. Equation (8.2.20) is applied to calculate U_o. Note that in this equation, A_o, A_i, and A_{lm} are defined from the heat transfer surface for a single passage for one of the fluids, such as a pipe or tube. In an industrial heat exchanger, there are many such passages. The three areas in Eq. (8.2.20) are usually known at the start of the design process; the total heat transfer area ($A_{o,\text{total}}$) is not. This latter is obtained from Eq. (9.5.6)

$$q = U_o A_{o,\text{total}} \, \phi \Delta T_{\text{lm}} \tag{9.5.6}$$

where
$\quad q = $ heat duty (Btu/lb)
$\quad U_o = $ overall heat transfer coefficient (Btu/h-ft^2-°F)
$\quad A_{o,\text{total}} = $ total heat transfer area (ft^2)
$\quad \phi = $ correction (dimensionless)
$\quad \Delta T_{\text{lm}} = $ log-mean temperature difference (°F)

The factor (ϕ) is a correction for the use of ΔT_{lm} as the average temperature difference driving force. If the flows are parallel (i.e., either cocurrent or countercurrent), this factor is unity. In industrial heat exchangers, this is usually not the case, and ϕ must be obtained either from the literature or the vendor for the particular flow patterns used in the exchanger.

9.7 ILLUSTRATIVE EXAMPLES

Example 9.7.1. For the system shown in Fig. 9.7.1, design an adiabatic quench spray tower to cool the combustion gases from the incinerator. Provide calculations based on (1) top, (2) bottom, and (3) average tower conditions. The composition of the combustion gas stream (Stream 1) is given in the following list:

	MW	(lb/h)	(lbmol/h)	Mole Fraction
CO_2	44	12,023	273.3	0.0536
H_2O	18	9,092	505.1	0.0990
N_2	28	106,783	3,813.7	0.7476
HCl	36.5	193	5.3	0.0010
O_2	32	16,115	503.6	0.0988
Total		144,206	5,101.0	1.0

Average MW = 28.27

For a reference temperature (T_r) of 32°F:
$c_{P,\text{BDA}}$ = 0.26 Btu/lb-°F between 32 to 2050°F
$c_{P,\text{BDA}}$ = 0.25 Btu/lb-°F between 32 to 200°F
$c_{P,\text{H}_2\text{O}}$ = 1.0 Btu/lb-°F
$c_{P,\text{wv}}$ = 0.5 Btu/lb-°F
$c_{P,\text{CG}}$ = 0.3 Btu/lb-°F (constant)
Δh_{vap} = 1075 Btu/lb at 32°F

(*Note:* The subscript BDA stands for bone dry air, wv for water vapor, and CG for combustion gases.)

Assume (a) the water makeup temperature (T_5) equal to 100°F and (b) a 2°F approach temperature for T_2.

SOLUTION. An enthalpy balance around the adiabatic quencher gives:

$$H_1 - H_2 + H_3 - H_4 + H_5 = 0 \tag{9.7.1}$$

The enthalpy of an air–water vapor mixture is defined as

$$H_{\text{air},T} = w_{\text{BDA}} c_{P,\text{BDA}}(T - T_r) + w_{\text{H}_2\text{O}}[\Delta h_{\text{vap},T_r} + c_{P,\text{wv}}(T - T_r)] \tag{9.7.2}$$

A water balance on the recycle line yields

$$w_3 = w_4 \tag{9.7.3}$$

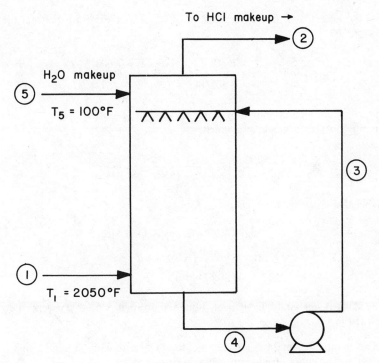

Figure 9.7.1. Quench spray tower.

Equation (9.7.1) now becomes

$$H_1 - H_2 + w_3 c_{P,H_2O}(T_3 - T_r) - w_4 c_{P,H_2O}(T_4 - T_r) + w_5 c_{P,H_2O}(T_5 - T_r) = 0 \quad (9.7.4)$$

By definition, for an adiabatic quench, $T_3 = T_4$. The makeup temperature is now set at $T_5 = 100°F$, and the exit vapor temperature is allowed to approach the inlet liquid by 2°F, that is, $T_2 = T_3 + 2$. The enthalpy balance becomes

$$H_1 - H_2 + w_5 c_{P,H_2O}(T_5 - T_r) = 0$$

$$H_{1,2050} - H_{2,(T_3+2)} + w_5 c_{P,H_2O}(100 - 32) = 0 \quad (9.7.5)$$

The enthalpy of the combustion gas is first calculated.

$$H_1 = (135,114)(0.26)(2050 - 32) + (9092)[1075 + 0.5(2050 - 32)]$$

$$= 89,893,340 \, \text{Btu/h}$$

T_3 (or T_2) is needed in order to calculate H_2. A trial-and-error calculation must be performed in order to obtain T_3. This is accomplished by assuming values of T_3 until Eq. (9.7.5) is satisfied. After several trials, a T_3 of 178.5°F is assumed so that $T_2 = 180.5°F$. Since Stream 2 is saturated with water, the mole fraction of water is given by the ratio of the vapor pressure of water at 178.5°F to the total pressure of the system.

$$y_{2, H_2O} = P'/P; \ P' \text{ at } 178.5°F = 7.603 \text{ psia}$$
$$= 7.603/14.7 = 0.5172$$

If z is the number of moles of water in the vapor outlet stream (Stream 2) and assuming no HCl removal in quench,

$$0.5172 = \frac{z}{z + (5101 - 505.1)}$$

Solving this equation for z gives

$$z = 4923.4 \text{ lbmol } H_2O \text{ is Stream 2}$$
$$w_{2, H_2O} = (4923.4)(18)$$
$$= 88,621 \text{ lb/h}$$

Note: This result could have been obtained from use of the psychrometric chart provided in Fig. 4.5.2.

The number of moles of water in Stream 5 is that in Stream 2 minus that in Stream 1. The mass of makeup water in Stream 5 is

$$w_5 = (4923.4 - 505.1)(18)$$
$$= 79,529 \text{ lb/h}$$

The enthalpy of Stream 2 is then

$$H_2 = (135,114)(0.25)(18.5 - 32) + 88,621[1075 + 0.5(180.5 - 32)]$$
$$= 106,863,400 \text{ Btu/h}$$

An enthalpy balance check on Eq. (9.7.5) is approximately accomplished for this assumed T_3 of 178.5°F. The composition of the effluent gas from the quencher is given in the following table:

	MW	(lb/h)	(lbmol/h)	Mole Fraction
CO_2	44	12,023	273.3	0.0287
H_2O	18	88,621	4923.4	0.5172
N_2	28	106,783	3813.7	0.4006
HCl	36.5	193	5.3	0.0005
O_2	32	16,115	503.6	0.0529
Total		223,735	9519.3	1.0
Average MW = 23.50				

One can now proceed to the heat duty (q) calculation. For adiabatic operation, this is approximately given by Eq. (9.7.6)

$$q = w_5(h_{2,\mathrm{H_2O}} - h_{5,\mathrm{H_2O}})$$
$$= 79,529[1075 + 0.5(180.5 - 32) - 1.0(100 - 32)] = (79,529)(1081.25)$$
$$= 85,990,731\,\mathrm{Btu/h} \tag{9.7.6}$$

The log-mean temperature difference across the quench unit is

$$\Delta T_{\mathrm{lm}} = \frac{[(T_1 - T_4) - (T_2 - T_3)]}{\ln[(T_1 - T_4)/(T_2 - T_3)]} = \frac{(1871.5 - 2.0)}{\ln(1871.5/2.0)}$$
$$= 273.3^\circ\mathrm{F}$$

With the material and energy balances complete, one can proceed to the physical design of the quencher. The recirculating liquid flow rate must be set. This information is also required to size the recirculating pump. Approximately 8–10% of the liquid entering at the top is vaporized. Select 9% liquid vaporization across the unit.

$$w_5 = 79,529\,\mathrm{lb/h}$$
$$w_3 = (79,529)(9) = 715,761\,\mathrm{lb/h}$$

The total liquid flow at the top of the tower is then 795,290 lb/h. At 178.5°F the liquid density and viscosity are 60.6 lb/ft^3 and 0.347 cP, respectively. The inlet vapor density at 2050°F (MW = 28.27) is 0.0154 lb/ft^3. The outlet vapor density at 180.5°F (MW = 23.50) is 0.0503 lb/ft^3.

Top Conditions (1)

The diameter of the spray tower is calculated from the superficial velocity (v_t) provided in Eq. (9.3.1).

$$v_t = 0.2\left(\frac{\rho_{\mathrm{L}} - \rho_{\mathrm{G}}}{\rho_{\mathrm{G}}}\right)^{0.5} \tag{9.3.1}$$

If this is evaluated at the gas exit (top),

$$v_t = 0.2\left(\frac{60.6 - 0.0503}{0.0503}\right)^{0.5} = 6.94\,\mathrm{ft/s}$$

The required area (A) and diameter (D) are then

$$A = \frac{223,735}{(3600)(0.0503)(6.94)}$$
$$= 178\,\mathrm{ft}^2$$
$$D \simeq 15\,\mathrm{ft}$$

At this diameter, the real area becomes $176.6\,\text{ft}^2$.

In order to calculate the tower height (Z), the volumetric heat transfer coefficient (U) is employed. If the liquid side resistance to heat transfer is small, as is the case with adiabatic humidification, U is equal to the hot gas film heat transfer coefficient (h) as given in Eq. (9.3.5).

$$h = \frac{0.043G^{0.8}L^{0.4}}{Z^{0.5}} \tag{9.3.5}$$

The terms G and L in Eq. (9.3.5) are superficial mass velocities for the gas and liquid, respectively, with units of pounds per hour per square foot (lb/hr-ft^2). Thus

$$G = 223{,}735/176.6 = 1267\,\text{lb/h-ft}^2$$

$$L = 795{,}290/176.6 = 4503\,\text{lb/h-ft}^2$$

Equation (9.3.5) is rearranged to solve for $hZ^{0.5}$

$$hZ^{0.5} = (0.043)(1267)^{0.8}(4503)^{0.4} = 377.7$$

The design equation for the spray tower volume (V) is given by Eq. (9.7.7).

$$V = AZ = \frac{qZ^{0.5}}{hZ^{0.5}\nabla T_{\text{lm}}}$$

$$= 176.6Z = \frac{85{,}990{,}731Z^{0.5}}{(377.7)(273.3)}$$

$$= 176.6Z = 833.1Z^{0.5} \tag{9.7.7}$$

$$Z^{0.5} = 4.72$$

$$Z = 22.25\,\text{ft}$$

$$V = 3930\,\text{ft}^3$$

Note: The volumetric heat transfer coefficient $(U$ or $h)$ is $79.6\,\text{Btu/h-ft}^3\text{-}°\text{F}$.

Bottom and Average Conditions (2,3)

The key calculated results for a design based on either bottom conditions or average (arithmetic) tower conditions—as well as the results for top conditions—are presented here in tabular form.

	Top	Bottom	Average
w_L (lb/h)	795,290	715,761	755,526
w_G (lb/h)	223,735	144,206	183,971
ρ_L (lb/ft^3)	60.5	60.5	60.5
ρ_G (lb/ft^3)	0.0503	0.0154	0.0329
μ_L (cP)	0.345	0.345	0.345

v_t (ft/s)	6.94	12.53	8.58
A (ft^2)	178	208	181
D (ft^2)	15(15)	16.3(17)	15.2(15)
A_{real} (ft^2)	176.6	227	176.6
L (lb/h-ft^2)	4503	3153	4278
G (lb/h-ft^2)	1267	635	1042
V (ft^3)	3930	12222	5598
Z (ft)	22.5	53.84	31.70
U (Btu/h-ft^3-°F)	79.6	25.7	56.2

A tower design based on average conditions, is recommended. Thus, a 15-ft-diameter column with approximately 30 ft between gas inlet and spray nozzles should be employed. Because of the assumptions made in the calculations, the design should allow for the installation of additional nozzles lower in the column and/or for the capability of increasing the water recirculation rate.

Example 9.7.2. With reference to Example 9.7.1, obtain an approximate design of this unit using the procedure outlined in Eqs. (9.3.13)–(9.3.15).

SOLUTION. The hot gas density (2050°F) is first calculated.

$$\rho = PM/RT = (1.0)(28.27)/(0.73)(2510)$$

$$= 0.0154 \, \text{lb/ft}^3$$

The tower area (A) and the diameter (D) are then

$$A = (0.2)(144{,}206)/(3600)(0.0154)$$

$$= 520 \, \text{ft}^2$$

$$D = 25.7 \, \text{ft}$$

The tower height is estimated from Eq. (9.3.15)

$$Z = 2.0D = 2.0(25.7)$$

$$= 51.4 \, \text{ft}$$

The results are in reasonable agreement with those in Example (9.7.1).

Example 9.7.3. It is proposed to cool the waste flue combustion gas in Example 9.7.1 by using ambient air at 70°F. Calculate the quantity (mass, mole, and volume basis) of air required to cool the gases to an acceptable temperature of 560°F.

SOLUTION. Under adiabatic conditions,

$$q_{flue} = q_{air}$$

$$q_{flue} = w_{flue} \overline{c}_P \Delta T = (144{,}206)(0.3)(2050 - 560)$$

$$= 64.5 \times 10^6 \, \text{Btu/h}$$

Since $q_{flue} = q_{air}$,

$$64.5 \times 10^6 = w_{air}(0.26)(560 - 70)$$

$$\text{or } w_{air} = 506{,}280 \,\text{lb/h}$$

$$= 17{,}458 \,\text{lb-mol/h}$$

$$= 6{,}743{,}800 \,\text{ft}^3/\text{h} \ (70°\text{F})$$

Note: The combined mass (and volume) flow of the gas is now significantly higher. This will adversely impact on the economics for the air pollution control equipment.

Example 9.7.4. Design the air quench tank in Example 9.7.3 if a 1.5-s residence time is required.

SOLUTION. The total mass flow rate of the gas is now

$$w_t = 144{,}206 + 506{,}280$$

$$= 650{,}486 \,\text{lb/h}$$

The volumetric flow at 560°F (assuming air) is

$$Q_a = w_t RT/PM = (650{,}486)(0.73)(1020)/(1.0)(29)$$

$$= 1.67 \times 10^7 \,\text{ft}^3/\text{h}$$

$$= 4640 \,\text{ft}^3/\text{s}$$

The volume of the tank is, therefore,

$$V_t = (4640)(1.5)$$

$$= 6960 \,\text{ft}^3$$

The physical dimensions of the tank are usually set by minimizing surface (materials) cost. The total surface is given by

$$S = 2(\pi D^2/4) + \pi DH$$

To minimize the total surface, the derivative of S with respect to D is set equal to zero.

$$\frac{dS}{dD} = \pi D + \pi H = 0 \quad \text{or} \quad D = H$$

The dimensions of the tank may now be calculated:

$$V_t = (\pi D^2/4)(H) = \pi D^3/4$$

$$D = 20.7\,\text{ft}$$

$$H = 20.7\,\text{ft}$$

Note: The size of the tank is excessive. This is another reason why air quenching is rarely used.

Example 9.7.5.   It is proposed to reduce the temperature of 10% of the 144,206 lb/h of flue gas described in Example 9.7.1 to 550°F, using a *solids contact* method that operates on a 1-h cooling cycle. Assume the average heat capacities of the flue gas and solid to be 0.26 and 0.88 Btu/lb-°F, respectively, over the temperature range in question. The initial temperature of the solids if 70°F. Assuming an approach temperature of 40°F, what mass of solid must be provided in order to cool the flue gas to the required temperature during each hour of operation.

SOLUTION. Set up an enthalpy balance on both the flue gas and solid.

$$\Delta H_{\text{flue}} = \Delta H_{\text{solid}}$$

For the flue gas, the enthalpy change for one hour of operation is

$$\Delta H_f = mc_P(T_2 - T_1) = (0.1)(144{,}206)(0.26)(2050 - 550)$$

$$= 5{,}623{,}800\,\text{Btu}$$

For the solids,

$$\Delta H_s = m_s(0.88)(510 - 70)$$

Since $\Delta H_s = \Delta H_f$,

$$m_s = 5{,}623{,}800/(0.88)(440) = 14{,}424\,\text{lb solids}$$

Example 9.7.6.   With reference to Example 9.7.1, design a radiative heat exchanger to cool the combustion gases from 2050 to 180°F. The ambient air temperature is 60°F and an overall heat transfer coefficient for the cooler of 1.5 Btu/h-ft²-°F may be assumed to apply.

SOLUTION. Based on the solution in Example 9.7.1,

$$q = (144{,}206)(0.3)(2050 - 180)$$

$$= 80.9 \times 10^6\,\text{Btu/h}$$

The log-mean temperature difference is given by

$$\Delta T_{lm} = [(2050 - 60) - (180 - 60)]/\ln[(2050 - 60)/(180 - 60)]$$
$$= 666°F$$

The radiative surface area required is

$$A = q/U\Delta T_{lm} = 80.9 \times 10^6/(1.5)(666)$$
$$= 80,980 \, ft^2$$

The volumetric flow (Q_a) at inlet conditions is given by

$$Q_a = 5100(379)(2050 + 460)/(60 + 460)$$
$$= 9,330,000 \, ft^3/h$$

Duct or pipe velocities of 60 ft/s for this type of application are typical. Assume this value to apply at inlet (2050°F) conditions. The duct area and diameter may now be calculated

$$A_d = (9,330,000)/(3600)(60)$$
$$= 43.2 \, ft^2$$
$$D = 7.4 \, ft$$

The length of required heat exchange ducting is then

$$L = A/\pi D$$
$$= 3475 \, ft$$

The practicality of this method of cooling is questionable.

PROBLEMS

1. For the system described in Example 9.7.1, propose a design if the quencher is a packed tower. Perform the calculation based on bottom conditions.
2. Using the simplified approximate procedure presented in Section 9.3, estimate the packing height and tower diameter if the waste flue gas flow rate in Example 9.7.1 is 166,500 lb/h.
3. Using ambient air at 60°F, calculate the quantity of dilution air required to cool the gases in Problem 2 to a temperature of 520°F.
4. Design the air quencher in Problem 2 if a 1.2-s residence time is recommended.
5. With reference to Problem 2 calculate the mass of 0.91 Btu/lb-°F heat capacity solids required to cool the gases to 180°F for one hour of operation. The initial temperature of the solids is 60°F, and an approach temperature of 20°F may be assumed.
6. With reference to Problem 2, design a radiative cooler to reduce the flue gas temperature from 2050 to 130°F. Ambient air is at 70°F.

REFERENCES

1. J. Fair, "Designing Direct Contact Coolers/Condensers," *Chem. Eng.*, 91–100, June 12 (1972).
2. J. Fair, "Process Heat Transfer by Direct Fluid-Phase Contact," *AIChE Heat Transfer Symposium*, Series No. 118, Vol. 68, 1–11, 1971.
3. R. Farrell, private communication, 1986.
4. J. Reynolds, personal notes, 1987.
5. T. Shen, U. Choi, and L. Theodore, *Hazardous Waste Incineration Manual*, U.S. EPA Air Pollution Training Institute, Research Triangle Park, NC, to be published.
6. W.H. McAdams, *Heat Transmission*, 2nd ed. McGraw-Hill, New York, 1942.
7. D.Q. Kern, *Process Heat Transfer*, McGraw-Hill, New York, 1950.

10

Air Pollution Control Equipment

10.1 INTRODUCTION

One major concern associated with hazardous waste incineration (HWI) is the emission of air pollutants. The greatest mass of air contaminants consists primarily of the following criteria pollutants: oxides of nitrogen, oxides of sulfur, and particulate matter. Trace levels of noncriteria pollutants such as chlorinated by-products, benzene, heavy metals, and acid gases are also of great concern. Available data indicate that criteria pollutant emissions from hazardous waste incinerators are about the same as those from various industrial combustion processes. The emission rates are dependent on the waste incineration rate, chemical composition of the waste, incinerator type, air pollution control equipment, and incinerator system operating parameters.

The primary end products in the combustion of hydrocarbons in air are mostly carbon dioxide (CO_2) and water vapor (H_2O). However, air pollutants may result from two sources when hazardous wastes are incinerated:

1. *Products of Incomplete Combustion (PICs)*. These result from the incomplete combustion of organic waste constituents and conversion of certain constituents present in the waste, auxiliary fuel, and/or combustion air. The PICs include carbon monoxide, carbon, hydrocarbons, aldehydes, amines, organic acids, polycyclic organic matter (POM), and any other waste constituents or their partially degraded products that escape complete thermal destruction in the incinerator. (In well-designed and well-operated incinerators, these incomplete combustion products are emitted only in trace amounts.)

2. *Combustion Products from Hazardous Wastes*. Hazardous waste compounds commonly contain other elements in addition to carbon, hydrogen, and oxygen. When such wastes or some auxiliary fuels are burned, those other elements form combustion products other than carbon dioxide and water vapor. These other products that are generally considered as air pollutants include:

330

- Hydrogen chloride (HCl) and small amounts of chlorine (Cl_2) from the incineration of chlorinated hydrocarbons.
- Acid halogens from the incineration of organic halogens.
- Sulfur oxides, mostly as sulfur dioxide (SO_2), but also including small amounts of sulfur trioxide (SO_3), formed from sulfur or sulfur compounds present in the waste and/or fuel mixture.
- Phosphorus pentoxide and/or phosphoric acid formed from the incineration of organophosphorus compounds.
- Nitrogen oxides (NO_x) from the nitrogen in the combustion air and/or from organic nitrogen compounds present in the waste.
- Particulate matter consisting of metal salts from the waste, metal oxides formed by combustion, and fragments of incompletely burned material (primarily carbonaceous).

If the incineration system is operated properly, 99.99% DRE of the hazardous wastes can be achieved. In addition, the PICs are usually not a problem and, of the combustion products in the second source just listed, only HCl and particulates are regularly encountered.

Regarding gaseous pollutants, Cl_2 is present in conjunction with HCl, but equilibrium favors HCl formation at the high temperatures employed in chlorinated waste incinerators (see Chapters 5 and 6). The NO_x emissions are not economically amenable to control by scrubbing or other post-generation removal techniques. If the waste does not contain nitrogen compounds, NO_x emissions can be minimized by low temperature combustion. The SO_x emissions are naturally a strong function of the sulfur content of the waste–fuel mixture. As with NO_x, this has not surfaced as an emission problem in HWI applications. Thus, the only gaseous emission of concern is HCl and the accompanying trace quantities of Cl_2. Federal regulation limits stack emission of HCl from an incinerator burning hazardous waste to no more than 4 lb/h, except that larger emission is permitted if the pollution control equipment removes 99% of the HCl in the combustion gas products. These gaseous pollutants are usually removed from the combustion gases by absorbers.

Regarding particulates, these emissions are strongly influenced by the chemical composition of the waste being incinerated and the auxiliary fuel, the type of incinerator and its operating parameters, and the air pollution control system. Particulate matter primarily consists of inert ash, various salts, and condensed gaseous contaminants. Most of the pollutants of concern, other than criteria pollutants such as heavy metals and organic toxic by-products, are collected as particulate matter. The metals and toxic by-products condense as or on fine particles as the exhaust gas stream cools. Concentrations of uncontrolled emissions of particulate matter from hazardous waste incinerators can normally be expected to be between 0.2 and 10 gr/dscf (grains per dry standard cubic feet) of air. Controlled concentrations typically range from 0.015 to 0.5 gr/dscf. Incinerators burning waste with high solids or ash content emit higher grain loadings and larger particle sizes. Most engineers in the HWI field define ash as the solid residue that remains after the waste–fuel mixture is incinerated. Ash produced during incineration is primarily inorganic and falls into two basic categories. (a) *Fly ash* consists of the ash that is entrained in exhaust gases leaving the incinerator, and which is usually captured in air pollution control equipment. (b) *Bottom ash* refers to the ash remaining in the combustion chamber after incineration and is normally associated with inerts. The composition of the ash depends on the composition of the waste being incinerated and can, therefore, vary greatly. Because hazardous waste incinerators are designed for complete destruction of these organic compounds, the ash normally contains very little carbonaceous material. Solid materials not susceptible to oxidation (e.g., glass or ceramic) constitute the major ash species. The amount of ash produced is very small in relation to total mass of waste incinerated. The relative proportion of fly ash to bottom ash is influenced by the waste composition and the incinerator design and operation. As expected, relatively little bottom ash and fly ash result when liquid or gaseous hazardous wastes are incinerated. It is common practice for purposes of emission calculations for liquid injection incinerators to assume that all the ash present in the waste–fuel mixture will *fly*, that is, will be carried out and appear in the flue gas.

There are four classes of air pollution equipment that are employed for particulate control: electrostatic precipitators, venturi scrubbers, ionizing wet scrubbers, and baghouses. The selection procedure for choosing one of these control devices depends on the inlet particulate loading, particle size distribution, potential acid removal capability, the gaseous control equipment employed (packed tower vs. dry scrubber), and (perhaps) regulatory requirements. In any event, the technologies for these devices are well established.[1]

In addition to discussing individual particulate and gaseous air pollution control equipment, this chapter will review dry scrubbing—a relatively new technology for emission control. This technique provides for capture of both gaseous and particulate pollutants.

Most HWI facilities today employ one of three possible process schemes for air pollution control:

1. Venturi scrubber (for particulates) followed by a packed tower absorber (for gases).
2. Ionizing wet scrubber (for particulates) combined with a packed tower (for gases).
3. Dry scrubber (for gases) followed by a baghouse—or possibly in the future, a dry electrostatic precipitator (for particulates).

Solid waste incinerators on the other hand, have traditionally employed either an electrostatic precipitator or a baghouse, although venturi scrubbers are occasionally used. When burning hazardous wastes the air pollution control equipment (APCE) is located between the incinerator and the induced fan or stack. If a quencher or waste heat boiler is employed, the APCE is located downstream of these units.

10.2 ELECTROSTATIC PRECIPITATORS

Electrostatic precipitator technology was developed in the United States by Dr. Fredrick Gardner at the turn of the century. Frederick Cottrell's precipitator was successfully applied in 1907 to the collection of sulfuric acid mist and shortly thereafter proven in a number of ore processing, chemical, and cement plants.

Electrostatic precipitators are satisfactory devices for removing small particles from moving gas streams at high collection efficiencies. They have been used almost universally in power plants for removing fly ash from the gases prior to discharge.

Although electrostatic precipitators have found limited application in treating gases from incinerators because of their high cost and the varying nature of the electrical resistivity of the particulate matter, they do have the capability of fine particulate control and may well find application in the future. Resistivity plays an important role in determining whether a particle can be readily collected in this device.

There are three classes of electrostatic precipitators:

- Dry electrostatic precipitators (ESP).
- Wet electrostatic precipitators (WEP).
- Ionizing wet scrubbers (IWS).

The ESP is by far the most popular; it has been used successfully to collect both solid and liquid particulate matter from many operations including smelters, steel furnaces, petroleum refineries, and utility boilers.

Dry High-Voltage Electrostatic Precipitators

Dry ESPs come in two varieties: (a) the high-voltage single-state unit and (b) the low-voltage two-stage unit. The low-voltage precipitator, where the particles are charged and collected in separate sections, is used only for small applications and is not generally employed at hazardous waste incineration facilities; this type will not be discussed here. The two major high-voltage ESP configurations currently used are tubular and plate. A tubular precipitator consists of a cylindrical collection electrode with the discharge electrode located on the axis of the cylinder, as schematically shown in Fig. 10.2.1. However, the vast majority of ESPs installed are of the plate type. Particles are collected on flat, parallel collecting surfaces spaced 8- to 12-in. apart, with a series of discharge electrodes spaced along the center line of adjacent plates, as shown schematically in Fig. 10.2.1. A typical arrangement of a commercial plate-type ESP is shown in Fig. 10.2.2. The gas to be cleaned passes horizontally between the plates or vertically up through the plates. Collected particles are usually removed by rapping and deposited in bins or hoppers at the base of the precipitator. Typical design parameters for plate-type precipitators are presented in Table 10.2.1.

Recently, for increased performance and reliability, precipitators have been divided into a number of independently energized bus sections. Each bus section has its own transformer rectifier, voltage-stabilization controls, and high-voltage conductors that energize the discharge electrodes within that section. The main reasons for sectionalization are to offset the dampening effects on corona power input of heavy flue dust loadings and to reduce the effect of bus section failures. These heavy flue gas dust loadings occur mainly in the inlet sections of a precipitator. By sectionalization, corona power input and particle charging can be increased in the inlet sections, thereby raising the overall precipitator efficiency.

The approach taken by industry to size ESPs for various applications makes use of the Deutsch–Anderson equation [Eq. (10.2.1)] or a modified form of it.

$$E = 1 - e^{-uA/Q_a} \tag{10.2.1}$$

where E = collection efficiency (dimensionless)
u = particle migration velocity (ft/s)
A = collecting area (ft^2)
Q_a = gas flow rate (ft^3/s)

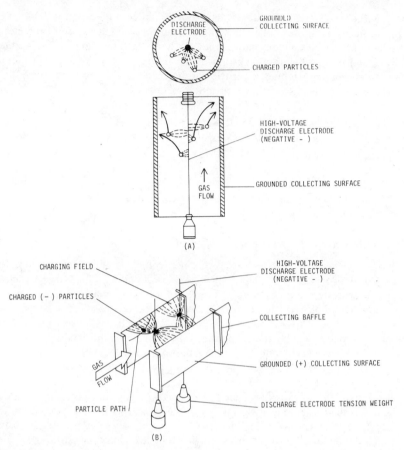

Figure 10.2.1. Electrostatic precipitator: (A) tubular-type precipitator; (B) plate-type precipitator.

In practice, the term u in Eq. (10.2.1) represents the *effective* migration velocity or precipitation rate parameter that is selected on the basis of experience with a particular dust. Since the desired collection efficiency and gas flow rate are usually specified, the required collecting area can be determined once an appropriate precipitation rate parameter has been chosen. Recently, better correlations with field data on high efficiency ESPs have been obtained by raising the exponential term in Eq. (10.2.1) to a power m using existing values of u. Typical values of m range between 0.4 and 0.7, with 0.5 as the norm.

The power required for a particular application is usually determined on an empirical basis, with past experience playing the key role. Performance curves for a given application typically relate power requirements to the efficiency and gas volume handled. The degree of sectionalization required is also determined on an empirical basis. The type and number of rappers for the collection and discharge electrodes depend on the properties of the dust and gas, the current densities, and

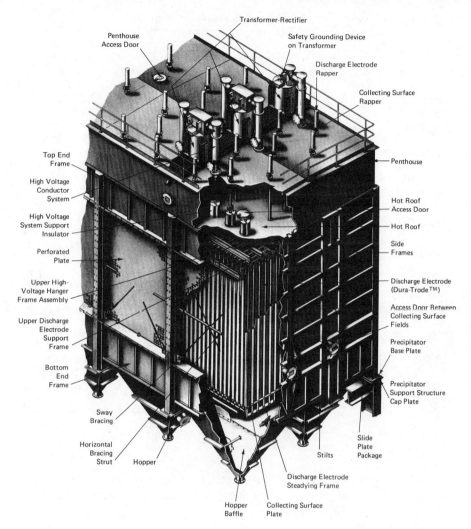

Figure 10.2.2. Plate-type electrostatic precipitator.

the configuration of the electrodes and electrode support structures. High-resistivity dust is usually more difficult to remove because of the increased electrical force holding it to the plate. Dust thickness buildup of 0.5 to 1.0 in. prior to dislodging is common. Typical rapper accelerations range from 10 to 100 times that due to gravity (g). With fly ash, for example, accelerations of 10 to 30 g's may be satisfactory, while highly resistive dust (difficult to remove) may require accelerations as high as 200 g's. Table 10.2.2 shows the number of rappers per unit area of collecting electrode and the number per unit length of discharge wire for various applications.

Depending on gas and dust conditions, and the required collecting efficiency, the gas velocities in an industrial ESP are between 2.5 and 8.0 ft/s. A uniform gas distribution is of prime importance for precipitator operation; it should be

TABLE 10.2.1. Typical Design Parameters for Electrostatic Precipitators

Parameter	Range of Values
Precipitation rate (effective migration velocity)	0.1–0.7 ft/s
Plate spacing	8–11 in.
Gas velocity	2–8 ft/s
Plate height	12–45 ft
Plate length	0.5–2.0 × height
Applied voltage	30–75 kV
Corona strength	0.01–1.0 mA/ft of wire
Field strength	7–15 kV/in.
Residence (treatment) time	2–10 s
Draft loss (pressure drop)	0.1–0.5 in. H_2O
Efficiency	to 99.9+%
Gas temperature	to 700°F (standard)
	1000°F (high temperature)
	1300°F (special)

TABLE 10.2.2. Typical Rapping Practices for Various Applications

Application	Collection Electrodes (rappers/1000 ft^2)	Discharge Electrodes (rappers/1000 ft)
Utilities	0.25–0.90	0.09–0.66
Pulp and paper	0.25–0.99	0.21–0.32
Metals	0.11–0.82	0.28–0.50
Cement	0.33–0.52	0.19–0.33

accomplished with a minimum expenditure of pressure drop. This is not always easy, since gas velocities in the duct ahead of the precipitator may be 30 to 100 ft/s in order to prevent dust buildup.

Wet Electrostatic Precipitators

The wet electrostatic precipitator (WEP) is a variation of the dry ESP design. The two major added features in a WEP system are

1. A preconditioning step, where inlet sprays in the entry section are provided for cooling, gas absorption, and removal of coarse particles.
2. A wetted collection surface, where liquid is used to continuously flush away collected materials.

Particle collection is achieved by the introduction of evenly distributed liquid droplets to the gas stream through sprays located above the electrostatic field sections, and migration of the charged particles and liquid droplets to the collection plates. The collected liquid droplets form a continuous downward-flowing film over the collection plates, and keep them clean by removing the collected particles.

The WEP overcomes some of the limitations of the dry ESP. Its operation is not influenced by the resistivity of the particles. Furthermore, since the internal components are continuously being washed with liquid, buildup of tacky particles is controlled and there is some capacity for removal of gaseous pollutants. In general, applications of the WEP fall into two areas: removal of fine particles and removal of condensed organic fumes. Outlet particulate concentrations are typically in the 10^{-3} to 10^{-2} gr/ft^3 range.[2] Data on the capability of the WEP to remove acid gases are very limited. This device has been installed to control HF emissions. Using a L/G ratio of 5 gal/1000 acf and a liquid pH between 8 and 9, fluoride removal efficiencies > 98% have been measured; outlet concentrations of HF were found to be < 1 ppm.[3]

At the present time, there are no WEP installations at HWI facilities. A potential application is the use of the WEP in conjunction with a low pressure-drop venturi scrubber upstream, where a major portion of the gaseous contaminants and heavy particles will be removed. The WEP will then serve as a second-stage control device for removal of the submicron particles and remaining gaseous pollutants. Because of its limited application history, extensive pilot testing prior to design and installation may be necessary.

Some of the advantages of a WEP include:

- Simultaneous gas absorption and dust removal.
- Low energy consumption.
- No dust resistivity problems.
- Efficient removal of fine particles.

Disadvantages of the WEP are the following:

- Low gas absorption efficiency.
- Sensitive to changes in flow rate.
- Dust collection is wet.

Ionizing Wet Scrubbers

The ionizing wet scrubber (IWS) is a relatively new development in the technology of removal of particulate matter from a gas stream. These devices have been incorporated in commercial incineration facilities.[4,5] In the IWS, high-voltage ionization in a charge section places a static electrical charge on the particles in the gas stream, which then passes through a crossflow packed-bed scrubber. The packing is polypropylene in the form of circular-wound spirals and gearlike wheel configurations, providing a large surface area. Particles with sizes of 3 μm or larger are trapped by inertial impaction within the packed bed. Smaller charged particles pass close to the surface of either the packing material or a scrubbing water droplet. An opposite charge in that surface is induced by the charged particle, which is then attracted to and attached to the surface. All collected particles are eventually washed out of the scrubber. The scrubbing water also functions to absorb gaseous pollutants. According to Ceilcote (the IWS vendor), collection efficiency of a two-stage IWS is greater than that of a baghouse or a conventional electrostatic precipitator for particles in the 0.2 to 0.6-μm range. For 0.8 μm and above, the efficiency of bag collectors is greater; at 1.8 μm and above, the ESP is as effective as the IWS.[2] Scrubbing water can include caustic soda or soda ash when needed for efficient absorption of acid gases. Corrosion resistance of the IWS is achieved by fabricating its shell and most internal parts from glass fiber reinforced plastic (FRP) and thermoplastic materials. Pressure drop through a single-stage IWS is around 5 in. H_2O (primarily through the wet scrubber section). All internal areas of the ionizer section are periodically deluge flushed with recycled liquid from the scrubber recycle system. The advantages of an IWS are those of the combination of a wet electrostatic precipitator and a packed-bed absorber. The disadvantages include the need for separation of particulates from the scrubbing medium, potential scaling and fouling problems (aggravated by recycle neutralizing solution), possible damage to the scrubber if scrubber solution pumps fail, and the need for a downstream mist eliminator. Despite some of these limitations, it is a particlate control device that has worked efficiently at a number of HWI facilities.

10.3   WET SCRUBBERS

Wet scrubbers have found widespread use in cleaning contaminated gas streams because of their ability to effectively remove both particulate and gaseous pollutants. Specifically, *wet scrubbing* describes the technique of bringing a contaminated gas stream into intimate contact with a liquid. In this section, the term *scrubber* will be restricted to those systems that utilize a liquid, usually water, to achieve or assist in the removal of particulate matter from a carrier gas stream.

The particulate collection mechanisms involved in the wet scrubbing operation may include some or all of the following:

- Inertial impaction
- Direct interception
- Diffusion (Brownian motion)
- Electrostatic forces
- Condensation
- Thermal gradients

The types of scrubbers for particulate control reviewed in this section are spray towers, packed-bed units, ionizing wet scrubbers, and venturi scrubbers. More extensive details are presented for the venturi since this control device is often used in HWI facilities for high efficiency particulate control.

The most common low-energy scrubbers are *gravity spray towers* in which liquid droplets are made to fall through rising exhaust gases and are drained at the bottom of the chamber (see Fig. 10.3.1). The droplets are usually formed by liquid atomized in spray nozzles. The spray is directed into a chamber shaped to conduct the gas through the finely divided liquid. In a vertical tower, the relative velocity between the droplets and the gas is eventually the terminal settling velocity of the droplets. To avoid spray droplet reentrainment, however, the terminal settling velocity of the droplets must be greater than the velocity of the rising gas stream. In practice, the

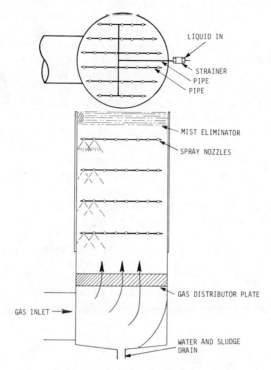

Figure 10.3.1. Spray scrubber.

vertical gas velocity typically ranges from 2–5 ft/s. For higher velocities, a mist eliminator must be used in the top of the tower.

Spray towers are suited for both particle collection and mass transfer (gas absorption), as are all wet scrubbers. Operating characteristics include low pressure drop (typically < 1–2 in. H_2O exclusive of mist eliminator and gas distribution plate), ability to handle liquids having a high solids content, and liquid requirements ranging from 3 to 30 gal/1000 ft^3 of gas treated. Spray towers are capable of handling large gas volumes and are often used as precoolers to reduce gas stream temperatures. Gas rates from 800 to 2500 lb/h-ft^2 are typical. Gas retention times within the tower typically range from 20 to 30 s. The chief disadvantage of spray towers is their relatively low scrubbing efficiency for particles in the 0 to 5-μm range.

Packed-bed scrubbers find normal application in the removal of pollutant gases and vapors. They will be considered in depth in a later section. To date, the major problems in using packed-bed scrubbers for particulate removal are plugging of the packing and the maintenance problems subsequently incurred. A general rule of thumb is to allow a maximum of 0.2 gr/ft^3 of dust to enter a packed tower.

To achieve high collection efficiency of particulates by *impaction*, a small droplet diameter and high relative velocity between the particle and droplet are required. In a *venturi scrubber* this is often accomplished by introducing the scrubbing liquid at right angles to a high-velocity gas flow in the venturi throat (vena contracta). Very small water droplets are formed, and high relative velocities are maintained until the droplets are accelerated to their normal velocity. Gas velocities through the venturi throat typically range from 12,000 to 24,000 ft/min. The velocity of the gases alone causes the atomization of the liquid. The energy expended in the scrubber is accounted for by gas stream pressure drop through the scrubber. Another factor important to the effectiveness of the venturi scrubber is the conditioning of the particulates by condensation. If the gas in the reduced-pressure region in the throat is fully saturated, or superheated (preferably), some condensation will occur on the particulates in the throat due to the cooling effect of the scrubbing liquid.

The venturi itself is only a gas conditioner and must be followed by a separating section for the elimination of entrained droplets (see Fig. 10.3.2). Water is injected into the venturi in quantities ranging from 6 to 15 gal/1000 ft^3 of gas. Very high collection efficiencies are achievable with operating pressure drops of 25 to 60 in. H$_2$O and are not at all uncommon. For example, a 10-in. H$_2$O pressure drop venturi can typically remove particles as small as a couple of microns with virtually 100% efficiency, while a 60-in.-H$_2$O pressure drop venturi is often required to remove particles as small as 0.3 to 0.4 μm. Since collection efficiency is directly related to pressure drop, variable-throat venturis have been introduced to maintain pressure drop with varying gas flows.

One of the more popular and widely used collection efficiency equations is that originally suggested by Johnstone et al.[6]

$$E = 1 - e^{-kR\psi^{0.5}}$$

(10.3.1)

where E = efficiency (fractional)
ψ = inertial impaction parameter (dimensionless)
R = liquid-to-gas ratio (gal/1000 acf or gpm/1000 acfm)
k = correlation coefficient, the value of which depends on the system geometry and operating conditions (typically 0.1–0.2 acf/gal)

The term ψ is given by

$$\psi = \frac{Cd_p^2\rho_p v_t}{9\mu_G d_1}$$

(10.3.2)

where d_p = particle diameter (ft)
ρ_p = particle density (lb/ft^3)
v_t = throat velocity (ft/s)
μ_G = gas viscosity (lb/ft-s)
d_1 = mean droplet diameter (ft)
C = Cunningham correction factor

Values for the Cunningham correction factor are provided in Table 10.3.1. This correction is usually neglected in scrubber calculations, but the effect becomes more pronounced as the particle size decreases, particularly below 1 μm. The mean droplet diameter, d_1, for standard air and water in a venturi scrubber is given by the Nukiyama and Tanasawa relationship:

$$d_1 = \frac{16,400}{v_t} + 1.45R^{1.5}$$

(10.3.3)

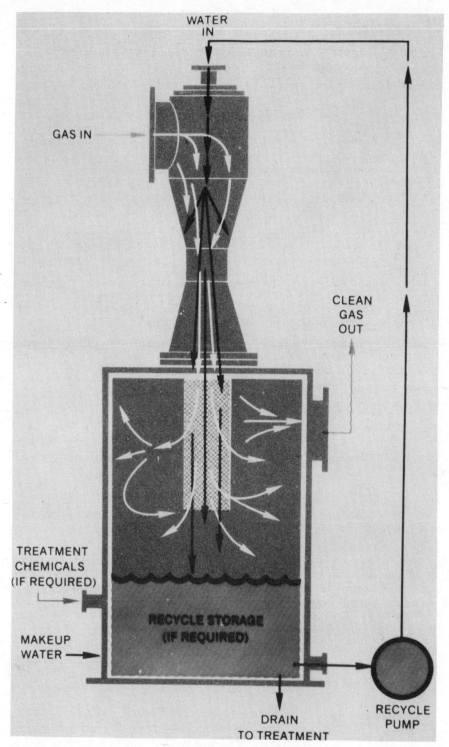

WATER IN

GAS IN

CLEAN GAS OUT

TREATMENT CHEMICALS (IF REQUIRED)

MAKEUP WATER

RECYCLE STORAGE (IF REQUIRED)

DRAIN TO TREATMENT

RECYCLE PUMP

Figure 10.3.2. Venturi scrubber.

TABLE 10.3.1. Cunningham Correction Factors for Air at Atmospheric Pressure

Particle Diameter (μm)	70°F	212°F	500°F
0.1	2.88	3.61	5.14
0.25	1.682	1.952	2.525
0.5	1.325	1.446	1.711
1.0	1.160	1.217	1.338
2.5	1.064	1.087	1.133
5.0	1.032	1.043	1.067
10.0	1.016	1.022	1.033

The pressure drop for gas flowing through a venturi scrubber can be estimated from knowledge of liquid acceleration and frictional effects along the wall of the equipment. Frictional losses depend largely on the scrubber geometry and are usually determined experimentally. The effect of liquid acceleration is, however, predictable. Equation (10.3.4) for estimating pressure drop through venturi scrubbers (given as a function of throat gas velocity and liquid-to-gas ratio) assumes that all the energy is used to accelerate the liquid droplets to throat velocity.

$$\Delta P' = 5 \times 10^{-5} v_t^2 R \tag{10.3.4}$$

where $\Delta P'$ = pressure drop (in. H_2O)

Another somewhat simpler equation that applies over a fairly wide range of R's is[7]

$$\Delta P^* = 0.8 + 0.12R \tag{10.3.5}$$

where ΔP^* is a dimensionless pressure drop equal to the pressure drop divided by a velocity head.

Studies by Hesketh[8] show that the pressure drop predictions obtained from throat velocity measurements may be subject to error at low velocities if Eq. (10.3.4) is applied for all ranges of velocities. Hesketh's equation for venturi scrubbers that have liquid injected before the throat[9] is given by

$$\Delta P' = \frac{v_t^2 \rho_G A_t^{0.133} R^{0.78}}{1270} \tag{10.3.6}$$

where ρ_G = gas density (lb/ft^3)
A_t = throat cross-sectional area (ft^2)

Contact power theory is an empirical approach relating particulate collection efficiency and pressure drop in wet scrubber systems. The concept is an outgrowth of the observation that particulate collection efficiency in spray-type scrubbers was mainly determined by pressure drop for the gas plus any power expended in atomizing the liquid. Contact power theory assumes that the particulate collection efficiency in a scrubber is solely a function of the total pressure loss in the unit. The total power loss (p_t) is assumed to be composed of two parts: the power loss of the gas passing through the scrubber (p_G) and the power loss of the spray liquid during atomization (p_L). These two terms are given here in equation form.

$$p_G = 0.157\Delta P' \tag{10.3.7}$$

where p_G = contacting power based on gas stream energy input (hp/1000 acfm)
$\quad \Delta P'$ = pressure drop across the scrubber (in. H_2O)

and

$$p_L = 5.83 \times 10^{-4} P_L R \tag{10.3.8}$$

where p_L = contacting power based on liquid energy input (hp/1000 acfm)
$\quad P_L$ = liquid inlet pressure (psia)
$\quad R$ = liquid-to-gas ratio (gal/1000 acf)

Then,

$$p_t = p_G + p_L$$

To correlate contacting power with scrubber collection efficiency, the latter is best expressed as the number of transfer units. The number of transfer units is defined by analogy to mass transfer and given by

$$N_t = \ln\left(\frac{1}{1 - E}\right) \tag{10.3.9}$$

where N_t = number of transfer units (dimensionless)
$\quad E$ = fractional collection efficiency (dimensionless)

The relationship between the number of transfer units and collection efficiency is by no means unique. The number of transfer units for a given value of contacting power (hp/1000 acfm) or vice versa, varies over nearly an order of magnitude. For example, at 2.5 transfer units ($E = 0.918$), the contacting power ranges from approximately 0.8 to 10.0 hp/1000 acfm, depending on the scrubber and the particulate.

For a given scrubber and particulate properties, there will usually be a very distinct relationship between the number of transfer units and the contacting power. Semrau plotted the number of transfer units (N_t) for a series of scrubbers and particulates against total power consumption; a linear relation on a log–log plot was obtained.[10] This relationship is independent of the type of scrubber and can be expressed by[11]

$$N_t = \alpha p_t^\beta \tag{10.3.10}$$

where α, β = parameters for the type particulates being collected (see Table 10.3.2)

TABLE 10.3.2. Constants for Use with Eq. (10.3.10)

Aerosol	Scrubber Type	α	β
Raw gas (lime dust and soda fume)	Venturi and cyclonic spray	1.47	1.05
Prewashed gas (soda fume)	Venturi, pipe line, and cyclonic spray	0.915	1.05
Talc dust	Venturi	2.97	0.362
	Orifice and pipe line	2.70	0.362
Black liquor recovery furnace fume			
Cold scrubbing water humid gases	Venturi and cyclonic spray	1.75	0.620
Hot fume solution for scrubbing (humid gases)	Venturi, pipeline, and cyclonic spray	0.740	0.861
Hot black liquor for scrubbing (dry gases)	Venturi evaporator	0.522	0.861
Phosphoric acid mist	Venturi	1.33	0.647
Foundry cupola dust	Venturi	1.35	0.621
Open-hearth steel furnace fume	Venturi	1.26	0.569
Talc dust	Cyclone	1.16	0.655
Copper sulfate	Solivore (A) with mechanical spray generator	0.390	1.14
	(B) with hydraulic nozzles	0.562	1.06
Ferrosilicon furnace fume	Venturi and cyclonic spray	0.870	0.459
Odorous mist	Venturi	0.363	1.41

When a combustion gas contains a high particulate loading, as well as one or more of the gaseous pollutants discussed previously, a venturi scrubber is often used in conjunction with a packed-bed or plate-tower scrubber. The venturi scrubber removes the particulates from the stream to prevent fouling of the packed-bed or plate-tower absorber, and may also remove a significant fraction of gases highly soluble in water. However, venturi scrubbers alone are not considered suitable for the removal of low solubility gases; when water is used as the scrubbing medium, estimated efficiencies are $<$ 50 to 70%.[5] Venturi scrubbers using water are not suitable for highly efficient (more than 99%) removal of either HCl or HF. Thus, although venturi scrubbers are usually designed for particulate collection, they can be used for simultaneous gas absorption as well. There is no satisfactory generalized design correlation for these types of scrubbers, especially when absorption with a chemical reaction is involved. Reliable design should be based on full-scale data or at least laboratory- or pilot-scale data.

Available data indicate that venturis at HWI facilities operate with pressure drops in the 30- to 60-in.-H_2O range. Liquid-to-gas ratios for venturi scrubbers are usually in the range of 5 to 20 gal/1000 ft^3 of gas. At existing HWI facilities, liquid-to-gas ratios ranging from 7 to 45 gal/1000 ft^3 of gas have been reported. In many cases, a minimum ratio of 7.5 gal/1000 ft^3 is needed to ensure that adequate liquid is supplied to provide good gas sweeping. Gas velocity data are not available at this time for venturi scrubbers operating at HWI facilities. Typical venturi throat velocities for other applications, however, are in the 100–400-ft/s range. The low end of this range, 100–150 ft/s, is typical of power plant applications, while the upper end of the range has been applied to lime kilns and blast furnaces.

If water is the scrubbing fluid for an incinerator burning chlorinated organic compounds, the wastewater effluent will contain suspended particulates, some dissolved HCl (hydrochloric acid), and other soluble constituents that may be present. The venturi scrubbing process also involves either a single pass or recirculation of the scrubbing fluid. If recirculation is used, scrubber fluid is recycled through the venturi scrubber until the TDS content reaches approximately 3%. When this occurs, a portion of the scrubbing fluid is removed (*blowdown*) and new scrubbing fluid is added to make up for the fluid lost as blowdown. The blowdown from the single pass or recirculation scrubbing system is neutralized (as needed) before delivery to on-site wastewater treatment processes, on-site storage facilities (e.g., vaporization ponds), municipal sewer, or a receiving water body.

10.4 BAGHOUSES

One of the oldest, simplest, and most efficient methods for removing solid particulate contaminants from gas streams is by filtration through fabric media. The fabric filter is capable of providing high collection efficiencies for particles as small as 0.5 μm and will remove a substantial quantity of those particles as small as 0.01 μm. In its simplest form the industrial fabric filter consists of a woven or felted fabric through which dust-laden gases are forced. A combination of factors results in the collection of particles on the fabric filters. When woven fabrics are used, a dust cake eventually forms; this, in turn, acts predominantly as a sieving mechanism. When felted fabrics are used, this dust cake is minimal or nonexistent and the primary filtering mechanisms are a combination of inertial forces, impingement, and so on. These are essentially the same mechanisms that are applied to particle collection on wet scrubbers, wherein the collection media is in the form of liquid droplets rather than solid fibers.

As particles are collected, the pressure drop across the fabric filtering media increases. Due in part to fan limitations, the filter must be cleaned at predetermined intervals. Dust is removed from the fabric by gravity and/or mechanical means. The fabric filters or bags are usually tubular or flat. The structure in which the bags hang is referred to as a *baghouse*; the number of bags in a baghouse may vary from a couple to several thousand. Quite often, when great numbers of bags are involved, the baghouse is compartmentalized so that one compartment may be cleaned while others are still in service.

The basic filtration process may be conducted in many different types of fabric filters in which the physical arrangement of hardware and the method of removing collected material from the filter media will vary. The essential differences may be related, in general, to the following:

- Type of fabric.
- Cleaning mechanism(s).
- Equipment geometry.
- Mode of operation.

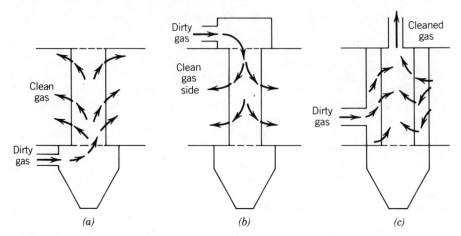

Figure 10.4.1. Baghouse filtration methods: (*A*) bottom feed; (*B*) top feed; (*C*) exterior filtration.

Depending on these factors the equipment will follow one of three systems as shown in Fig. 10.4.1. Bottom-feed units are characterized by the introduction of dust-laden gas through the baghouse hopper and then to the interior of the filter tube. In top-feed units, dust-laden gas enters the top of the filters to the interior or clean-air side. When the gas flow is from inside the bag to the outside, by virtue of the pressure differential, the internal area of the filter element will be open and self-supporting; unsupported filter elements are tubular. When the filtration process is reversed, with the gas flow from outside the bag to inside, it is necessary to support the media against the developed pressures so that the degree of collapse is controlled. Supported filter elements are either of the tubular or envelope shape.

Baghouse collectors are available for either intermittent or continuous operation. Intermittent operation is employed where the operational schedule of the dust-generating source permits halting the gas cleaning function at periodic intervals (regularly set by time or by pressure differential) for removal of collected material from the filter media (*cleaning*). Collectors of this type are primarily utilized for the control of small-volume operations such as grinding and polishing, and for aerosols of a very coarse nature. For most air pollution control installations and major dust control problems, however, it is desirable to use collectors that allow for continuous operation. This is accomplished by arranging several filter areas in a parallel flow system and cleaning one area at a time according to some preset mode of operation (see Fig. 10.4.2).

Baghouses may also be characterized and identified according to the method used to remove collected material from the bags. Particle removal can be accomplished in a variety of ways, including shaking the bags, blowing a jet of air on the bags from a reciprocating manifold, or rapidly expanding the bags by a pulse of compressed air. In general, the various types of bag-cleaning methods can be divided into those involving flexing and those involving a reverse flow of clean air. Figure 10.4.3 illustrates some of these fabric flexing cleaning methods.

There are two basic types of filtration that occur in commercial fabric filters; the first is referred to as *media* or *fiber filtration* and the second is *layer* or *cake filtration*. In fiber filtration the dust is retained on the fibers themselves by settling, impaction,

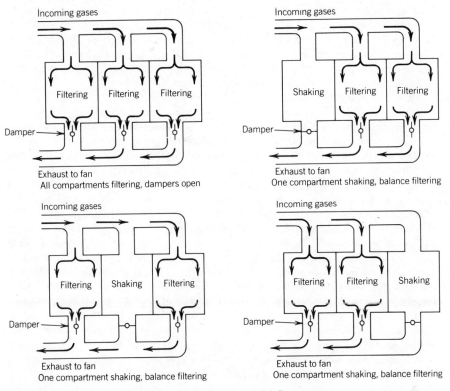

Figure 10.4.2. Baghouse parallel flow system.

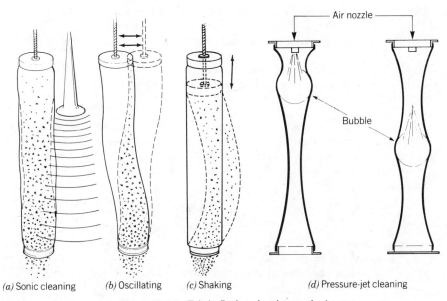

Figure 10.4.3. Fabric flexing cleaning methods.

interception, and diffusion. In cake filtration the fiber acts as a support on which a layer of dust is deposited to form a microporous layer capable of removing additional particles by sieving as well as other basic filtration mechanisms (impaction, interception, diffusion, settling, and electrostatic attraction). In practical industrial cloth filters, both methods occur, but cake filtration is the more important process after the new filter cloth becomes thoroughly impregnated with dust.

Perhaps the outstanding characteristic of fabric filters is their inherent very high collection efficiency with even the finest particles. These units usually have the capability of achieving efficiencies of 99% almost automatically—provided they are properly constructed and maintained in satisfactory operating condition.

The number of variables necessary to design a fabric filter is very large. Since fundamentals cannot treat all of these factors in the design and/or prediction of performance of a filter, this determination is basically left up to the experience and judgment of the design engineer. In addition, there is no one formula that can determine whether or not a fabric filter application is feasible. A qualitative description of the filtration process is possible, although quantitatively the theories are far less successful. Theory, coupled with some experimental data, can help predict the performance and design of the unit. The state of the art of engineering process design is the selection of filter medium, superficial velocity, and cleaning method that will yield the best economic compromise. Industry relies on certain simple guides and calculations, which are usually considered proprietary information, to achieve this. Despite the progress in developing pure filtration theory, and in view of the complexity of the phenomena, the most common methods of correlation are based on predicting a form of a final equation that can be verified by experiment.

Pressure drop will be considered first. The pressure drop equation for a fluid flowing through a fabric in a baghouse is given by

$$\Delta P = \Delta P(\text{fabric}) + \Delta P(\text{cake}) = K_1 v + K_2 W_i v^2 \theta \tag{10.4.1}$$

where ΔP = total pressure drop across both fabric and cake (in. H_2O)
v = superficial velocity through the bag/cake (ft/min)
W_i = inlet particulate loading in gas (lb/ft^3)
θ = elapsed time in filtering cycle (min)
K_1 = resistance coefficient for the bag (fabric) [(in. H_2O)-ft/min]
K_2 = resistance coefficient for the deposited dust (cake) [(in. H_2O)/(lb dust/ ft^2 cloth area)-(ft/min)]

Thus, a plot of ΔP versus θ yields a straight line. At $\theta = 0$, the only resistance to flow is that of the filter medium; the pressure drop, however, is a linear function of time, and as time increases, the resistance due to the dust cake may predominate.

An equation that can be used for determining the collection efficiency of a baghouse is[1]

$$E = 1 - e^{-(\psi L + \phi \theta)} \tag{10.4.2}$$

where ψ = constant based on fabric (ft^{-1})
ϕ = constant based on cake (s^{-1})
θ = time of operation to develop the cake thickness (s)
L = fabric thickness (ft)
E = collection efficiency (dimensionless)

The exit concentration (W_e) for the combined resistance system (the fiber and the cake) is

$$W_e = W_i e^{-(\psi L + \phi \theta)} \qquad (10.4.2a)$$

where W_e = exit concentration (lb/ft^3)
$\quad W_i$ = inlet concentration (lb/ft^3)

Design Practice

The design of industrial baghouse equipment requires consideration of many factors. The most important design considerations include the operational pressure drop, cloth area, cleaning mechanism, fabric and fabric life, baghouse configuration, and costs. Exhaust volume through the usual single-compartment fabric collector will not be constant because of the increasing resistance to air flow as the dust cake accumulates. The reduction in flow rate will be a function of the system pressure relationships, the exhaust fan characteristics, and the point of rating. Drop-off of exhaust volume is usually not severe in practice because pressure losses for the system of ducts and hoods usually equal or exceed that of the fabric collector; therefore a reduction in exhaust volume will cause a corresponding reduction in the pressure needs for that portion of the system. Most fabric collectors have the centrifugal fan on the clean-air side of the collector, where the more efficient backward-curved blade fan can be applied. This fan construction has a steeply rising pressure–volume characteristic, providing a significant increase in available static pressure as exhaust volume is reduced. This fan characteristic prevents overloading when collectors with new, clean media are first placed in service and where, for a short time, the pressure loss through the clean fabric is low. When fabric collectors are designed for continuous operation, pressure relationships become more complex, but approach a constant value. The smaller the fraction of the elements taken out of service at any one time for dust-cake removal, the more uniform the pressure drop of the fabric collector. For reverse-jet types of continuous collectors, minor pressure variations occur, depending on the position and direction of travel of the blow-back device. Pressure variations are minimal in pulse-jet designs because the cleaning cycle is extremely short and relatively few elements are taken out of service at any one time. Typical pressure drops for woven cloths range between 3 and 6 in. H$_2$O. However, the use of high-velocity filtration, felted fabrics, or the presence of a sticky or low-porosity dust cake often requires pressure drops on the order of 6 to 10 in. H$_2$O.

The size of a baghouse is primarily determined by the area of filter cloth required to filter the gases. The choice of a filtration velocity (or its equivalent, the *air-to-cloth ratio* (ACR) in cubic feet per minute of gas filtered per square foot of filter area) must take certain factors into consideration. Although the higher velocities are usually associated with the greater pressure drops, they also reduce the filter area required. Practical experience has led to the use of a series of ACR ratios for various materials collected and types of equipment. Ratios in current use range from < 1:1 to > 20:1. The choice depends on cleaning method, fabric, and characteristics of the particles. The rule of thumb for ACR ratios for conventional fabric filter baghouses with woven cloth is 1.5 to 3.0 ft^3/min-ft^2. Remember that during the cleaning cycle, the ACR is increased if compartmentalization is used with one compartment always off-stream. This increased value is then given by (ACR) $(N - 1)/N$, where N is the total number of compartments. When dust is fine or loadings are high, it is recommended that the filtering velocity be < 3 acfm/ft². Experience has also demonstrated that more cloth is required with increased temperature.

It is best to select the cleaning mechanism and the filter fabric together, since both items are closely related. For example, felted fabrics are almost exclusively cleaned by pulse or reverse-jet air, whereas most woven fabrics are cleaned by other means. By the process of elimination, a review of past successful filter operations will usually show that only a few cleaning mechanism–fabric combinations are compatible and sufficiently attractive to warrant economic evaluation.

Fabric deterioration often results from the combined assault of several factors (rather than from any single effect) such as thermal erosion, mechanical stress through repeated flexing, chemical attack, and abrasion. All possible modes of failure should be considered and again, previous experience is usually the best guide. For example, extrapolating from experience might show that the reduction in fiber life through thermal erosion might double for a 20°F rise in temperature or that the mechanical attrition rate doubles when the frequency of cleaning is doubled.

Selecting the number of compartments for a fabric filter installation requires information such as the allowable variation in gas flow, the availability of sizes of commercial units, and the expected frequency of maintenance. A single compartment unit is the most foolproof, predictable, and least expensive fabric collector design. It may be necessary, however, to provide additional compartments for emergencies, extended maintenance, or unexpected increases in process effluent.

The use of baghouses for particulate control is expected to increase in the near future. Many states are now requiring a dry scrubber (see Section 10.6) to control flue gas emissions. Although an ESP may be used downstream from the dry scrubber, the choice in all installations thus far has been the baghouse. Pulse-jet units with some type of acid-resistant glass bags is the preferred choice.

10.5 ABSORBERS

The process of absorption conventionally refers to the intimate contacting of a mixture of gases with a liquid so that part of one or more of the constituents of the gas will dissolve in the liquid. In a HWI facility, the contact usually takes place in some type of packed column. (The use of the packed column as a quenching device was discussed in Chapter 9. In this section, its use as a gas-cleaning device will be covered.)

Gas absorption as applied to the control of air pollution is concerned with the removal of one or more pollutants from a contaminated gas stream by treatment with a liquid. Consider, for example, the process taking place when a mixture of air and gaseous HCl is brought into contact with H_2O. The HCl is soluble in H_2O, and those molecules that come into contact with the water surface dissolve immediately. However, the HCl molecules are initially dispersed throughout the gas phase, and they can only reach the water by gas phase diffusion. When the HCl at the water surface has dissolved, it is distributed throughout the water phase by a second diffusional process. Consequently, the rate of absorption is determined by the rates of diffusion in both the gas and the liquid phases. Equilibrium is another important factor to be considered in controlling the operation of absorption systems. The rate at which the pollutant will diffuse into an absorbent liquid will depend on the

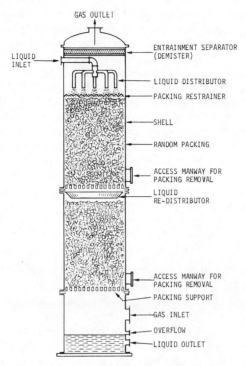

GAS OUTLET

LIQUID INLET

ENTRAINMENT SEPARATOR (DEMISTER)

LIQUID DISTRIBUTOR

PACKING RESTRAINER

SHELL

RANDOM PACKING

ACCESS MANWAY FOR PACKING REMOVAL

LIQUID RE-DISTRIBUTOR

ACCESS MANWAY FOR PACKING REMOVAL

PACKING SUPPORT

GAS INLET

OVERFLOW

LIQUID OUTLET

Figure 10.5.1. Countercurrent packed column.

departure from equilibrium that is maintained. The rate at which equilibrium is established is then essentially dependent on the rate of diffusion of the pollutant through the nonabsorbed gas and through the absorbing liquid.

Packed columns, as described in Chapter 9, are used for the continuous contact between liquid and gas. The countercurrent packed column (see Fig. 10.5.1) is the most common type of unit encountered in gaseous pollutant control for the removal of the undesirable gas, vapor, or odor. This type of column has found widespread application in both the chemical and pollution control industries. The gas stream containing the pollutant moves upward through the packed bed against an absorbing or reacting liquid that is injected at the top of the packing. This results in the highest possible efficiency. Since the pollutant concentration in the gas stream decreases as it rises through the column, there is constantly fresher liquid available for contact. This provides maximum average driving force for the diffusion process throughout the bed.

Liquid distribution plays an important role in the efficient operation of a packed column. A good packing from the process viewpoint can be reduced in effectiveness by poor liquid distribution across the top of its upper surface. Poor distribution reduces the effective wetted packing area and promotes liquid channeling. The final selection of the mechanism of distributing the liquid across the packing depends on the size of the column, type of packing, tendency of packing to divert liquid to column walls, and materials of construction for distribution. For stacked packing, the liquid usually has little tendency to cross distribution and thus moves down the column fairly uniformly in the cross-sectional area that it enters. In the dumped condition, most packings follow a conical distribution down the column, with the apex of the cone at the liquid impingement point. For well-distributed liquid flow and reduced channeling of gas and liquid to produce efficient use of the packed bed, the impingement of the liquid onto the bed must be as uniform as possible. The liquid coming down through the packing and on the inside wall of the column should be redistributed after a bed depth of ~ 3 column diameters for Rasching rings and 5 to 10 column diameters for saddle packing. As a guide, Raschig rings usually have a maximum of 10 to 15 ft of packing per section, while saddle packing can use a maximum of 12 to 20 ft. As a general rule of thumb, however, the liquid should be redistributed every 10 ft of packed height. The redistribution brings the liquid off the wall and outer portions of the column and directs it toward the center area of the column. Redistribution is usually not necessary for stacked bed packings, as the liquid flows essentially in vertical streams.

Packed columns are characterized by a number of features to which their widespread popularity may be attributed:

- *Minimum Structure*. The packed column needs only a packing support and liquid distributor about every 10 ft along its height.
- *Versatility*. The packing material can be changed by simply dumping it and replacing it with a type giving better efficiency, lower pressure drop, or higher capacity. The depth of packing can also be easily changed if efficiency turns out to be less than anticipated, or if feed or product specifications change.
- *Corrosive-Fluids Handling*. Ceramic packing is common and often preferable to metal or plastic because of its corrosion resistance. It may also be preferred when handling hot combustion gases, as in a hazardous waste incineration facility.
- *Low Pressure Drop*. Unless operated at very high liquid rates, where the liquid becomes the continuous phase as its films thicken and merge, the pressure drop per linear foot of packed height is relatively low.
- *Low Investment*. When plastic packings are satisfactory or when the columns are < 3 to 4 ft in diameter, cost is relatively low.

The packing is the heart of the performance of this type of equipment. Its proper selection entails an understanding of packing operational characteristics. The main points to be considered in choosing the column packing include:

- *Durability and Corrosion Resistance*. The packing should be chemically inert to the fluids being processed.
- *Free Space per Cubic Foot of Packed Space*. This controls the liquid hold-up in the column, as well as the pressure drop across it.
- *Wetted Surface Area per Unit Volume of Packed Space*. This affects the pressure drop across the column.
- *Packing Resistance to the Flow of Gas*. This affects the pressure drop across the column.
- *Packing Stability and Structural Strength to Permit Easy Handling and Installation*.
- *Weight per Cubic Foot of Packed Space*.
- *Cost per Square Foot of Effective Surface*.

The pressure drop through a packed column for any combination of liquid and gas flows in the operable range is an important economic consideration in the design of such columns. For most random packings, the pressure drop suffered by the gas is influenced by the gas and liquid flow rates. At constant gas rate, an increase in liquid throughput—which takes up more room in the packing (increased holdup) and so leaves less room for the gas (greater restriction)—is accompanied by an increase in pressure drop until the liquid flooding rate is reached. At this point, any slight excess that cannot pass through remains at the top of the packing, building up a deeper and deeper head (or pressure drop), hypothetically reaching an infinite value. Similarly, at constant liquid downflow, increasing gas flow is again accompanied by rising pressure drop until the flooding rate is reached, whereupon the slightest increase will cause a decline in permissible liquid throughput. This causes the remainder to again accumulate above the packing, so that pressure again increases. For a particular packing, the most accurate pressure drop data will be those available directly from the manufacturer. However, for the purposes of estimation, Fig. 10.5.4 is simple to use and usually gives reasonable results.

Some general rules of thumb in the design of packed columns do exist, but these must be applied discriminately because there are other considerations that might have to be taken into account, for example, allowable pressure drop, possible column height restrictions, and so on. For approximation purposes, if the gas rate is greater than about 500 acfm, a nominal packing size < 1 in. would probably not be practical; similarly, at about 2000 acfm, sizes < 2 in. would likely also be impractical. The nominal size of the packing should never exceed about $\frac{1}{20}$th of the column diameter. The usual practice is to design so that the operating gas rate is approximately 60% of the rate that would cause flooding. If possible, column dimensions should be in readily available sizes (i.e., diameters to the nearest $\frac{1}{2}$-ft and heights to the nearest foot). If the column can be purchased "off the shelf" as opposed to being specially made, a substantial savings can be realized.

Plate columns may also be employed as absorbers, although they are rarely used. These devices are essentially vertical cylinders in which the liquid and gas are contacted in stepwise fashion on plates or trays, in a manner shown schematically in Fig. 10.5.2. The liquid enters at the top and flows downward via gravity. On the way, it flows across each plate and through a downspout to the plate below. The gas passes upward through openings of one sort or another in the plate, then bubbles through the liquid to form a froth, disengages from the froth, and passes on to the next plate above. The overall effect is a multiple countercurrent contact of gas and liquid. Each plate of the column is a stage since on the plate the fluids are brought into intimate contact, interphase diffusion occurs, and the fluids are separated. The number of theoretical plates is dependent on the difficulty of the separation to be carried out and is determined solely from material balances and equilibrium considerations. The diameter of the column, on the other hand, depends on the quantities of liquid and gas flowing through the column per unit time. The *actual* number of plates required for a given separation is greater than the *theoretical* number due to plate inefficiency.

In bubble-cap plates, the vapor moves upward through *risers* into the bubble cap, out through the slots as bubbles, and into the surrounding liquid on the plates. Figure 10.5.3 demonstrates the vapor–liquid action for a bubble-cap plate. The bubble-cap plate design is the most flexible of plate designs for high and low vapor and liquid rates. On the average, plates are usually spaced approximately 24-in. apart.

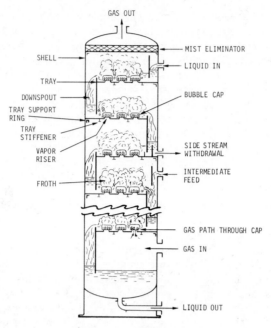

Figure 10.5.2. Plate column.

The *design* of the absorption tower is similar in some aspects to that for the water-quenching packed tower discussed in Section 9.3. Assuming that the liquid and gas superficial mass velocities (again as measured near the bottom of the packing where these quantities are at their maximum values) are known, Equation (9.3.7) can be used to determine the column diameter. The pressure drop may be evaluated directly from Fig. 10.5.4. The column height is given by Eq. (10.5.1).

$$Z = N_{OG} H_{OG} \qquad\qquad (10.5.1)$$

where N_{OG} = number of overall transfer units (dimensionless)
H_{OG} = height of overall transfer unit (ft)
Z = height of packing (ft)

In actual gas absorption design practice, the number of transfer units (N_{OG}) is obtained experimentally or calculated using any of the methods to be explained later in this section. The height of a transfer unit (H_{OG}) is also usually determined experimentally for the system under consideration. Information on many different systems using various types of packings has been compiled by the manufacturers of gas absorption equipment and should be consulted prior to design. The data are usually in the form of graphs depicting, for a specific system and packing, the H_{OG} versus the gas rate (in lb/h-ft²) with the liquid rate (in lb/h-ft²) as a parameter. The packing height Z is then simply the product of the H_{OG} and the N_{OG}.

In incineration operations, the pollutant to be absorbed (usually HCl) is in the very dilute range. If Henry's law applies (see Section 4.5), the number of transfer units is given by

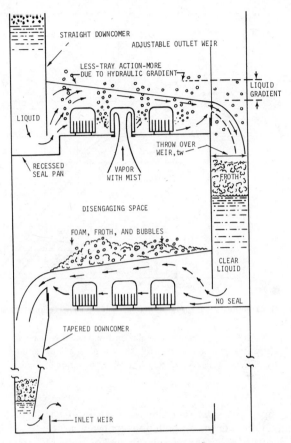

Figure 10.5.3. Bubble-cap plates.

$$N_{OG} = \ln\left[\left(\frac{y_1 - mx_2}{y_2 - mx_2}\right)\left(1 - \frac{1}{A}\right) + \frac{1}{A}\right]\bigg/\left(1 - \frac{1}{A}\right) \qquad (10.5.2)$$

where y_1 = inlet mole fraction of pollutant in vapor stream (dimensionless)
   y_2 = outlet mole fraction of pollutant in vapor stream (dimensionless)
   m = Henry's law constant (dimensionless)
   x_2 = outlet mole fraction of pollutant in liquid stream (dimensionless)
   A = absorption factor = L_m/mV_m, (dimensionless)
   L_m = liquid superficial mole velocity (lbmol/h-ft^2)
   V_m = vapor superficial mole velocity (lbmol/h-ft^2)

The solution to this equation can conveniently be found graphically from Fig. 10.5.5.

The term m represents the slope of the *equilibrium* line, which according to Henry's law should appear as a straight line on an x–y plot. A line describing the actual (x, y) operating points within the absorber may also be graphed on the same coordinates; this line is determined by means of a material (mol) balance around the

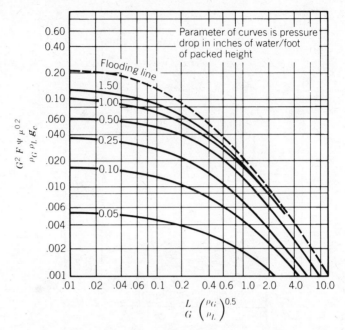

Figure 10.5.4. Generalized pressure drop correlation for packed towers: L = liquid rate (lb/s-ft^2); G = gas rate (lb/s-ft^2); ρ_L = liquid density (lb/ft^3); ρ_G = gas density (lb/ft^3); F = packing factor; μ = viscosity of liquid (cP); ψ = ratio of densities of water to liquid; and g_C = constant = 32.2 lb-ft/lb$_f$-s^2.

column. (The slope of the *operating* line is given by the ratio of the liquid-to-gas molar flow rates.)[1] For most air pollution applications where small quantities of pollutants are absorbed, both the equilibrium and operating lines may be assumed to be straight. When a highly soluble gas like HCl is absorbed or when a chemical reaction consumes the absorbate (pollutant), the value of m approaches zero, and Eq. (10.5.2) reduces to

$$N_{OG} = \ln\left(\frac{y_1}{y_2}\right) = \ln\left(\frac{w_1}{w_2}\right) \qquad (10.5.2a)$$

where w_1 = inlet mass flow rate of solute in vapor stream (lb/h)
$\quad\ w_2$ = outlet mass flow rate of solute in vapor stream (lb/h)

The pressure drop through a packed column, which is a function of tower size, flow rates, type of packing, and so on, is also an important design consideration for packed columns. The reader is again referred back to Section 9.3 for a discussion on this topic.

The most important design considerations for plate columns include calculation of the column diameter, type and number of plates to be used, actual plant layout and physical design, and plate spacing (which in turn, determines the column height).

The column diameter and consequently its cross section must be sufficiently large to handle the gas and liquid at velocities that will not cause flooding or excessive

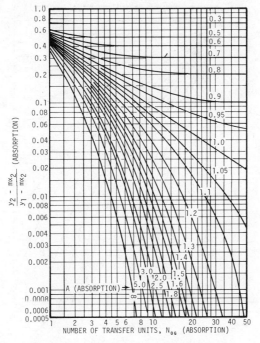

Figure 10.5.5. N_{OG} for absorption columns with constant absorption factor (A).

entrainment. The superficial gas velocity for a given type of plate at flooding is given by the relation

$$v_f = C_f[(\rho_L - \rho_G)/\rho_G]^{0.5} \qquad (10.5.3)$$

where v_f = gas velocity through the net column cross-sectional area for gas flow (ft^3/s-ft)

ρ_L, ρ_G = liquid and gas densities, respectively (lb/ft^3)

C_f = empirical coefficient that depends on the type of plate and operating conditions

The net cross section is the difference between the column cross section and the area taken up by downcomers and unused plate space. In actual design, some percentage of v_f is usually used—for nonfoaming liquids, 80 to 85% of v_f, and 75% or less for foaming liquids.

For cases where both the operating line and the equilibrium line curve may be considered straight (dilute solutions), the number of theoretical plates may be determined directly without recourse to graphical techniques. This will frequently be the case for a relatively dilute gas (as usually encountered in air pollution control) and liquid solutions where, more often than not, Henry's law is usually applicable. Since the quantities of gas absorbed are small, the total flows of liquid and gas entering and leaving the column remain essentially constant. Hence the operating line will be substantially straight. For such cases, the Kremser–Brown–Souders equation for determining the number of theoretical plates (N_p) applies.

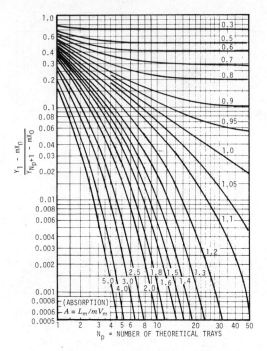

Figure 10.5.6. Number of theoretical stages for absorption columns with constant absorption factor (A).

$$N_p = \left[\left(\frac{y_{N_p+1} - mx_0}{y_1 - mx_0} \right) \left(1 - \frac{1}{A} \right) + \frac{1}{A} \right] \bigg/ \ln A \qquad (10.5.4)$$

where y_{N_p+1} = inlet mole fraction of solute in vapor stream
$\quad\quad\quad y_1$ = outlet mole fraction of solute in vapor stream
$\quad\quad\quad x_0$ = inlet mole fraction of solute in liquid stream
$\quad\quad\quad A$ = absorption factor = L_m/mV_m

Here mx_0 is the gas composition in equilibrium with the entering liquid (m is Henry's law constant = slope of equilibrium curve). If the entering liquid contains no solute gas, then $x_0 = 0$ and this equation can be simplified further. The solute concentration in the gas stream, y_{N_p+1} and y_1 represent inlet and outlet conditions, and L_m and V_m, the total molar flow rates of liquid and gas per unit column cross-sectional area. Equation (10.5.4) has been plotted in Fig. 10.5.6 for convenience and may be used for solution to this equation.

Water is often employed in absorbers to control emissions. When water is used as the scrubbing medium, an acidic stream is produced that must be neutralized prior to discharge, unless the HCl concentration in the incinerator off-gases is high enough to make HCL recovery economically attractive.[5] Hydrochloric acid concentration in the scrubbing liquor is normally limited to 1 to 2% by adjusting the make-up water and gas flow rates. However, higher acid concentrations can be achieved through some modest innovative design procedures that can include stage operations and liquid recycling. The scrubber must also be lined with or made of acid-resistant material.

Caustic solution (typically 18–20 wt% caustic soda in water) is also commonly used in packed-bed absorbers to control HCl and HF emissions. Because these compounds react with caustic, the driving force for mass transfer is increased and more efficient removal is achieved at the same liquid-to-gas ratio and packing depth. Neutralization is also achieved *in situ* if sufficient caustic is supplied for complete conversion of HCl to NaCl. Unlike water scrubbing, caustic scrubbing can also achieve high removal efficiencies for SO_2 and HBr, which are less soluble in water than HCL or HF. When gases such as SO_2 are being scrubbed, the caustic addition rate is adjusted to maintain an alkaline scrubbing medium. Alternatively, the caustic addition rate can be adjusted to substoichiometric levels. This reduces the scrubber water makeup rate needed to maintain a specified acid concentration in the scrubber liquor.

Lime slurry, typically 10–32 wt% $Ca(OH)_2$ in water, can also be used to control emissions of HCl, HF, and SO_2. However, lime slurries are not often used as the scrubbing liquid in packed-bed designs because of plugging problems. The use of lime slurries can lead to plugging of the spray nozzles and cause scale formation on the surfaces of the scrubber equipment, particularly scrubber internal and mist eliminator surfaces. The magnitude of the scaling problem will depend on the levels of HCl, HF, and SO_x in the incinerator exhaust gases. Lime solutions are used in plate tower scrubbers, however, because lime is less expensive than caustic. At several HWI facilities, venturi scrubbers with lime slurry injection as well as absorbers are used to control emissions of HCl.

Materials of construction are selected, in part, on the basis of the degree of acidity or alkalinity provided by the scrubbing liquid during operation. Deviation from the design pH condition or range may result in deterioration of the scrubber structure in contact with the liquid. Furthermore, maintenance of the pH design condition is important to scrubber liquid absorption efficiency when removing gaseous contaminants.

Quench water and scrubber (venturi and/or absorber) effluents are normally combined for treatment and ultimate disposal. Depending on the scrubbing liquids used and gaseous contaminants removed, the wastewater may contain chlorides, fluorides, sulfites, sulfates, phosphates, bromides, and bromates, as well as particulate matter. Liquid waste streams containing sodium fluoride can be treated with lime or limestone slurry to yield insoluble calcium fluoride. Sulfates and phosphates can also be readily removed from the wastewater stream because of the low solubility of their calcium salts. Therefore, treatment normally includes clarification (to remove particulates), neutralization (to take care of any residual acid or base that may still be present), and dilution (to help control TDS levels). Particulates that are insoluble in the scrubber fluid become suspended solids in the scrubber wastewater. If the particulates dissolve in the scrubber fluid, they contribute to the wastewater's TDS level. Suspended solids in scrubber wastewater generally present few, if any, problems[5] because their concentrations are usually < 2.2 gr/ft^3. Suspended solids are usually removed by on-site settling ponds. Overflow from settling ponds can be recycled to the scrubber. Wastewater with either high or low pH levels is usually neutralized by adding acid or base, as necessary, prior to the final discharge to a municipal sewer or receiving stream. A high concentration of total dissolved solids (NaCl, $CaCl_2$, and in some cases, the excess NaOH not used to neutralize HCl) is commonly reduced by piping scrubber effluents to in-plant treatment systems or by diluting with other plant process streams and storing in a holding pond or lagoon.

10.6 DRY SCRUBBERS

The success of fabric filters in removing fine particles from flue gas streams has encouraged the use of combined dry-scrubbing/fabric-filter systems for the dual purpose of removing both particulates and acid gases simultaneously. Dry scrubbers offer potential advantages over their wet counterparts, especially in the areas of energy savings and capital costs. Furthermore, the dry-scrubbing process design is relatively simple, and the product is a dry waste, rather than a wet sludge.

There are two major types of so-called dry scrubber systems: spray drying and dry injection. The first process is often referred to as a *wet–dry* system. When compared to a conventional wet scrubber, it uses significantly less scrubbing liquid. The second process may be referred to as a *dry–dry* system since no liquid scrubbing is involved. The spray drying system is predominantly used in utility and industrial applications. However, either process holds great promise in the HWI field. As discussed earlier, many states are now requiring spray drying units on proposed HWI facilities. Since there are essentially no meaningful data presently available on spray dryers in the HWI industry, the presentation to follow will focus on data and information available on flue gas desulfurization (FGD) applications in the utility industry.

The method of operation of the spray dryer is relatively simple, requiring only two major equipment items—a spray dryer similar to those used in the chemical food-processing and mineral-preparation industries, and a baghouse (or ESP) to collect the fly ash and entrained solids. In the spray dryer, the sorbent solution, or slurry, is atomized into the incoming flue gas stream to increase the liquid–gas interface and to promote the mass transfer of the SO_2 from the gas to the slurry droplets, where it is absorbed. Simultaneously, the thermal energy of the gas evaporates the water in the droplets to produce a dry powdered mixture of sulfite–sulfate and some unreacted alkali. Since the flue gas is not saturated and contains no liquid carry-over, potentially troublesome mist eliminators are not required. The SO_2 reacts with the lime slurry as follows:

$$SO_2 + Ca(OH)_2 \longrightarrow CaSO_3 + H_2O$$

$$2CaSO_3 + O_2 \longrightarrow 2CaSO_4$$

After leaving the spray dryer, the solid-bearing gas passes through the fabric filter (or ESP) where the dry product is collected, and where a percentage of the un-reacted alkali reacts with SO_2 for further removal. The cleaned gas is then discharged through the fabric-filter plenum to an induced draft (ID) fan and the stack.

Among the inherent advantages that the spray dryer enjoys over wet scrubbers are

- Lower capital costs.
- Lower draft losses.
- Reduced auxiliary power.
- Reduced water consumption, with liquid-to-gas ratios (L/G) significantly lower than those of wet scrubbers.
- Continuous, two-stage operation, from liquid feed to dry product.

Because the spray dryer operates near the flue-gas saturation point, no reheat is required, and there is little effect of water-vapor plume problems.

The sorbent of choice for most spray-dryer systems is a lime slurry. One system under development uses a sodium carbonate solution. Although the latter will generally achieve a higher level of SO_2 removal than a lime slurry at similar operating conditions, the significant cost advantage that lime has over sodium carbonate makes it the overwhelming favorite. Also, when sodium alkalis are used, the products are highly water soluble and may create disposal problems.

When compared to limestone, lime has the higher SO_2-removal efficiency (85 versus 60% for limestone, at a stoichiometric ratio of 1.0). Hand in hand with removal efficiency is the concept of alkali utilization; increased utilization allows for lower stoichiometric ratios and generates less waste product. While 100% utilization is attainable with lime by using performance-enhancing modifications, such as sorbent recycle and hot-gas bypass, limestone typically gives 60–70% utilization.

Spray-dryer design can be affected by the choice of the particulate-collection device. Here, fabric filters have an inherent advantage over ESPs in that a percentage of unreacted alkali in the collected waste on the bag surface can react with the remaining SO_2 in the flue gas. This percentage depends mainly on the temperature of the gas entering the fabric filter; in some cases it can be as much as 10%. One disadvantage of using a fabric filter is that, since the fabric is somewhat sensitive to wetting, a margin above saturation temperature (about 25–35°F) must be kept for bag protection. Increased utilization allows for lower stoichiometric ratios, and generates less waste products.

Limitations to the spray-dryer system do exist, of course. Efficiencies of 85% can be obtained with good sorbent utilization, but anything > 85% removal efficiency requires too high a consumption of expensive alkali to be economical. Also, the higher L/G ratios required at higher sulfur content can result in cooling of the gas below the acid dew point.

Dry injection processes generally involve pneumatic introduction of a dry, powdery alkaline material, usually a sodium-base sorbent, into the flue gas stream with subsequent fabric-filter collection. The injection point in such processes can vary from the boiler-furnace area all the way to the flue gas entrance to the baghouse, depending on operating conditions and design criteria.

The first U.S. studies of the dry-injection method were made with a dry-powdered nahcolite (a mineral form of sodium bicarbonate) injected both into the flue gas stream and onto the filter bags for SO_2 removal. The system chemistry is based on the following reaction:

$$Na_2CO_3 + SO_2 \longrightarrow Na_2SO_3 + CO_2$$

Efficiencies of 90% with 60% sorbent utilization were achieved in pilot-plant tests on flue gas from a lignite-fired boiler. This was encouraging, but the outlook for ready availability of nahcolite was not. Also, as was the case with spray-dryer systems, the waste solids from this sodium-based process are water soluble, posing potential disposal problems. There are some data available on HCl emission control with dry scrubbers. Interestingly, ETS, Inc., an air pollution consulting firm with headquarters in Roanoke, Virginia, has successfully installed a dry reactor system as a pilot plant on a slip stream of flue gas from a PCB incinerator, removing more than 98% of the HCl emissions. The reactor system consists of a dry reactor and baghouse. Flue gas is diverted cyclonically into the reactor where the rotating slinger unit delivers the dry reactant (either hydrated lime or a sodium-based compound) perpendicularly by to the flue gas flow, thus creating maximum mixing and intimate contact with the gaseous pollutants. The gaseous pollutants combine with the alkaline reactant forming dry particles. This dry material is then collected in a baghouse. This system was originally designed to remove SO_2 emissions from boiler flue gas. (ETS, Inc., has also installed a spray cooler and baghouse pilot plant on a glass melting furnace to remove both gaseous and particulate arsenic emissions.)

10.7 ILLUSTRATIVE EXAMPLES

Example 10.7.1.   A horizontal parallel-plate ESP consists of a single duct 24 ft high and 20 ft deep with an 11-in. plate-to-plate spacing. A collection efficiency of 88.2% is obtained with a flow rate of 4200 acfm. The inlet loading is 2.82 gr/ft^3. Calculate the following:

1. The bulk velocity of the gas (assume a uniform distribution.)
2. The outlet loading.
3. The drift velocity for this system.
4. A revised collection efficiency if the flow rate is increased to 5400 acfm.
5. A revised collection efficiency if the plate spacing is decreased to 9 in.

SOLUTION

1. The bulk velocity is given by $v = Q_a/A_d = 4200/24(\frac{11}{12}) = 191$ ft/min, where $A_d =$ duct cross-sectional area (ft^2).
2. The outlet loading is $2.82(1 - 0.882) = 0.333$ gr/ft^3.
3. ϕ is first calculated from Eq. (10.2.1).

$$E = 1 - e^{-\phi}$$
$$0.882 = 1 - e^{-\phi}$$

Solving for ϕ,

$$\phi = 2.14$$

Since

$$\phi = uA/Q_a$$
$$2.14 = (u)(24)(20)(2)/(4200/60)$$
$$u = 0.156 \text{ ft/s}$$

4. If $Q_a = 5400$ acfm, then a new ϕ can be calculated assuming the same drift velocity.

$$\phi = (0.156)(24)(20)(2)/(5400/60) = 1.67$$

Calculate the revised collection efficiency.

$$E = 1 - e^{-1.67} = 0.812$$

5. Since Q_a, u, and A are all constant, the Deutsch–Anderson equation predicts that the efficiency does not change if the plate spacing is 9 in.

Example 10.7.2. A duct-type electrostatic precipitator is to be used to clean 100,000 acfm of a gas stream containing particulates from a hazardous waste incinerator. The proposed precipitator consists of three bus sections (fields) arranged in series, each with the same collection surface. The inlet loading has been measured as 40 gr/ft^3 and a maximum outlet loading of 0.08 gr/ft^3 is allowed by local EPA regulations. The drift velocity for the particulates has been experimentally determined in a similar incinerator installation with the following results:

First section (inlet)   0.37 ft/s.
Second section (middle)   0.35 ft/s.
Third section (outlet)   0.33 ft/s.

1. Calculate the total collecting surface required based on the *average* drift velocity and the required total efficiency.
2. Find the total mass flow rate (lb/min) of particulates captured by each section using these drift velocities.

SOLUTION

1. Calculate the required total collection efficiency based on the given inlet and outlet loading.

$$E = 1 - [(\text{outlet loading})/(\text{inlet loading})] = 1 - [(0.08)/(40)] = 0.998$$

$$= 99.8\%$$

Calculate the average drift velocity (u).

$$u = (0.37 + 0.35 + 0.33)/3$$

$$= 0.35 \text{ ft/s}$$

Calculate the total surface area required using the following equation for the electrostatic precipitator [from Eq. (10.2.1)]

$$A = -\ln(1 - E)/(\mu/Q_a)$$

where A = total surface area
E = collection efficiency
u = drift velocity (average)
Q_a = gas volumetric flow rate

$$A = -\ln(1 - 0.998)/(0.35 \times 60/100{,}000)$$

$$= 29{,}593 \text{ ft}^2$$

2. Calculate the collection efficiencies of each section. Assume that each section has the same surface area.

$$E_1 = 1 - \exp(-Au_1/3Q_a) = 1 - \exp[-(29{,}593)(0.37)/3(100{,}000/60)] = 0.888$$

$$E_2 = 1 - \exp(-Au_2/3Q_a) = 0.874$$

$$E_3 = 1 - \exp(-Au_3/3Q_a) = 0.858$$

Calculate the mass flow rate of particulates captured by each section using the collection efficiencies calculated previously.

$$w_1 = (E_1)(\text{inlet loading})(Q_a)$$
$$= 3.552 \times 10^6 \, \text{gr/min}$$
$$= 507.4 \, \text{lb/min}$$

$$w_2 = (1 - E_1)(E_2)(\text{inlet loading}) \, (Q_a)$$
$$= 3.915 \times 10^5 \, \text{gr/min}$$
$$= 55.9 \, \text{lb/min}$$

$$w_3 = (1 - E_1)(1 - E_2)(E_3)(\text{inlet loading}) \, (Q_a)$$
$$= 4.843 \times 10^4 \, \text{gr/min}$$
$$= 6.9 \, \text{lb/min}$$

Example 10.7.3. A venturi scrubber is employed to reduce the discharge of particulates from the stack of a hazardous waste incineration facility to the atmosphere. The unit is presently treating 80,000 acfm of gas with a concentration of 8.2 gr/ft^3 of particulates and operating at a pressure drop of 5 in. H$_2$O. Experimental studies have yielded the following particle-size collection efficiency data:

Particle Diameter (μm)	Collection Efficiency (%)
20	58
30	76
50	91
75	94
100	98.9
150	99.4
200	99.9+

Determine the overall collection efficiency of the venturi scrubber assuming the following particle size distribution: 2% by weight of 20-μm particles; 3%, 30μm; 7%, 50μm; 10%, 75μm; 13%, 100μm; 19%, 150μm; 46%, 200μm.

SOLUTION. The calculations are presented in tabular form:

Particle Diameter (μm)	Weight Fraction (x_i)	(E_i)	Collection Efficiency ($x_i E_i$)
20	0.02	58	1.16
30	0.03	76	2.28
50	0.07	91	6.37
75	0.10	94	9.40
100	0.13	98.9	12.85
150	0.19	99.4	18.89
200	0.46	99.9	46.00

$$E = 96.95\%$$

The overall collection efficiency is 96.95%.

Example 10.7.4. A fly ash laden gas stream is to be cleaned by a venturi scrubber using a L/G ratio of 8.5 gal/1000 ft^3. The efficiency can be calculated from Eq. (10.3.1).

$$E_i = 1 - e^{-kR\psi^{0.5}}$$

where E_i = fractional collection efficiency of particle size d_{pi}
    R = liquid-to-gas ratio (gal/1000 acf)
    k = a constant (1000 ft^3/gal)
    ψ = inertial impaction number (dimensionless)

The fly ash has a particle density of 43.7 lb/ft^3 and $k = 200$ ft^3/gal. The throat velocity is 272 ft/s and the gas viscosity is 1.5×10^{-5} lb/ft-s. The particle size distribution is

d_{pi} (μm)	% by Weight
< 0.10	0.01
0.1– 0.5	0.21
0.6– 1.0	0.78
1.1– 5.0	13.0
6.0–10.0	16.0
11.0–15.0	12.0
16.0–20.0	8.0
> 20.0	50.0

Make use of the Nukiyama and Tanasawa relationship and neglect the Cunningham correction factor.

SOLUTION. Use the Nukiyama and Tanasawa relationship [Eq. (10.3.3)] to calculate to the mean droplet diameter (d_1).

$$d_1 = (16{,}400/v_t) + 1.45R^{1.5}$$

where v_t = throat velocity (ft/s)
R = liquid-to-gas ratio (gal/1000 ft^3)
d_1 = mean droplet diameter (μm) = $(16{,}400/272) + 1.45(8.5)^{1.5}$
= 96.2 μm

Express the inertial impaction number (ψ) in terms of d_p using Eq. (10.3.2) (assuming $C = 1$)

$$\psi = \frac{d_{pi}^2 \rho_{pi} v_t}{9 \mu_G d_1}$$

$$= d_{pi}^2 (43.7)(272)/[(9)(1.5 \times 10^{-5})(96.2)(3.048 \times 10^5)]$$

$$= 3.003 d_{pi}^2 \ (d_{pi} \ \text{in} \ \mu\text{m})$$

where d_{pi} = particle diameter (average for the ith size range)
ρ_{pi} = particle density
v_t = throat velocity
μ_G = gas viscosity
d_1 = mean droplet diameter

Make sure to use consistent units. There are $3.048 \times 10^5 \ \mu$m/ft.
Obtain the individual collection efficiency in terms of d_{pi} using Eq. (10.3.1).

$$E_i = 1 - e^{(-kR\psi_i^{0.5})}$$

$$= 1 - e^{-(2.94)d_{pi}}$$

Note: The numerical value of k in this equation is 0.2.

Comment: The previously given collection efficiency is referred to in industry as the size efficiency or grade efficiency equation for the system. One can use this type of an equation to calculate efficiencies for different particle sizes. (Similar equations also are available for other particulate control devices). These individual collection efficiencies are used to calculate an overall efficiency. This is demonstrated in the next step.
Calculate the overall collection efficiency by completing the following table.

d_p (μm)	E_i	x_i (%)	$x_i E_i$ (%)
0.05	0.1367	0.01	0.001367
0.30	0.586	0.21	0.123
0.80	0.905	0.78	0.706
3.0	0.9998	13.0	12.998
8.0	0.9999	16.0	15.999
13.0	0.9999	12.0	12.000
18.0	0.9999	8.0	8.000
20.0	0.9999	50.0	50.000
		$E = \Sigma x_i E_i = $	99.928%

Comment: The Johnstone equation assumes particle capture to occur by inertial impaction, that is, the particles, because of their inertia, strike the liquid droplets and are captured. This does not apply to submicron particulates. However, another collection mechanism defined as *molecular diffusion* comes into play. The Cunningham correction factor should be included in the analysis if a significant quantity of these very small particles are in the gas stream. Generally, venturi scrubbers are not overly effective with submicron particles.

Example 10.7.5.   A vendor proposes to use a spray tower in a hazardous waste incineration operation to reduce the discharge of solids to the atmosphere. The inlet loading of the gas stream from the kiln is 5.0 gr/ft^3 and is to be reduced to 0.05 in order to meet state regulations. The vendor's design calls for a water pressure drop of 80 psi and a pressure drop across the tower of 5.0 in. H$_2$O. The gas flow rate is 10,000 acfm and a water rate of 50 gal/min is proposed. Assume the contact power theory to apply.

1. Will the spray tower meet regulations?
2. What total power loss is required to meet regulations?
3. Propose a set of operating conditions that will meet the standard. The maximum gas and water pressure drop across the unit are 15 in. H$_2$O and 100 psi, respectively.
4. What conclusions can be drawn concerning the use of a spray tower for this application?

SOLUTION

1. Using Eqs. (10.3.7)–(10.3.9),

$$N_t = \ln(1/1 - E)$$

The term p_t is calculated as follows:

$$p_t = p_G + p_L$$
$$p_G = 0.157\Delta P' = 0.157(5) = 0.785$$
$$p_L = 5.83 \times 10^{-4} \, P_L R$$
$$= 5.83 \times 10^{-4} \, (80)(50/10) = 0.233$$

Therefore,

$$p_t = 1.018 \, \text{hp}/1000 \, \text{acfm}$$

For a HWI dust and/or ash, Table 10.3.2 gives $\alpha = 1.47$ and $\beta = 1.05$. Thus,

$$N_t = 1.47(0.18)^{1.05} = 1.50$$

$$1.5 = \ln(1/1 - E)$$

$$E = 77.7\%$$

Calculate the required efficiency.

$$E = (5.0 - 0.05)/5.0 = 0.99 = 99\%$$

Since the regulations require 99% efficiency, the spray tower will not meet the regulations.

2. Calculate p_t at the required efficiency

$$E = 0.99$$
$$N_t = \ln(1 - 0.99) = 4.605$$
$$4.605 = 1.47(p_t)^{1.05}$$
$$p_t = 2.96 \, hp/1000 \, acfm$$

3. Assume the maximum gas and water pressure drops across the unit to be 15 in. H_2O and 100 psi, respectively. Calculate p_G and p_L.

$$p_G = 2.36$$
$$p_L = 0.60$$

Calculate R in gal/1000 acf.

$$R = p_L/[5.83 \times 10^{-4} \, P_L] = 0.6/[5.83 \times 10^{-4}(100)] = 10.3 \, gal/1000 \, acf$$

Determine the new water flow rate.

$$(10.3 \, gal/1000 \, acf)(10,000 \, acfm) = 103 \, gal/min$$

4. The unit has limited applicability, at best, for high collection efficiency operations.

Example 10.7.6. It is proposed to install a pulse-jet fabric filter system to clean a 10,000 scfm (32°F) air stream at 250°F, containing 4.0 gr/ft^3 of pollutant from a hazardous waste incinerator facility. To reduce the outlet loading to below 0.08 gr/dscf, an average ACR of 2.5 cfm/ft^2 cloth is required. The following information, given by filter bag manufacturers, is available at the beginning of the selection process:

	Filter Bag			
	A	B	C	D
Tensile strength	Excellent	Above average	Fair	Excellent
Recommended maximum operation temperature (°F)	260	275	260	220
Resistance factor	0.9	1.0	0.5	0.9
Relative cost/bag	2.6	3.8	1.0	2.0
Standard size	8 in. × 16 ft	10 in. × 16 ft	1 ft × 16 ft	1 ft × 20 ft

1. Determine the filtering area required for this operation.
2. Based on the required area and this information, select the most suitable filter bag and calculate the number of bags that should be used.

SOLUTION

1. $Q_a = 10,000(250 + 460)/(460 + 32) = 14,430\,\text{acfm}$

Filtering area $= 14,430/2.5 = 5772\,\text{ft}^2$
2. D is not good since the maximum temperature is 220°F.
C is not good because we need either excellent or above average tensile strength
for pulse units. Therefore only A or B is acceptable.
Calculate the area/bag for bag A and bag B.

Area/bag for A $= (\pi)(8/12)(16) + (18/2)^2(\pi/4) = 34\,\text{ft}^2$
Area/bag for B $= (\pi)(10/12)(16) + (10/12)^2)(\pi/4) = 42\,\text{ft}^2$
Compare the relative cost of bags A and B.

Bag Type	Area/Bag	No. of Bags	Relative Cost
A	34.0	170	442
B	42.0	138	524

Type A bags should be used.

Example 10.7.7. A hazardous waste incinerator plant has an exhaust flow of
175,000 acfm following a dry scrubber with a loading of 6.0 gr/ft³. The ACR is 8 and
the efficiency is 99.3%. The allowable pressure drop is 10 in. H₂O. How many
18 in. × 20 ft bags are needed? What is the frequency of cleaning if the vendor's
equation for the pressure drop is

$$\Delta P = 0.3v + 4.0W_i v^2 \theta$$
$$v = \text{ft/min}$$
$$W_i = \text{lb/ft}^3$$
$$\theta = \text{min}$$

where ΔP = pressure drop (in. H₂O)
v = filtration velocity (ft/min)
W_i = dust concentration (lb/ft³ of gas)
θ = time after bags were cleaned (min)

SOLUTION
Bag area $- (18/12)(\pi)(20) + (18/12)^2(\pi/4) = 96\,\text{ft}^2$
Filtering area required $= 175,000/8 = 21,875\,\text{ft}^2$
Number of bags $= 21,875/96 = 230$ bags
Frequency of cleaning, $\theta = (\Delta P - 0.3v)/4W_i v^2$ where

$$W_i = (6.0\,\text{gr/ft}^3)(\text{lb}/7,000\,\text{gr}) = 8.57 \times 10^{-4}\,\text{lb/ft}^3$$

$$= [10 - (0.3)(8)]/4(8.57 \times 10^{-4})(8^2) = 34.6\,\text{min}$$

Example 10.7.8. 50,000 acfm of gas with a dust loading of 5.0 gr/ft³ flows through a
baghouse with an average filtration velocity of 10 ft/min. The pressure drop is given
by

$$\Delta P = 0.20v + 5.0W_i v^2 \theta$$

The fan can maintain the volume flow rate up to a pressure drop of 8 in. H_2O. Show that the baghouse can be operated for 16.8 min between cleanings.

In an attempt to determine the efficiency of this unit at "terminal" conditions, both the fabric and deposited cake (individually) were subjected to laboratory experimentation with the same gas and same flow velocity. The following data were recorded:

Fabric alone: thickness = 0.1 in.

concentration at the inside surface of the cloth = 1.0 g/cm^3
concentration at the outside surface of the cloth = 0.1245 g/cm^3

Cake alone: time = 16.8 min (end of cleaning cycle)

concentration at the inside surface of the cake = 1.0 g/cm^3
concentration at the outside surface of the cake = 0.0778 g/cm^3

Using the "Theodore" collection efficiency model, determine the overall efficiency at the start and end of a cleaning cycle.

SOLUTION. First show that the baghouse can be operated for 16.8 min.

$$(\Delta P - 0.2v)/5W_i v^2 = [8 - (0.2)(10)]/[(5)(5/7000)(10^2)]$$

$$= 16.8 \, min$$

Consider the collection efficiency of the fabric alone.

$$\ln(W_o/W_i) = -\psi L$$

$$\ln(0.1245/1.0) = -\psi(0.1/12)$$

$$\psi = 250 \, ft^{-1}$$

For cake alone,

$$\ln(W_o/W_i) = -\phi\theta$$

$$\ln(0.0778/1.0) = -\phi(16.8)$$

$$= 0.152 \, min^{-1}$$

Calculate the collection efficiency with the cake and fabric combined. [Eq. (10.4.1)].

$$E = 1 - e^{-[\psi L + \phi\theta]}$$

$$E = 1 - e^{-[(250)(0.1/12)+(0.152)(16.8)]}$$

$$= 0.9903 = 99.03\% \, (\text{at the end of cleaning cycle})$$

$$E = 87.55\% \, (\text{at the start of the cycle})$$

Example 10.7.9. A waste incinerator emits 300 ppm HCl with peak values of 500 ppm 15% of the time. The air flow is a constant 25,000 acfm at 75°F and 1 atm. Only sketchy information was submitted with the scrubber permit application for a spray tower. Determine if the spray unit is satisfactory.

DATA

Emission limit = 25 ppm HCl
Maximum gas velocity allowed through the tower = 3 ft/s
Number of sprays = 6
Diameter of the tower = 14 ft

The plans show a countercurrent water spray tower. For a very soluble gas (Henry's law constant approximately zero), the number of transfer units (N_{OG}) can be determined by Eq. (10.5.2a).

$$N_{OG} = \ln(y_1/y_2)$$

where y_1 = concentration of inlet gas
   y_2 = concentration of outlet gas

In a spray tower, the number of transfer units (N_{OG}) for the first (or top) spray will be ~ 0.7. Each tower spray will have only ~ 60% of the N_{OG} of the spray above it. The final spray, if placed in the inlet duct, has a N_{OG} of 0.5. The spray sections of a tower are normally spaced at 3-ft intervals. The inlet duct spray adds no height to the column.

SOLUTION

$$v = Q_a/A = Q_a/(\pi D^2/4) = (25{,}000)(4)/(60)(3.14)(14)^2$$

$$= 2.7\,\text{ft/s}$$

The gas velocity meets the requirement, since the maximum allowable gas velocity is 3 ft/s.

The number of transfer units (N_{OG}) to meet the regulations is

$$N_{OG} = \ln(y_1/y_2) = \ln(500/25) = 2.99$$

The total number of transfer units provided by a tower with six spray sections is

Spray Section	N_{OG}
Top	0.7
2nd	0.7 × 0.6 = 0.42
3rd	0.42 × 0.6 = 0.25
4th	0.25 × 0.6 = 0.15
5th	0.15 × 0.6 = 0.09
Inlet	0.5
	Total = 2.11

$$2.11 = \ln(500/y_2)$$
$$y_2 = 60.6\,\text{ppm}$$

The spray tower does not meet the HCl regulation that limits the emission of HCl to 25 ppm.

Example 10.7.10. A packed column is to be installed in a hazardous waste incinerator facility as an air pollution control device. It is designed to absorb HCl from a gas stream. Given the operating conditions and type of packing, calculate the height of packing and column diameter.

OPERATING DATA AND ASSUMPTIONS

Ga mass flow rate = 5,000 lb/h

HCl concentration in inlet gas stream = 2.0 mol%

Scrubbing liquid = pure H_2O

Packing type = 1-in. Raschig rings

Packing factor (F) for 1-in. Raschig rings = 160[1]

H_{OG} of the column = 2.5 ft

Henry's law constant, $m = 0.20$

Density of gas (air) = 0.075 lb/ft^3

Density of liquid = 62.4 lb/ft^3

Viscosity of liquid = 1.8 cP

The unit operates at 60% of the flooding gas mass velocity, the actual liquid flow rate is 500% more than the minimum, and 99% of HCl is to be collected based on state regulations.

SOLUTION

According to Henry's law $= x_1^* = y_1/m$

where x_1^* = equilibrium outlet liquid composition

$\quad\quad y_1$ = inlet gas composition

$\quad\quad m$ = Henry's constant

$$x_1^* = 0.02/0.20 = 0.10$$

Since it is required to remove 99% of the HCl, there will be 1% of HCl remaining in the outlet gas stream.

$$y_2 = (0.01 y_1)/[(1 - y_1) + (0.01 y_1)]$$

where y_2 = outlet gas composition

$$y_2 = 0.01(0.02)/[(1 - 0.02) + 0.01(0.02)] = 0.0002$$

The minimum ratio of molar liquid flow rate to molar gas flow rate, $(L_m/G_m)_{min}$, by a material balance, is

$$(L_m/G_m)_{min} = (y_1 - y_2)/(x_1^* - x_2)$$

where x_2 = inlet liquid composition

$$(L_m/G_m)_{min} = (0.02 - 0.0002)/(0.1 - 0) = 0.198$$

For a liquid flow rate 500% more than the minimum,

$$(L_m/G_m) = 0.198 \times 6.00 = 1.188$$

N_{OG} is determined from Eq. (10.5.2) or Fig. 10.5.5.

$$(y_1 - mx_2)/(y_2 - mx_2) = 0.02/0.0002 = 100$$
$$A = L_m/mG_m = 1.188/0.2 = 5.94$$
$$N_{OG} = 5.3$$

The height of packing is

$$Z = H_{OG} \times N_{OG} = (2.5)(5.3) = 13.2 \text{ ft}$$

The generalized flooding correlation [Eq. (9.3.7)] is applied to determine the cross-sectional area.

$$(L/G)(\rho_G/\rho_L)^{0.5} = (L_m/G_m)(18/29)(\rho_G/\rho_L)^{0.5}$$

where 18/29 = ratio of MW of H_2O to air
$\quad\quad L$ = liquid mass velocity (lb/s-ft^2)
$\quad\quad G$ = gas mass velocity (lb/s-ft^2)

From Eq. (9.3.7),

$$X = \ln\left[\frac{L}{G}\left(\frac{\rho_G}{\rho_L}\right)^{0.5}\right] = \ln[(1.188)(18/29)(0.075/62.4)^{0.5}]$$

$$= -3.67$$

$$Y = -3.84 - 1.06X - 0.119X^2 = -3.85 - 1.06(-3.67) - 0.119(-3.67)^2$$

$$= -1.55$$

Equation (9.3.8) may be used to calculate the cross-sectional area for flooding (A_f).

$$e^Y = e^{-1.55} = 0.212$$

$$A_f = \left[\frac{w_G^2 F \psi \mu^{0.2}}{\rho_G \rho_L g_c e^Y}\right]^{0.5}$$

$$= \left[\frac{(5000)^2(160)(1)(1.8)^{0.2}}{(0.075)(62.4)(4.173 \times 10^8)(0.212)}\right]^{0.5} = 3.3 \text{ ft}^2 \quad\quad (9.3.8)$$

The cross section for 60% flooding is given by Eq. (9.3.9)

$$A = \frac{A_f}{f} = \frac{3.3}{0.60} = 5.5 \text{ ft}^2$$

The column diameter is then

$$D = \left(\frac{4A}{\pi}\right)^{0.5} = 2.6\,\text{ft}$$

Approximately 40 ft of packing in a 2.6-ft-i.d. column is required. Four 10-ft packed columns with redistributors is recommeneded. The liquid flow rate should be checked with the vendor to determine if it is above the minimum wetting rate.

PROBLEMS

1. A horizontal parallel-plate ESP consists of a single duct 24 ft high and 20 ft deep with an 11-in. plate-to-plate spacing. A collection efficiency of 88.2% is obtained with a flow rate of 4200 acfm. The inlet loading is 2.82 gr/ft³. Calculate the following using a modified form of the Deutsch–Anderson equation, with the exponent $m = 0.5$ (i.e., exponent on ϕ).
   (a) The outlet loading.
   (b) The drift velocity for this system.
   (c) A revised collection efficiency if the flow rate is increased to 5400 acfm.
   (d) A revised collection efficiency if the plate spacing is decreased to 9 in.

2. A horizontal parallel-plate ESP consisting of four passages 24 ft high and 20 ft deep with an 11-in plate spacing is used in an incineration process to treat 41,800 acfm of air containing 4.3 gr/ft³ of fly ash. Using an average drift velocity of 36.3 ft/min for the system, calculate the collection efficiency of the unit.

3. As a recently hired engineer in a hazardous waste incineration facility, you have been assigned the job of submitting a preliminary design of an ESP to treat 180,000 acfm of gas laden with solids from the incinerator burning dry sewage sludge. The inlet loading of 3.47 gr/ft³ is to be reduced to 0.07. A previous (similar) design employed a drift velocity, throughput velocity, and plate height of 0.2 ft/s, 4.0 ft/s, and 25 ft, respectively. Submit your design.

4. A fly ash laden stream is to be cleaned by a venturi scrubber using a L/G ratio of 8.5 gal/1000 ft³. The fly ash has a particle density of 0.7 g/cm³. Use a throat velocity of 272 ft/s and a gas viscosity of 1.5×10^{-5} lb/ft-s. The particle size distribution is

Particle Size Range (μm)	% by Weight
0.10	0.01
0.10– 0.50	0.21
0.60– 1.00	0.78
1.10– 5.00	18.00
6.00–10.00	16.00
11.00–15.00	12.00
16.00–20.00	8.00
20.00	45.00

Assume a correlation coefficient (k) of 250 ft³/gal. Determine the overall collection efficiency.

5. A vendor proposes to use a spray tower in a hazardous waste incineration operation to reduce the discharge of solids to the atmosphere. The inlet loading of the gas stream from the kiln is 5.0 gr/ft³ and is to be reduced to 0.08 gr/ft³ in order to meet state regulations. The vendor's design calls for a water pressure drop of 80 psi and a pressure drop across the tower of 5.0 in. H₂O The gas flow rate is 10,000 scfm and a water rate of 50 gal/min is proposed. Assume the contact power theory to apply.

   (a) Will the spray tower meet regulations?
   (b) What total pressure loss is required to meet regulations?
   (c) What conditions can be drawn concerning the use of a spray tower for this application?

6. A hazardous waste incineration facility plans to install a fabric filter as an air cleaning device.

   (a) How many bags, each 8 in. in diameter and 12 ft long, must be used to treat the exhaust gas that has a particulate loading of 2 gr/ft³; the exhaust fan is rated at 7000 acfm? Assume the filtering velocity is 2 ft/min.
   (b) Estimate the pressure drop after 4 h of operation if the resistance coefficients of the filter and dust cake are, respectively, $K_1 = 0.8$ in. H₂O/(ft/min) and $K_2 = 3.0$ in. H₂O/(lb dust/ft² cloth area)-(ft/min). [See Eq. (10.4.1).]
   (c) Calculate the fan horsepower (fan/motor efficiency of 63%) at the maximum pressure drop.

7. A hazardous waste incineration plant has an exhaust flow of 175,000 acfm following a dry scrubber with a loading of 8.0 gr/ft³. The ACR ratio is 5.0 and the efficiency is 99.0%. The allowable pressure drop is eight in. H₂O. How many 18-in. × 20-ft bags are needed? What is the frequency of cleaning if the vendor's equation for the pressure drop is

$$\Delta P = 0.2v + 5.0W_i v^2 \theta$$
$$v = \text{ft/min}$$
$$W_i = \text{lb/ft}^3$$
$$\theta = \text{min}$$

where ΔP = pressure drop (in. H₂O)
   v = filtration velocity (ft/min)
   W_i = dust concentration (lb/ft³ of gas)
   θ = time after bags were cleaned (min)

8. 50,000 acfm of gas with a dust loading of 5.0 gr/ft³ flows through a baghouse with an average filtration velocity of 10 ft/min. The pressure drop is given by

$$\Delta P = 0.20v + 5.0W_i v^2 \theta$$

The fan can maintain the volumetric flow rate up to a pressure drop of 8 in. H₂O. Show that the baghouse can be operated for 3.4043 min between cleanings. In an attempt to determine the efficiency of this unit at "terminal" conditions, both the fabric and deposited cake (individually) were subjected to laboratory experimentation. The following data were recorded:

Fabric alone: thickness = 0.05 in.
concentration at inlet side of cloth = $1.0 \, g/cm^3$
concentration at outlet side of cloth = $0.185 \, g/cm^3$

Cake alone: time = 4.0 min (end of cleaning cycle)
concentration at inlet side of cake = $1.0 \, g/cm^3$
concentration at outlet side of cake = $0.08 \, g/cm^3$

Using the "Theodore" collection efficiency model, determine the overall efficiency at the start and end of a cleaning cycle.

9. A waste incinerator emits 300 ppm HCl with peak values of 600 ppm. The air flow is a constant 5000 acfm at 75°F and 1 atm. Only sketchy information was submitted with the scrubber permit application for a spray tower. You are requested to determine if the spray unit is satisfactory.

DATA

Emission limit = 30 ppm HCl
Maximum gas velocity allowed through the tower = 3 ft/s
Number of sprays = 6
Diameter of the tower = 10 ft

The plans show a countercurrent water spray tower.

10. A packed column is to be installed in a hazardous waste incineration facility as the gaseous pollution control equipment. It is designed to absorb HCl from a gas stream. Given the operating conditions and type of packing, calculate the height of packing and column diameter.

OPERATING DATA AND ASSUMPTIONS

Gas mass flow rate = 4000 lb/h
HCl concentration in inlet gas stream = 2.5 mol%
Scrubbing liquid = pure water
Packing type = 1-in Raschig rings
H_{OG} of the column = 2.5 ft
Henry's law constant, $m = 0.20$
Density of gas (air) = $0.075 \, lb/ft^3$
Density of water = $62.4 \, lb/ft^3$
Viscosity of water = 1.8 cP

The unit operates at 60% of the flooding gas mass velocity, the actual liquid flow rate is 200% more than the minimum, and 99% of the HCl is to be collected based on state regulations.

REFERENCES

1. L. Theodore, and L. Buonicore, *Air Pollution Control Equipment*, CRC Press, Boca Raton, FL, 1987.
2. The Ceilcote Corporation, IWS Bulletin, Berea, OH, 1983.
3. J.W. MacDonald, private communication, 1986.
4. U.S. EPA, *Engineering Handbook for Hazardous Waste Incineration*, prepared by Monsanto Research Corporation, Dayton, OH, EPA Contract No. 68–03–3025, September, 1981.
5. Revised U.S. EPA, *Engineering Handbook for Hazardous Waste Incineration*, unpublished, 1985.
6. H.F. Johnstone, R.B. Feild, and M.C. Tassler, *Ind. Eng. Chem.*, **46**, 1601 (1954).
7. L. Theodore, C. LoPinto, L. Llamas, and C. Murray, *Proc. Environ. Eng. Sci. Conf.*, **4**, 365 (1974).
8. H.E. Hesketh, "Fine Particle Collection Efficiency Related to Pressure Drop, Scrubbant and Particle Properties and Contact Mechanism," *J. Air Pollut. Control Assoc.*, **24**, 938–942, (1974).
9. H.E. Hesketh, "Atomization and Cloud Behavior in Wet Scrubbers," US–USSR Symposium on Control of Fine Particulate Emissions, January 15–18, 1974.
10. K.T. Semrau, "Correlation of Dust Scrubber Efficiency," *J. Air Pollut. Control Assoc.*, **10** 200–207 (1960).
11. W. Strauss, *Industrial Gas Cleaning*, Pergamon, New York, 1966.

11

Ancillary Equipment

Contributing author: Abdool Jabar

11.1 INTRODUCTION

While the previous chapters of Part III covered the major pieces of equipment used in the hazardous waste incineration industry—incinerators, waste heat boilers, quenchers, and air pollution devices—this chapter deals with some of the ancillary equipment. Most of this equipment is involved with the transporting of material such as wastes, combustion gases, air, water, and so on, to, from, or between the major units. Although the movement of dry solid material, either waste or combustion products, is a part of the total picture, it is a rather minor part when compared to the transport through the facility of gases, liquids, and liquid–solid mixtures in the form of suspensions, slurries, sludges, and the like. All of these, unlike the dry solids, are pumpable to some extent and the pieces of equipment discussed in this chapter have for their main function the movement of these materials. Some of these devices are simply conduits for the moving material—pipes, ducts, fittings, stacks; others control the flow of material—valves; and others provide the mechanical driving force for the flow—fans, pumps, compressors. (It should be noted that there are many other devices that could appropriately be considered under the title "Ancillary Equipment" for HWI, for example, blenders and evaporators used in the pretreatment of waste and activated carbon cannisters used in process vents for capturing fugitive gases. These are too numerous to include here and the interested reader is directed to the literature for information on these devices.)

One of the more critical parameters in fluid flow is pressure. Three pressure terms should be defined before proceeding to the body of this chapter. These are the static pressure (P_s), the velocity pressure (P_v), and the total of the two (P_t). Any fluid confined in a stationary enclosure has *static* pressure simply because the molecules of that fluid are in constant random motion and are continually colliding with the container walls. The bulk velocity of this fluid is zero, and the total pressure (P_t) is equal to the static pressure. If the same fluid is flowing and the temperature has not changed, it possesses the same static pressure, since its molecules still have the same degree of random motion. Its total pressure is now higher, however, because it also possesses the second pressure component, *velocity* pressure. If the fluid flow were to suddenly change direction because of a solid obstruction, for example, a flat plate perpendicular to the flow direction, an extra pressure on the plate (over and above the static pressure) would be exerted because of the momentum of the bulk flow against the plate. This extra pressure is the *velocity* pressure (P_v) and the total fluid pressure is the sum of the *static* and *velocity* pressures. Static pressure is therefore the result of motion on the molecular level, while velocity pressure is due to motion on the macroscopic or bulk level.

The difference in total pressure between two different points along the stream is called the *pressure loss* or the *pressure drop*. Pressure losses from fluid flow are due to any factors that can change fluid momentum on either the molecular or macroscopic levels; the two main contributing factors are skin friction and form friction. Skin friction losses are caused by fluid moving along (parallel to) a solid surface such as a pipe or duct wall. The layers of fluid immediately adjacent to the wall are in laminar flow and moving much slower than the bulk of the fluid. The pressure drop caused by the drag effect of the wall on the fluid is due to skin friction. Form friction losses are due to acceleration or deceleration of the fluid. These include changes in bulk fluid velocity that occur because of changes in either flow direction or flow speed. An example of a change in flow direction is fluid flowing through a 90° elbow; a change in flow speed occurs when the cross section of the conduit changes. Besides changes in bulk fluid velocity, form friction losses also include changes in velocity that occur locally, that is, internal to the bulk of the fluid. This occurs in turbulent flow, which is characterized by rapidly swirling masses of fluid called *eddies*.

In a straight length of pipe or duct, skin friction is usually the main contributor to pressure loss; when the fluid travels through the various fittings and valves discussed in Section 11.2, form friction predominates.

11.2 PIPES, DUCTS, FITTINGS, AND VALVES

Pipes

The most common conduits for fluids are pipes and tubing. Both have circular cross sections but pipes tend to have larger diameters and thicker walls. Because of the heavier walls, pipes can be threaded, while tubing cannot. Because HWI systems usually handle large flow rates that require the larger diameters associated with pipes, the use of tubing in these systems is minimal and will not be covered in any detail in this section.

Tubing and pipes are manufactured from many construction materials. The selection of the material depends on the corrosive nature of the fluid and the flow (system) pressure. If corrosive liquids or high standards of purity require special piping, stainless steel, nickel alloys, or materials of high resistance to heat and mechanical damage are used. Steel pipe can also be lined with tin, plastic, rubber, lead, or other coatings for special purposes. If problems of corrosion or contamination are the controlling factors, the use of a nonmetallic pipe such as glass, porcelain, thermosetting plastic, or hard rubber is often acceptable.

There are several techniques used to join pipe sections. For small pipes

TABLE 11.2.1. Nominal Pipe Diameters for Liquid Waste Flows

Capacity (gpm)	Nominal Pipe Diameter (in.)
0–15	1
15–70	2
70–150	3
150–250	4

(diameters > 2 in.), threaded connectors are the most common; for larger pipes, flanged fittings or welded connections are normally employed.

Because pipes are manufactured with so many different diameters and wall thicknesses, the following method of standardization has been established by the American National Standards Institute (ANSI). By convention, pipe and fitting sizes are characterized by nominal diameters and wall thicknesses. For pipes over 12 in. in diameter, the nominal diameter is the actual outside diameter. For pipes from 3 to 12 in. in nominal diameter, the nominal value is approximately equal to the inside diameter. For pipes < 3 in. the nominal size can be considerably smaller than the actual inside diameter. Typical pipe diameters for various liquid flow capacities in HWI applications are given in Table 11.2.1.

The wall thickness of a pipe is specified by a schedule number, which is a function of the internal pressure and allowable stress, as follows:

$$\text{schedule number} = 1000P/S \qquad (11.2.1)$$

where P = internal working pressure (lb/in.^{2})
$\quad\ S$ = allowable stress (lb/in.^{2})

There are 10 schedule numbers in use and these range from 10 to 160; the thickness of the pipe wall increases with the schedule number. Schedule 40 pipe is commonly used in many industrial applications.

Piping requirements for HWI facilities include liquid waste feeding and transfer within the storage complex, fuel oil and caustic soda solution feeding, and nitrogen hookup to storage tanks. The total length of pipe depends on a number of factors, including overall facility size, number of storage tanks, and storage facility layout.

Ducts

Pipes and tubing are used as conduits for the transporting of liquids or gases; ducts are used only for gases. By using thicker walls, pipes can be used for flows at higher pressures; ducts are always thin walled and are generally used for gas flows close to ambient pressure. Pipes are usually circular in cross section; ducts come in many shapes: circular, oval, rectangular, and so on. In general, ducts are larger in cross section than pipes because they carry fluids with low densities and high volumetric flow rates. In HWI applications, ducts are used primarily to move exhaust gases to the stack.

Ducts are most often constructed of field-fabricated galvanized sheet steel, although other materials such as fibrous glass board, factory-fabricated round fibrous glass, spiral sheet metal, and flexible duct materials are becoming increasingly popular. Other duct construction materials include black steel, aluminium, stainless steel, plastic and plastic-coated steel, cement asbestos, and copper. In HWI applications, the corrosion resistance of the duct materials deserves special consideration. Material costs generally increase along with corrosion resistance, so the selection of material must be determined by the desired life span in the anticipated environment; this environment is a function of the characteristics of the waste being incinerated and the operating conditions of the incinerator. For maximum resistance to moisture or corrosive gases, stainless steel and copper are used where their cost can be justified. Aluminium sheet is used where lighter weight and superior resistance to moisture are needed. For the same applications, nonferrous metals must normally be larger in wall thickness, since they generally lack the high strength and rigidity of steel.

Gases traveling through a duct impose two loads on the duct's structure: one related to pressure, the other to velocity. The dominant load is normally the difference in static pressure across the duct wall; the second load is that caused by turbulent air flow causing relatively small but pulsating forces on the duct wall. Duct construction is classified in terms of operating pressure. The low pressure range extends up to 2 in. H_2O (0.072 psig) and the high pressure range up to 10 in. H_2O (0.36 psig). Recommended wall thickness depends on the material of construction, operating pressure, and the cross-section shape. Typical wall thicknesses for a 3-ft-diameter round duct made of galvanized steel are about 0.003 in. for low pressure and 0.004 in. for high pressure operation.

The most common cross sections for ducts are rectangular, round, and oval. The versatility of a *rectangular* cross section allows convenient adaptation of the duct system to the available space. The *round* cross section is the most economical and efficient; it is used more extensively than other shapes because it gives higher strength and rigidity for a given wall thickness and allows air-tight connections to be made more conveniently. The *oval* cross section combines some advantages of round and rectangular duct; it can be joined using the same techniques of round duct assembly and permits economical use of space, though not nearly to the same extent as rectangular duct.

Fittings

A fitting is a piece of equipment that has for its function one or more of the following:

- The joining of two pieces of straight pipe; for example, couplings and unions.
- The changing of pipeline direction; for example, elbows and T's.
- The changing of pipeline diameter; for example, reducers and bushings.
- The terminating of a pipeline; for example, plugs and caps.
- The joining of two streams; for example, T's and Y's.

A *coupling* is a short piece of pipe threaded on the inside and used to connect straight sections of pipe. A *union* is also used to connect two straight sections but differs from a coupling in that it can be opened conveniently without disturbing the rest of the pipeline. When a coupling is opened, considerable amount of piping must usually be dismantled. An *elbow* is an angle fitting used to change flow direction, usually by 90°. A T (shaped like the letter T) can also be used to change flow direction, but is more often used to combine two streams into one. A *reducer* is a coupling for two pipe sections of different diameter. A *bushing* is also a connector for pipes of different diameter, but, unlike the reducer coupling, is threaded on both inside and outside. The larger pipe screws onto the outside of the bushing and the smaller pipe screws into the inside of the bushing. To terminate a pipeline, *plugs*, which are threaded on the outside, and *caps*, which are threaded on the inside, are used. A Y (shaped like the letter Y) is similar to the T and is used to combine two streams.

Valves

Valves have two main functions in a pipeline: to control the amount of flow, or to stop the flow completely. There are many different types of valves, the most commonly used are the gate valve and the globe valve. The *gate* valve contains a disk that slides at right angles to the flow direction. This type of valve is used primarily for *on–off* control of a liquid flow. Because small lateral adjustments of the disk cause extreme changes in the flow cross-sectional area, this type of valve is not suitable for adjusting flow rates. As the fluid passes through the gate valve, only a small amount of turbulence is generated; the direction of flow is not altered and the flow cross-sectional area inside the valve is only slightly smaller than that of the pipe. As a result, the valve causes only a minor pressure drop. Problems with abrasion and erosion of the disk arise when the valve is used in positions other than fully open or fully closed.

Unlike the gate valve, the *globe* valve is designed for flow control. In this type of valve, the liquid passes through the valve in a somewhat circuitous route. In one form, the seal is a horizontal ring into which a plug with a slightly beveled edge is inserted when the stem is closed. Good control of flow is achieved with this type of valve, but at the expense of a higher pressure loss than a gate valve.

Some other types of valves are the *check* valve, which permits flow in one direction only; the *butterfly* valve, which operates in damperlike fashion by rotating a flat plate to either a parallel or perpendicular position relative to the flow; the *plug* valve, in which a rotating tapered plug provides on–off service; the *needle* valve, which is a variation of the globe valve that gives improved flow control; and the *diaphragm* valve, a valve specially designed to handle fluids such as very viscous liquids, slurries, or corrosive liquids that may tend to clog the moving parts of other valves.

In modern HWI facilities, many of the flow rates are controlled automatically using more sophisticated electrically or pneumatically driven valves.

11.3 FANS, PUMPS, AND COMPRESSORS

To move material (either a fluid or solid–liquid mixture) through the various pieces of equipment at the incinerator facility including piping and duct work, requires mechanical energy, not only to impart an initial velocity to the material, but more importantly to overcome pressure losses that occur throughout the flow path. This energy may be imparted to the moving stream in one or more of three modes; an increase in the stream's velocity, an increase in the stream pressure, or an increase in stream height. In the first case, the additional energy takes the form of increased *external* kinetic energy as the bulk stream velocity increases. In the second, the *internal* energy (mainly potential, but usually some kinetic as well) of the stream increases. *Internal energy* refers to energy exhibited on the molecular level. In this case, the higher pressure shortens the distance between molecules, which represents an increase in internal potential energy in the electrical force fields between neighboring molecules (bonds); this pressure increase may also cause stream temperature rise, which represents an internal kinetic energy increase. In the third case, which is relatively unimportant for HWI operations, the bulk fluid experiences an increase in external potential energy in the earth's gravitational field.

In this chapter, three devices that convert electrical energy into the mechanical energy that is to be applied to the various streams are discussed. These devices are *fans*, which move low pressure gases; *compressors*, which move high pressure gases; and *pumps*, which move liquids and liquid–solid mixtures such as suspensions, slurries, and sludges. For HWI operations, fans are used to move the gases, and pumps to move the liquids and semiliquids. These two devices will be discussed in some detail here. Compressors are rarely used in these operations and only passing attention will be given to them in this text.

Fans

The term *fans* and *blowers* are often used interchangeably, and no distinction will be made between the two in the following discussion. Whatever is stated about fans equally applies to blowers. Strictly speaking, however, fans are used for low pressure (drop) operation, generally below 2 psi. Blowers are generally employed when generating pressure heads in the 2.0 to 14.7-psi range. Higher pressure operations require compressors.

Fans are usually classified as the centrifugal or the axial-flow type. In centrifugal fans, the gas is introduced into the center of the revolving wheel (the eye) and discharges at right angles to the rotating blades. In axial-flow fans, the gas moves directly (forward) through the axis of rotation of the fan blades. Both types are used in industry, but it is the centrifugal fan that is employed at incineration facilities.

The gas in a centrifugal fan is subjected to centrifugal forces. These forces compress the gas giving it additional static pressure. Centrifugal fans are enclosed in a scroll-shaped housing that helps convert kinetic energy to static pressure. Gas rotating between the fan blades is compressed in the fan scroll, which increases the static pressure. Centrifugal fans are classified by blade configuration as radial, forward curved, backward curved, air foil, and radial tip-forward curved heel.

Radial or straight blade fans physically resemble a paddle wheel with long radial blades attached to the rotor and are the simplest design of all centrifugal fans. This enables most radial blade fans to be built with great mechanical strength and to be easily repaired. These fans can be used in a variety of situations, especially heavy duty applications. This fan can handle erosive and corrosive gases as well as very sticky gases. It is particularly well suited for high static pressure operations and can generate pressures in excess of 50 in. H_2O. When operated properly, the horsepower efficiency range is 55–69%, with 65% as a typical value. *Forward curved* fans are the most popular for general ventilation purposes (high flow rates and low static pressures). These fans have both the heel and the tip of the blade curved forward in the direction of rotation. Blades are smaller and spaced much closer together than in other blade designs. They are generally not used with dirty gases when dust or sticky materials are present because contaminants easily accumulate on the blades and cause imbalance. Efficiencies range from 52 to 71%, with 65% being typical. *Backward curved* or backward inclined fans have blades inclined in a direction opposite to that of the direction of rotation. This feature causes the gas to leave the tip of the blade at a lower velocity than the wheel-tip speed, a factor that improves the mechanical efficiency. These types of fans are not suitable for a heavily particulate-laden gas, sticky material, or abrasive dust. Centrifugal forces tend to build up particulate matter on the backside of the fan blades. These are rarely used in air pollution control despite typical efficiencies of 70%. The airfoil is similar to the backward curved fan, except that the blade has been contoured to increase stability and operating efficiency. These fans are more expensive to construct than backward curved fans, but have lower power requirements. They are used only occasionally in pollution control. The gas must be clean and noncorrosive. Since the blades are hollow, abrasion and wear could allow dust, water vapor, and so on, to enter the blade and cause imbalance. This is the most efficient of the various fan types, with typical efficiencies at the 85% level. A modification of the radial fan is the *radial tip-forward curved heel*. The blades are curved forward with this unit. This fan is reportedly more dependable than the radial or high tip speed applications (i.e., high static pressures) due to better vibrational characteristics and its ability to resist fatigue. It finds application in the processing of large flow rates ($>$ 200,000 acfm) with light to medium particulate loadings. However, sticky material and particulates, in general, can accumulate in the slight curvature of the blades and cause imbalance. Efficiencies here typically range from 52 to 74%, with 70% being common.

Generally, centrifugal fans are easier to control, more robust in construction, and less noisy than axial units. They have a broader operating range at their highest efficiencies. Centrifugal fans are better suited for operations in which there are flow variations and they can handle dust and fumes better than axial fans.

Fan laws are equations that enable the results of a fan test (or operation) at one set of conditions to be used to calculate the performance at another set of conditions, including differently sized but geometrically similar models of the same fan design. The fan laws can be written in many different ways. The three key laws are provided in the following equations:

$$Q_a = k_1(\text{RPM})D^3 \tag{11.3.1}$$

$$P_s = k_2(\text{RPM})^2 D^2 \rho \tag{11.3.2}$$

$$\text{HP} = k_3(\text{RPM})^3 D^5 \rho \tag{11.3.3}$$

where Q_a = volumetric flow rate
  P_s = static pressure
  HP = horsepower
  RPM = revolutions per minute
  D = wheel diameter
  ρ = gas density
  k_1, k_2, k_3 = proportionality constants

Thus, these three laws may be used to determine the effect of fan speed, fan size, and gas density on flow rate, developed static pressure head, and horsepower. For two conditions, where the constants k remain unchanged, Eqs. (11.3.1) through (11.3.3) become:

$$\left(\frac{Q'_a}{Q_a}\right) = \left(\frac{RPM'}{RPM}\right)\left(\frac{D'}{D}\right)^3 \tag{11.3.4}$$

$$\left(\frac{P'_s}{P_s}\right) = \left(\frac{RPM'}{RPM}\right)^2 \left(\frac{D'}{D}\right)^2 \left(\frac{\rho'}{\rho}\right) \tag{11.3.5}$$

$$\left(\frac{HP'}{HP}\right) = \left(\frac{RPM'}{RPM}\right)^3 \left(\frac{D'}{D}\right)^5 \left(\frac{\rho'}{\rho}\right) \tag{11.3.6}$$

Note: The *prime* refers to the new condition. It is also important to note that the fan laws are approximations and should not be used over wide ranges or changes of flow rate, size, and so on.

A rigorous, extensive treatment on fan selection is beyond the scope of this text. It is common practice among fan vendors to publish voluminous data in tabular form providing flow rate, static pressure, speed, and horsepower at a standard temperature and gas density. These are often referred to as *multirating tables*.

Note: These tables should not be used for fan selection except by those who have experience in this area. For those who do not, the proper course of action to follow is to provide the fan manufacturer with a complete description of the system and allow the manufacturer to select and guarantee the optimum fan choice.

To help in the actual selection of fan size, a typical fan rating table is given in Table 11.3.1. The fan size and dimensions are usually listed at the top of the table. Values of static pressure are arranged as columns that contain the fan speed and brake horsepower required to produce various volume flows. The point of maximum efficiency at each static pressure is usually underlined or printed in special type. In order to select a fan for the exact condition desired, it is sometimes necessary to interpolate between values presented in the multirating tables. Straight-line interpolation can be used with negligible error for multirating tables based on a single fan size. Some multirating tables attempt to show ratings for a whole series of geometrically similar (homologous) fans in one table; in this case, interpolation is not advised.[1]

The selection procedure is, in part, an examination of the fan curve and the system curve. A fan curve, relating static pressure with flow rate, is provided in Fig. 11.3.1. Note that each type of fan has its own characteristic curve. Also note that fans are usually tested in the factory or laboratory with open inlets and long smooth straight discharge ducts. Since these conditions are seldom duplicated in the field, actual operation often results in lower efficiency and reduced performance. A system curve is also shown in Fig. 11.3.1. This curve is calculated prior to the purchase of the fan and provides a best estimate of the pressure drop across the system through which the fan must deliver the gas. (This curve should approach a straight line with an approximate slope of 1.8 on log–log coordinates.) The system pressure (drop) is

TABLE 11.3.1. Typical Fan Rating Table[a-e]

Q_a (acfm)	Static Pressure (in. H_2O)											
	$\frac{1}{2}$		1		$1\frac{1}{2}$		2		$2\frac{1}{2}$		3	
	rpm	bhp	rpm	bhp	rpm	bhp	rpm	bhp	rpm	bhp	rpm	bhp
5,727	216	0.74	278	1.34	330	2.01	378	2.75	421	3.50	460	4.28
6,873	236	0.97	291	1.67	339	2.42	384	3.24	424	4.06	462	4.91
8,018	250	1.27	305	2.05	352	2.87	393	3.76	430	4.69	467	5.66
9,164	271	1.63	320	2.49	366	3.42	405	4.35	441	5.36	475	6.40
10,309	293	2.12	338	3.05	381	4.06	419	5.06	453	6.14	486	7.26
11,455	315	2.72	356	3.65	396	4.76	432	5.88	468	7.03	499	8.19
12,600	337	3.46	377	4.39	413	5.58	448	6.81	482	8.04	514	9.31
13,746	360	4.39	399	5.25	430	6.48	465	7.82	496	9.16	527	10.53
14,891	382	5.43	421	6.25	451	7.52	481	8.93	512	10.35	542	11.80
16,037	405	6.66	442	7.48	473	8.67	501	10.16	529	11.69	557	13.25
17,182	429	8.08	463	8.97	496	10.05	521	11.54	547	13.18	574	14.81
18,328	451	9.64	486	10.61	517	11.61	543	12.99	566	14.78	591	16.49
19,473	474	11.46	510	12.47	539	13.47	565	14.85	587	16.56	610	18.35
20,619	497	13.51	532	14.55	560	15.60	587	16.82	610	18.54	630	20.40
21,764	520	15.82	556	16.82	584	17.98	608	19.17	632	20.77	652	22.59

Source: Bayler Blower Co.

[a] Wheel style: backward inclined.
[b] Wheel diameter: 50½ in.
[c] Maximum fan speed: 1134 rpm.
[d] Performances underlined are those at maximum efficiency.
[e] Brake horsepower = bhp.

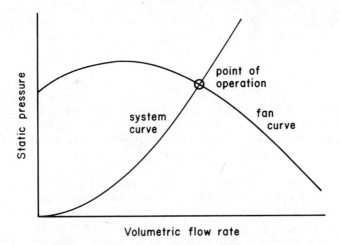

Figure 11.3.1. System and fan characteristic curves.

defined as the resistance through ducts, fittings, equipment, contractions and expansions, and so on. A number of methods are available to estimate the total system pressure change. These vary from very crude approximations to detailed, rigorous calculations. The simplest procedure is to obtain estimates of the pressure change associated with the movement of the gas through all of the resistances described previously. The sum of these pressure changes represents the total pressure drop across the system, and represents the total pressure (change) that must be developed by the fan. This calculation becomes more complex if branches (i.e., combining flows) are involved. However, the same stepwise procedure should be employed. When the two curves are superimposed, the intersection is defined as the *point of operation*. The fan should be selected so that it operates just to the right of the peak on the fan curve. The fan operates most efficiently and with maximum stability at this condition. If a fan is selected for operation too close to its peak, it will surge and oscillate. Thus, the point of intersection of the two curves determines the actual volumetric flow rate. If the system resistance has been accurately specified and the fan properly selected, the two performance curves will intersect at the design flow rate. If system pressure losses have not been accurately specified or if undesirable inlet and outlet conditions exist, design conditions will not be obtained. Dampers and fan speed changes can provide some variability on operating conditions.

There are a number of process and equipment variables that are classified as part of a fan specification. These include:

Flow rate (acfm)
Temperature
Density
Gas stream characteristics
Static pressure that needs to be developed
Motor type

Drive type

Materials of construction

Fan location

Noise controls

With respect to drive type, belt drives are usually employed if the power is < 200 hp. Direct drives are preferred for large horsepower systems. Direct drive units have lower maintenance costs and lower power transmission losses. However, the speed cannot be varied; if a speed change is required, the cost can be expensive. For materials of construction, mild carbon steel is commonly employed when treating dry air up to temperatures approaching 1000°F. Fiberglass may be used for corrosive conditions but the temperature should not exceed 250°F. With respect to fan location, the fan may be located downstream or upstream of a particular piece of equipment in the process. Fans located downstream are referred to as *induced draft* or *negative pressure* fans. Leakage occurs into, rather than out of, the unit. Possible lower flow rates and lower temperatures may lead to a smaller fan and reduced operating cost. Equipment costs could be higher because of the need of heavier construction material when operating under negative (vacuum) pressure. Fans located upstream may have erosion problems due to particulate loading; in addition, there may be an accumulation of particulates. A larger volumetric flow rate is possible, which would require a larger fan and high horsepower costs. With respect to noise controls, fans can create objectionable noise problems in the work area or neighboring residences. To minimize noise effects, the fan should, of course, be properly designed and properly operated. A fan should therefore be operated close to the point of maximum efficiency to reduce noise levels. Putting a fan in proper balance can also be an effective solution. If noise problems are expected or are already present, acoustical insulation should be applied to the fan housing. If the noise level is severe, insulation should also be added to the ductwork.

Gas (or air) horsepower (GHP) and brake horsepower (BHP) are the two terms of interest. These may be calculated from Eqs. (11.3.7) and (11.3.8).

$$GHP = 0.0001575 Q_a \Delta P = Q_a \Delta P / 6356 \tag{11.3.7}$$

$$BHP = 0.0001575 Q_a \Delta P / E_f \tag{11.3.8}$$

where Q_a = volumetric flow rate (acfm)

ΔP = SP = static pressure head developed (in. H_2O)

E_f = fractional fan efficiency

For cold start ups, the static pressure can be significantly higher than the normal operating pressure drop, and the fan must be able to handle the increased resistance. The correction for the cold static pressure may be calculated from Eq (11.3.9).

$$\Delta P_c = \Delta P(\rho_c / \rho) \tag{11.3.9}$$

where ρ = gas density (lb/ft³)

c = subscript denoting cold gas

For HWI systems, the ID fan (i.e., located downstream) is almost always used so that any leakage will be into rather than out of the unit. Either a single ID fan, or two ID fans in series, are used to create *draft* through the waste heat boiler (if there is one), quencher, and the air pollution control equipment, and to maintain a slight negative pressure in the incinerator. Typical pressure drops through the incinerator, waste heat boiler, quencher, and adsorber are 4, 4, 1, and 7 in. H_2O, respectively. The pressure drop across a venturi scrubber may range from 10 to 60 in. H_2O. Pressure drops across dampers and ductwork may be neglected. Typical fan–motor efficiencies are in the 50–60% range.[2]

All cooling and combustion air, excluding atomizing air, is supplied to the incinerator via one or more forced-draft fans or blowers. This assumption is somewhat conservative for certain designs, particularly rotary kiln incinerators in which a variable fraction of the combustion air flow is induced through negative pressure operation. The inlet air is generally assumed to be at 60°F and 14.7 psia.

Pumps

In a HWI facility, pumps are required to transport liquid wastes, wastes in the form of liquid–solid mixtures such as slurries and sludges, plus auxiliary fuel, if any is required, to the incinerator; pumps are also needed to transport water to and/or from such peripheral devices as waste heat boilers, quenchers, scrubbers, and so on.

Pumps may be classified as reciprocating, rotary, or centrifugal. The reciprocating and rotary types are referred to as *positive displacement* pumps, because, unlike the centrifugal type, the liquid or semiliquid flow is broken up into small portions as it passes through the pump.

Reciprocating pumps operate by the direct action of a piston on the liquid contained in a cylinder. As the liquid is compressed by the piston, the higher pressure forces it through discharge valves to the pump outlet. As the piston retracts, the next batch of low pressure liquid is drawn into the cylinder and the cycle is repeated. The piston may be either directly steam driven or moved by a rotating crankshaft through a crosshead. The rate of liquid delivery is a function of the volume swept out by the piston and the number of strokes per unit time. A fixed volume is delivered for each stroke but the actual delivery may be less because of both leakage past the piston and failure to fill the cylinder when the piston retracts. The volumetric efficiency of the pump is defined as the ratio of the actual volumetric discharge to the pump displacement. For well maintained pumps, the volumetric efficiency is at least 95%. Reciprocating pumps are used for many applications related to waste incineration, including pumping of boiler feed water, oil, waste slurries, and so on.

Reciprocating pumps can deliver the highest pressure of any type of pump (20,000 psig); however, their capacities are relatively small compared to the centrifugal pump. Also, because of the nature of the operation of the reciprocating pump, the discharge flow rate tends to be somewhat pulsating. Liquids containing abrasive solids can damage the machined surfaces of the piston and cylinder. Because of its positive displacement operation, reciprocating pumps can be used to measure liquid volumetric flow rates.

The *rotary* pump combines rotation of the liquid with positive displacement. The rotating elements mesh with elements of the stationary casing in much the same way that two gears mesh. As the rotating elements come together, a pocket is created that first enlarges, drawing in liquid from the inlet or suction line. As rotation continues, the pocket of liquid is trapped, reduced in volume, and then forced into the discharge line at a higher pressure. The flow rate of liquid from a rotary pump is a function of size and speed of rotation and is slightly dependent on the discharge pressure. Unlike reciprocating pumps, rotary pumps deliver nearly constant flow rates. Rotary pumps are used on liquids of almost any viscosity as long as the liquids do not contain abrasive solids. For this reason, they are very effective with many high viscosity waste mixtures. They operate in moderate pressure ranges (5000 psig), have small-to-medium capacities, and like the reciprocating pump, can be used for metering liquids.

Centrifugal pumps are widely used in the process industry because of simplicity of design, low initial cost, low maintenance, and flexibility of application. Centrifugal pumps have been built to move as little as a few gallons per minute against a small head, and as much as several thousand gallons per minute against a pressure of several hundred pounds per square inch. In its simplest form, this type of pump consists of an impeller rotating within a casing. Fluid enters the pump near the center of the rotating impeller and is thrown outwards by centrifugal force. The kinetic energy of the fluid increases from the center of the impeller to the tips of the impeller vanes. This high velocity is converted to a high pressure as the fast-moving fluid leaves the impeller and is driven into slower moving fluid in the volute or diffuser.[3]

Not all centrifugal pumps produce the radial flow (directed away from the axis of rotation) described previously; many produce an axial flow (directed along the axis of rotation) and others a combination of the two. The turbine type of centrifugal pump has smaller, straighter vanes, is driven at high speed, and generates a highly radial flow to produce higher pressures at lower flow rates. The axial-flow type employs multibladed propellers that generate a highly axial flow, resulting in large flow rates at lower pressures. For a specific balance between flow rate and pressure, impellers can be shaped to provide results between that of the turbine and axial-flow type; these are referred to as *mixed-flow* impellers.

The impeller is the heart of the centrifugal pump. It consists of a number of curved vanes or blades that are shaped in such a way as to give smooth fluid flow between the blades. In the *straight-vane, single-suction closed* impeller, the surfaces of the vanes are defined by straight lines parallel to the axis of rotation. The *double-suction* impeller is, in fact, two single-suction impellers arranged back to back in a single casing. Centrifugal pump casings may be of several designs but their main function is to convert kinetic energy imparted to the fluid by the impeller into a higher pressure. In addition, the casing provides an inlet and an outlet for the pump and contains the fluid. Casings may be either the *volute* or the *diffuser* type. The *volute* type has a continuous flow area that allows the velocity to decrease gradually, thereby reducing eddy formation; this minimizes the loss of energy due to turbulence. The *diffuser*-type casing has stationary guides that offer the liquid a widening path from impeller to casing; this also keeps turbulence at a minimum.[3]

In pumping fluids at a waste incineration facility, the operation of a pump may result in the release of air contaminants. Both the reciprocating and centrifugal pump can be sources of hazardous emissions. The opening in the cylinder through which the connecting rod drives the piston is the major source of contaminant release from a reciprocating pump. In centrifual pumps, normally the only potential source of leakage occurs where the drive shaft passes through the casing. Several methods have been devised for sealing the clearance between the pump shaft and fluid casing. For most applications, packed seals and mechanical seals are widely used. Packed seals can be used on both positive displacement and centrifugal pumps. A typical packed seal consists of a stuffing box filled with a sealing material or packing that encases the moving shaft. The stuffing box is fitted with a takeup ring that compresses the packing and causes it to tighten around the shaft. Materials used for packing vary with the fluid's temperature, physical and chemical properties, pressure, and pump type. Some commonly used materials are metal, rubber, leather, and plastics. For cases where the use of mechanical seals is not feasible, specialized pumps such as canned-motor, diaphragm, or electromagnetic pumps are used. These specialty pumps are used where no leakage can be tolerated and are available in a limited range of sizes; most are for low flow rates and all are of single- or two-stage construction. They have been used to handle both high temperature and very low temperature liquids. These pumps follow the same hydraulic principles as the traditional centrifugal pump. Because of their small size, they operate with rather low efficiencies, but in dangerous applications, efficiency must be sacrificed for safety.[2]

Pumps are employed at HWI sites for one and/or a combination of the following reasons:

1. The transportation of wastes from trucks or delivery vehicles to storage tanks.
2. The transportation of liquid wastes from storage tanks to incinerator.
3. The transportation of slurries and/or sludges from storage tanks to incinerators.

Pumps discharge pressures for reasons (2) and (3) are roughly 75 and 50 psig, respectively. Actual discharge pressures will vary, however, with waste viscosity, line pressure drop (which is a function of site-specific logistics), and atomization technique employed (particularly with liquid injection incinerators).

4. The delivery and transportation of fuel oil (where applicable) to the incinerator.
5. The delivery of water to the quench unit and waste heat boiler (where applicable).
6. The delivery and circulation of water in absorbers and scrubbers (where applicable).
7. The delivery and circulation of caustic solutions in absorbers (where applicable).
8. The transportation of liquids in reasons (6) and (7) to neutralization tanks or ponds or to water treatment facilities.

A general relationship that may be used to determine pump HP requirements is[2]

$$HP = \frac{7.27 \times 10^{-5} \, w \Delta P}{\rho E} \tag{11.3.10}$$

where w = liquid flow rate (lb/h)
   ΔP = developed pressure drop (psi)
   ρ = liquid density (lb/ft^3)
   E = pump–motor efficiency (fraction basis)

Compressors

In HWI operations, compressors, unlike fans and pumps, find only limited and specialized application. In some types of incinerators, liquid wastes are broken up into tiny droplets (atomized) before entering the burner. This is accomplished through the use of a high pressure stream of air or steam that impinges on the liquid waste stream and atomizes it. The pressurizing of the air or steam is accomplished through the use of compressors. Compressors are also used for atomization on certain types of air pollution control devices such as venturi scrubbers, which depend on fine water droplets to remove particulates from the flue gas stream. Compressors operate in a similar fashion to pumps and have the same classifications: rotary, reciprocating, and centrifugal. An obvious difference between the two operations is the large decrease in volume resulting from the compression of a gaseous stream compared to the negligible change in volume caused by the pumping of a liquid stream.

Centrifugal compressors are employed when large volumes of gases are to be handled at low to moderate pressure increases (0.5–50 psi). *Rotary* compressors have smaller capacities and can achieve discharge pressures up to 100 psi. *Reciprocating* compressors are the most common type used in industry and are capable of compressing small gas flows to as much as 3500 psig. With specially designed compressors, discharge pressures as high as 25,000 psig can be reached, but these devices are capable of handling only very small volumes, and do not work well for all gases. For the applications mentioned earlier, atomizing of liquid waste for combustion or of venturi scrubber water for gas cleaning, reciprocating compressors are normally used.

The following equation may be used to calculate compressor power requirements when compressor operation is adiabatic and the gas (usually air) follows ideal gas behavior.

$$W_s = \left(\frac{\gamma RT}{\gamma - 1}\right)\left[\left(\frac{P_2}{P_1}\right)^{(\gamma-1)/\gamma} - 1\right] \qquad (6.3.11)$$

where W_s = compressor work required per lbmol of air
R = 1.987 Btu/lbmol-°R
T = air temperature at compressor inlet conditions (°R)
P_1, P_2 = air inlet and discharge pressures
γ = ratio of the heat capacity at constant pressure to that at constant volume—typically 1.3 for air.

11.4 STACKS

Several types of stacks (referred to as *chimneys* by some in industry) are used to discharge incinerator flue gases into the ambient atmosphere. *Stub* or short stacks are usually fabricated of steel and extend a minimum distance upward from the discharge of an induced draft fan. These are constructed either of unlined or refractory-lined steel plate, or entirely of refractory and structural brick. Tall stacks are constructed of the same material as short stacks and are used to provide a greater pressure difference driving force (draft) than that resulting from the shorter stacks and to obtain more effective dispersion of the flue gas effluent into the atmosphere. Some chemical and utility applications use metal stacks that are made of a double wall with an air space between the metal sheets. This double wall provides an insulating air pocket to prevent condensation on the inside of the stack and thus avoid corrosion of the metal. As indicated in Section 11.3, a HWI system commonly employs an induced draft fan. The need for a fan, as opposed to the use of natural draft, arises from the high pressure drop from the air pollution equipment, the depressed stack temperatures that occur when venturi scrubbers and/or absorbers are used, and the control advantages realized with mechanical draft systems.

Whenever stack gas temperatures are higher than the ambient air temperature, draft will be produced by the bouyancy of the hot flue gases. The theoretical natural draft may be estimated by

$$\Delta P_{nd} = 0.019 PL \left(\frac{1}{T_o} - \frac{1}{T_s} \right) \tag{11.4.1}$$

where ΔP_{nd} = theoretical natural draft (psi)
   P = barometric (ambient) pressure (psia)
   L = height of stack above the breeching (ft)
   T_o = absolute ambient temperature (°R)
   T_s = absolute temperature of stack gases (°R)

The value for T_s should be an average value that takes into account heat losses and, for tall stacks, the temperature drop on expansion as the atmospheric pressure declines with increasing altitude. As the flue gases pass up the stack, frictional losses reduce the fraction of the theoretical draft that is available. Frictional losses may be estimated from the following equation:

$$\Delta P_f = 4.01 \times 10^{-5} (Lu^2 p/S) \tag{11.4.2}$$

where ΔP_f = frictional loss (psi)
   u = mean stack gas velocity (ft/s)
   p = stack perimeter (ft)
   S = stack cross-sectional area (ft^2)

In addition, some of the draft is lost to the velocity head in the rising gases. This *loss* is referred to as the *expansion loss* and may be estimated by

$$\Delta P_e = 2.59 \times 10^{-3} (u^2/S) \tag{11.4.3}$$

where ΔP_e = the expansion loss (atm).

Typical natural draft stacks[4] operate with gas velocities of about 30 ft/s.

The two process design variables for the stack are the height and diameter. The stack height is primarily dictated by atmospheric dispersion considerations. The diameter (D) may be determined by the equation[4]

$$D = (0.03 Q_a)^{0.5} \tag{11.4.4}$$

where D = stack diameter (ft)
   Q_a = flue gas flow rate (acfs)

An important factor in handling acid gases in a stack involves maintaining a high internal temperature. This often retards the detrimental effect on the masonry without the necessity for other precautions. If the flue gases are such that high temperatures alone are not sufficient, it may be necessary to protect the main walls by using an independent lining for the full height of the stack, and with a 3- to 4-in. air space between the lining and the main walls. The independent lining must be built of impervious brick with a low lime content and acid-proof mortar; very thin joints should be used. The mortar should be carefully chosen for its resistance not only to the particular acid involved, but to moisture as well. In addition, the top of the chimney should be protected by a cap covering both the lining and main walls and made of a material not affected by the flue gas. While room should be allowed for expansion, fumes and moisture must not be allowed to penetrate under the cap.

11.5   ILLUSTRATIVE EXAMPLES

Example 11.5.1.   Fan A has a blade diameter of 46 in. It is operating at about 1575 rpm while transporting 16,240 acfm of flue gas and requires 47.5 bhp. Fan B is to replace Fan A. It is to operate at 1625 rpm with a blade diameter of 42 in. and is of the same homologous series as Fan A. What is the power requirement of Fan B?

SOLUTION. Using the fan law [Eq. (11.3.6)], the power requirement for Fan B is calculated.

$$HP'/HP = (RPM'/RPM)^3 (D'/D)^5 (\rho'/\rho) = (1625/1575)^3 (42/46)^5 (1)$$

$$= 0.697$$

$$HP' = (0.697)(47.5) = 33.1 \, bhp$$

Example 11.5.2.   A fan operating at a speed of 1694 rpm delivers 12,200 acfm of flue gas at 5.0 in. H_2O static pressure and requires 9.25 bhp. What will be the new operating conditions if the fan speed is increased to 2100 rpm?

SOLUTION. The new fan flow rate is calculated using Eq. (11.3.4).

$$Q'_a = Q_a(RPM'/RPM)(D'/D)^3 = 12,200(2100/1694)(1)$$

$$= 15,124 \, acfm$$

Using Eq. (11.3.5), the new static pressure is calculated.

$$P'_s = P_s(RPM'/RPM)^2 (D'/D)^2 (\rho'/\rho) = (5.0)(2100/1694)^2 (1)(1)$$

$$= 7.68 \, in. \ H_2O$$

The required HP is calculated using Eq. (11.3.6).

$$HP' = HP(RPM'/RPM)^3 (D'/D)^5 (\rho'/\rho) = (9.25)(2100/1694)^3 (1)(1)$$

$$= 17.62 \, bhp$$

Example 11.5.3.   Calculate the HP required to process a 6500-acfm gas stream from an incinerator. The pressure drop across various pieces of equipment has been estimated to be 6.4 in. H_2O. The pressure loss for duct work, elbows, valves, and so

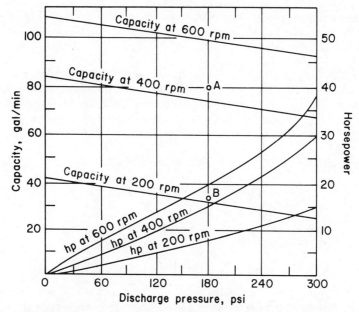

Figure 11.5.1. Performance characteristics of an external gear pump.

on, and expansion–contraction losses are estimated at 4.4 in. H_2O. Assume an overall fan–motor efficiency of 63%.

SOLUTION. The total pressure drop, ΔP (in. H_2O), is

$$\Delta P = (6.4 + 4.4) = 10.8 \text{ in. } H_2O$$

The brake horsepower required is calculated by Eq. (11.3.8).

$$\text{BHP} = Q_a(\text{acfm}) \; \Delta P(\text{in. } H_2O)(1.575 \times 10^{-4})/(\text{fan efficiency})$$

$$= (6500)(10.8)(1.575 \times 10^{-4})/(0.63)$$

$$= 17.55 \text{ bhp}$$

Note: The term 1.575×10^{-4} is a conversion factor to obtain units of horsepower (hp).

Example 11.5.4. It is necessary to deliver a liquid waste with density and viscosity similar to that of water to an incinerator at a rate of 80 gal/min. The pump must operate against a pressure of 180 psi. A pump with characteristics shown in Fig. 11.5.1 is available with variable speed drive. At what speed should the pump be operated? What horsepower is needed to maintain flow?

SOLUTION. Plotting this information on Figure 11.5.1. indicates that the speed is between the 400- and 600-rpm lines (point A). Interpolation gives a speed of about

425 rpm. Interpolating again on the HP curve gives a value of about 17 hp (point B). Using this data with Eq. (11.3.10) with an efficiency of 50% results in a pump horsepower of approximately 15.

Example 11.5.5. The exhaust flow rate from a small hazardous waste incineration facility is 1200 scfm (70°F). All of the gas is discharged through a small stack, which has an inlet area of 1.2 ft^2. The exhaust gas has a temperature of 300°F. What is the velocity of the gas through the stack inlet in ft/s? Neglect the pressure drop across the stack.

SOLUTION. Using Charles' Law to calculate the flow rate,

$$Q_a = Q_s(T_a/T_s) = 1200(760/530)$$

$$= 1721 \, \text{acfm}$$

The velocity is calculated from

$$v = Q_a/A = 1721/1.2 = 1434 \, \text{ft/min}$$

$$= 23.9 \, \text{ft/s}$$

Example 11.5.6. Compressed air is to be employed in the nozzle of a liquid injection incinerator to assist the atomization of a liquid hazardous waste. The air requirement for the nozzle is 7.5 lb/min at 40 psia. If atmospheric air is available at 60°F and 1.0 atm, calculate the power requirement.

SOLUTION. The ideal gas law will apply at these conditions. Set the coefficient γ for air equal to 1.3. The compressed energy requirement (delivered to the air) is given by Eq. (6.3.11).

$$W_s = -\frac{\gamma R T_1}{\gamma - 1}\left[\left(\frac{P_2}{P_1}\right)^{(\gamma-1)/\gamma} - 1\right] = -\frac{(1.3)(1.987)(520)}{1.3 - 1}\left[\left(\frac{40}{14.7}\right)^{(1.3-1)/1.3} - 1\right]$$

$$= -1163 \, \text{Btu/lbmol of air}$$

The power is

$$\text{HP} = -(1163)(7.5/29)(778)$$

$$= -234,000 \, \text{ft-lb}_f/\text{min}$$

This result may be divided by 33,000 to yield a HP requirement of 7.1 hp. The power required has the opposite sign. The reader should also note that this represents the *minimum* power required to accomplish this job.

PROBLEMS

1. Fan A has a blade diameter of 48 in., operates at 1650 rpm, delivers 18,450 acfm, and requires 50.0 bhp. Fan B operates at 1590 rpm and has a blade diameter of

45 in. If Fan B is moving a gas of the same density and is of the same homologous series as Fan A, what is its power requirement?

2. Calculate the HP needed to process a 8500 acfm gas stream from an incinerator. The pressure drop across the various pieces of equipment has been estimated to be 8.4 in. H_2O. The pressure loss for duct work, elbows, valves, and so on, and expansion–contraction losses are estimated at 5.8 in. H_2O. Assume an overall fan-motor efficiency of 58%.

3. A hazardous waste incinerator facility consists of a quench tower, scrubber, demister, and fans. The pressure drop due to bends, valves, elbows, and expansion–contraction losses is estimated to be 12.5 in. H_2O. The combustion gas to be treated has a flow rate of 23,000 acfm. The pressure drop across each device is given as follows:

Location	ΔP(in. H_2O)
Quench tower	5.1
Scrubber	6.8
Demister	3.9

Calculate the overall pressure drop across the facility.

4. A fan operating at a speed of 1750 rpm delivers 13,250 acrm at 5.5 in. H_2O static pressure and requires a BHP of 10.5. What will be the new operating conditions if the fan speed is increased to 2000 rpm?

5. It is necessary to pump a constant-flow stream of liquid with a density and viscosity similar to that of water in a reactor at a rate of 70 gal/min. The pump must operate against a pressure of 200 psi. A pump with characteristics shown in Fig. 11.5.1 is available with variable speed drive. At what speed should the pump be operated? What HP is needed to maintain this flow?

6. The exhaust gas flow rate from a facility is 1500 scfm. All of the gas is vented through a small stack, which has an inlet area of 1.3 ft². The exhaust gas temperature is 350°F. What is the velocity of the gas through the stack inlet in ft/s. Assume standard conditions to be 70°F and 1 atm. Neglect the pressure drop across the stack.

7. Compressed air is to be employed in the nozzle of a liquid injection incinerator to assist the atomization of a liquid hazardous waste. The air requirement for the nozzle is 0.5 lbmol/min at 50 psia. If atmospheric air is available at 50°F and 1.0 atm, calculate the power requirement.

REFERENCES

1. T. Shen, U. Choi, and L. Theodore, *Hazardous Waste Incineration Manual*, U.S. EPA Air Pollution Training Institute, Research Triangle Park, NC, to be published.

2. U.S. EPA, *Engineering Handbook for Hazardous Waste Incineration*, prepared by Monsanto Research Corporation, Dayton, OH, EPA Contract No. 68–03–3025, Sept., 1981.

3. A.S. Foust, L.A. Wenzel, C.W. Clump, L. Maus, and L. B. Anderson, *Principles of Unit Operations*, Wiley, New York, 1980.

4. W.R. Niessen, *Combustion and Incineration Processes*, Dekker, New York, 1978.

IV

FACILITY DESIGN

The purpose of Part IV is to expose the reader to some aspects of the design of an overall hazardous waste incineration facility. The complete engineering design of such an installation is a complex task involving not only the application of engineering principles of the type covered earlier in this text, but also the consideration of economic, legal, O&M (operation and maintenance), environmental, and often political and social factors. Most of these factors are too involved and too installation specific to be treated in this text. The authors feel, however, that an exposure to the basics of engineering design, even if only a rather sketchy one, can provide a better understanding of many of the technical (engineering) aspects associated with a hazardous waste facility.

The mechanics of the design process—the collecting of required information, preparation of schematics, application of material and energy balances plus other calculations, writing of a design report, and so on—are presented in Chapter 12. Economic analysis, an obviously important facet of any design procedure, is discussed in Chapter 13. Since illustrative examples often paint a clearer picture for the reader than textual presentation, the final chapter of this text, Chapter 14, consists mainly of two rather extensive plant design problems and their solutions. Since both problems involve overall facility design, these examples serve the function of merging many of the equipment design procedures presented earlier in the text into a more complete plant design package.

12

Design Principles

12.1 INTRODUCTION

Current design practices for hazardous waste incineration systems usually fall into the category of state of the art and pure empiricism. Past experience with similar applications is commonly used as the sole basis for the design procedure. The vendor maintains proprietary files on past installations; these files are periodically revised and expanded as new installations are evaluated. In designing a new unit, the files are consulted for similar applications and old designs are heavily relied on.

By contrast, the engineering profession in general, and the chemical engineering profession in particular, has developed well-defined procedures for the design, construction, and operation of chemical plants. These techniques, tested and refined for better than a half-century, are routinely used by today's engineers. These same procedures may also be used in the design of HWI facilities.

The purpose of this chapter is to introduce the reader to some of these process design fundamentals. Such an introduction to design principles, however sketchy, can provide the reader with a better understanding of the major engineering aspects of a HWI facility, including some of the operational, economic, and environmental factors associated with the process. No attempt is made in the sections that follow to provide extensive coverage of this topic; only general procedures and concepts are presented and discussed.

12.2 PRELIMINARY STUDIES

A process engineer is usually involved in one of two activities: building a plant or deciding whether to do so. The skills required in both cases are quite similar, but the money, time, and detail involved are not as great in the latter situation. It has been estimated that only one out of 15 proposed new processes ever achieves the construction stage. Thus, knowledge at the preliminary stage is vital to prevent financial loss on one hand and provide opportunity for success on the other. In well-managed process organizations, the engineer's evaluation is a critical activity that usually involves considerable preliminary research on the proposed process. Successful process development consists of a series of actions and decisions, the most significant of which takes place well before plant construction.

It is important to determine whether a project has promise as early in the development stage as possible. This section discusses some of the preparatory work required before the design of a HWI facility can be formally initiated. In the chemical process industry, there may be an extended period of preparatory work required if the proposed plant/facility is a unique or first-time application. This can involve bench-scale work by chemists to develop and better understand the process chemistry. This is often followed by pilot experimentation by process and/or development engineers to obtain scale-up and equipment performance information. However, these two steps are usually not required in the design of an established system. This is presently the situation with most HWI designs although some bench scale or pilot work may be necessary and deemed appropriate by management.

In general design practice, there are usually five levels of sophistication for evaluating and estimating. Each is discussed in the following list.

1. The first level requires little more than identification of products, raw materials, and utilities. This is what is known as an *order of magnitude* estimate and is often made by extrapolating or interpolating from data on similar existing processes. The evaluation can be done quickly and at minimum cost, but with a probable error exceeding ±50%.

2. The next level of sophistication is called a *study estimate* and requires a preliminary process flow sheet (to be discussed in the next section) and a first attempt at identification of equipment, utilities, materials of construction, and other processing units. Estimation accuracy improves to within ±30% probable error, but more time is required and the cost of the evaluation can escalate to over $10,000 for a million dollar plant. Evaluation at this level usually precedes expenditures for site selection, market evaluation, pilot plant work, and detailed equipment design. If a positive evaluation results, pilot plant and other activities may also begin.

3. A *scope* or *budget authorization*, the next level of economic evaluation, requires a more defined process definition, detailed process flow sheets, and prefinal equipment design (discussed in a later section). The information required is usually obtained from pilot plant, marketing, and other studies. The scope authorization estimate could cost upwards of $25,000 for a $1 million facility with a probable error exceeding ±20%.

4. If the evaluation is positive at this stage, a *project control* estimate is then prepared. Final flow sheets, site analyses, equipment specifications, and architectural and engineering sketches are employed to prepare this estimate. The accuracy of this estimate is about ±10% probable error. A project control estimate can serve as the basis for a corporate appropriation, for judging contractor bids, and for determining construction expenses. Due to increased intricacy and precision, the cost for preparing such as estimate for the process can approach $100,000.

5. The final economic analysis is called a *firm* or *contractor's* estimate. It is based on detailed specifications and actual equipment bids. It is employed by the contractor

to establish a project cost and has the highest level of accuracy, ±5% probable error. A cost of preparation results from engineering, drafting, support, and management/labor expenses. Because of unforeseen contingencies, inflation, and changing political and economic trends, it is impossible to assure actual costs for even the most precise estimates.

In the case of a HWI design, data on similar existing processes are normally available and economic estimates or process feasibility are determined from these data. It should be pointed out again that most processes in real practice are designed by duplicating or *mimicking* similar existing systems. Simple algebraic correlations that are based on past experience are the rule rather than the exception. This stark reality is often disappointing and depressing to students and novice engineers involved in design. For HWI facilities, the only preparatory work normally required is the gathering of all existing physical and chemical property data of the waste(s) and auxiliary materials and chemicals. Process chemistry information may also be needed but this too may be obtained directly from the literature, company files, or a similar type of application.

12.3   PROCESS SCHEMATICS

To the practicing engineer, particularly the chemical engineer, the process flow sheet is the key instrument for defining, refining, and documenting a chemical process. The process flow diagram is the authorized process blueprint, the framework for specifications used in equipment designation and design; it is the single, authoritative document employed to define, construct, and operate the chemical process.[1]

Beyond equipment symbols and process stream flow lines, there are several essential constituents contributing to a detailed process flow sheet. These include equipment identification numbers and names, temperature and pressure designations, utility designations, volumetric or molar flow rates for each process stream, and a material balance table pertaining to process flow lines. The process flow diagram may also contain additional information such as energy requirements, major instrumentation, and physical properties of the process streams. When properly assembled and employed, a process schematic provides a coherent picture of the overall process. It can point up some deficiencies in the process that may have been overlooked earlier in the study, for example, by-products (undesirable or otherwise) and recycle needs. But basically, the flow sheet symbolically and pictorially represents the interrelation between the various flow streams and equipment, and permits easy calculations of material and energy balances. These two topics are considered in the next sections.

As one might expect, a process flow diagram for a chemical or petroleum plant is usually significantly more complex than that for a HWI facility. As described earlier, a typical facility consists of the incinerator, followed by a waste heat boiler and/or a quench. This in turn is followed by the air pollution control equipment that may consist of separate devices for particulate and gaseous pollutant control (e.g., electrostatic precipitator–absorber) or a dry scrubber system (e.g., spray dryer–baghouse). Discharge gases from the control equipment are directed to an induced draft fan and then released to the atmosphere through a stack. There is usually no bypassing. There may be recycling and blowdown of the scrubbing liquid in the absorber or the combined quench–absorber system. The flow sequence and determinations thus approach a "railroad" or sequential type of calculation that does not require iterative calculations.

There are a number of symbols that are universally employed to represent equipment, equipment parts, valves, piping, and so on. These are depicted in the schematic in Fig. 12.3.1. Although a significant number of these symbols are used to describe some of the chemical and petrochemical waste processes, only a few are needed for even the most complex hazardous waste facilities. These symbols obviously reduce, and in some instances replace, detailed written descriptions of the process. Note that many of the symbols are pictorial, which helps in better describing process components, units, and equipment.

The degree of sophistication and details of a flow sheet usually vary with time. The flow sheet may initially consist of a simple free-hand block diagram with limited information that includes only the equipment; later versions may include line drawings with pertinent process data such as overall and componential flow rates, utility and energy requirements, and instrumentation. During the later stages of the design project, the flow sheet will consist of a highly detailed P&I (piping and instrumentation) diagram; this aspect of the design procedure is beyond the scope of this text; for information on P&I diagrams, the reader is referred to the literature.[2]

In a sense, flow sheets are the international language of the engineer, particularly the chemical engineer. Chemical engineers conceptually view a (chemical) plant as consisting of a series of interrelated building blocks that are defined as *units* or *unit operations*. The plant essentially ties together the various pieces of equipment that make up the process. Flow schematics follow the successive steps of a process by indicating where the pieces of equipment are located and the material streams entering and leaving each unit.

12.4   MATERIAL AND ENERGY BALANCES

Overall and componential material balances have already been described in rather extensive detail in Part II. Material balances may be based on mass, moles, or volume, usually on a rate (time rate of change) basis. Care should be exercised here since moles and volumes are *not* conserved, that is, the quantities change during the course of the combustion reaction. Thus, the initial material balance calculation should be based on *mass*. Mole balances and molar information are important in chemical reaction and phase equilibria calculations. Volume rates play a role in some equipment sizing calculations.

The units (feet, kilogram, etc.) employed may also create a problem. Despite earlier efforts by EPA and other government agencies, industry still primarily employs the British and/or engineering units in practice; there has been a reluctance to accept Standard International Units, commonly referred to as *SI*, that employ a modifed metric set of units. As indicated earlier, this text uses primarily engineering units. However, for those individuals who are more comfortable with the SI system, a short writeup on the conversion of units has been prepared by the authors and included in the appendix.

Most design calculations in the chemical process industry today include transient effects that can account for process upsets, startups, shutdowns, and so on. The describing equations for these time-varying (unsteady-state) systems are differential. The equations usually take the form of a first-order derivative with respect to time,

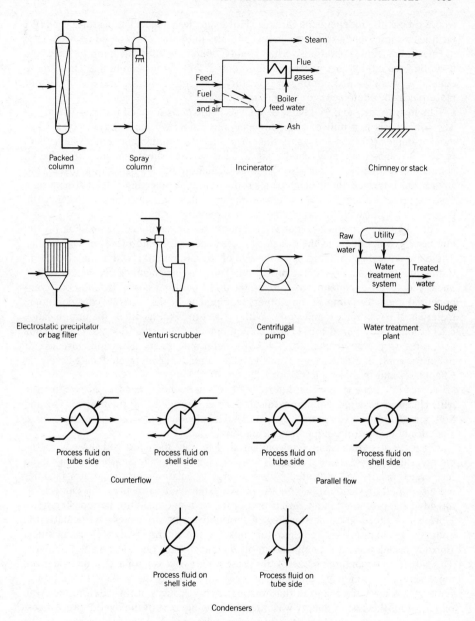

Figure 12.3.1. Flow sheet symbols.

where time is the independent variable. However, design calculations for most HWI facilities assume steady-state conditions, with the ultimate design based on worst case or maximum flow conditions. This greatly simplifies the calculations since the describing equations are no longer differential, but rather algebraic. Thus, these equations provide an accounting or inventory of all mass entering and leaving one or more pieces of equipment, or the entire process.

The heart of any material balance analysis is the *basis* selected for the calculation. The usual basis is a unit of time, for example, minutes, hours, days or year. For more complex calculations, and this may include multicomponent systems and recycle streams, one may choose as a basis a convenient amount of a "key" component or element. Note also that the calculation may be based on either a feed stream, an intermediate stream, or a product stream. Selecting a feed stream as a basis is preferred since it often allows one to follow the calculations through the process in a "railroad" manner, that is, in sequential order.[2]

The number of material balance equations can be significant depending on the number of components in the system, process chemistry, and pieces of equipment. These are critical calculations since the size of the equipment is often linearly related to the quantity of material being processed. This can then significantly impact—often linearly or even exponentially—on capital and operating costs. In addition, componential rates can impact on (other) equipment needs, energy considerations, materials of construction, and so on. With respect to HWI facilities, material balance calculations are required to: (1) size all equipment, (2) predict flue gas components and rates from the incinerator, and (3) determine particulate and gaseous air pollution control efficiencies. Topic (2) has already been treated extensively in Chapters 5 and 6; and topic (3), in Chapter 10.

Topics (2) and (3) are also interrelated. Componential balances in conjunction with chemical reaction equilibria calculations can provide information on Cl_2 and SO_3 generation (see Chapters 5 and 6). These individual flow rates must be treated analytically in control equipment calculations.

Once the material balance is completed, one may then proceed directly to the energy calculations, some of which play a significant role in the design of an incinerator facility. As indicated earlier, energy calculations for incinerator facilities are also based on steady-state conditions. An extensive treatment of this subject has already been presented in an earlier part of the text, and need not be repeated here. However, a thorough understanding of thermodynamic principles—particularly the enthalpy calculations—is required for most of the energy (balance) calculations. Entropy calculations are employed in all meaningful energy conservation analysis. This subject is beyond the scope of this presentation, but is a topic that may be given more serious attention by the engineering community in the not-to-distant future. Industry has always recognized that wasting energy reduces profits. Before the Arab oil embargo, the cost of energy was often a negligible part of the overall process cost and immense operational inefficiencies were tolerated. The sudden decrease in the availability of natural gas and oil resulting from the embargo raised the cost of energy and encouraged the elimination of unnecessary energy consumption. These energy-saving measures, once established, will probably continue to be employed even when energy costs are low.

One of the principal jobs of an engineer involved in the design of an incinerator facility is to account for the energy that flows into and out of each process unit, and to determine the overall energy requirement for the process. This is done by performing energy balances on each process unit and on the overall process. These balances can play an important role in HWI facility design. They find particular application in determining incinerator fuel requirements, in heat exchanger design, in heat recovery systems, in specifying materials of construction, and in calculating fan and pump power requirements.

The heat energy calculation is that associated with the incinerator. Both the conservation laws for mass and energy and the principles of thermodynamics must be applied in order to determine fuel requirements, operating temperatures, and air (excess) requirements for proper operation of the incinerator. This calculational procedure can be extremely tedious and detailed. An overall and componential material balance is applied around the incinerator that is then followed by an enthalpy balance. The enthalpy balance takes into account waste–fuel mixture, stoichiometric and excess air, product (flue) gas, latent (phase), and combustion (reaction) enthalpy effects. Fortunately, algorithms have been developed and are readily available to perform these calculations (see Chapter 6) and provide information on the interrelationship among incinerator operating temperature, waste–fuel heating value, and excess air requirements.

12.5 EQUIPMENT AND PROCESS DESIGN

Equipment design calculations are usually based on maximum throughput capacities for a new process. The reason for this is to enable the system to perform satisfactorily under the most extreme conditions. Material and energy balances based on these conditions are required before the individual equipment design. Each piece of equipment is designed individually and independently of all others. Equipment design is the first major step toward assessing the capital costs or purchase price of the entire plant.

As noted previously, chemical engineers describe the application of any piece of equipment that operates on the basis of mass, energy and/or momentum transfer as a *unit operation*. A combination of two or more of these unit operations is defined as a *unit process*. A whole chemical process can be described as a coordinated set of unit operations and unit processes. This subject matter has received much attention in recent years and, as a result, is well covered in the literature[3–5]. From details on these unit operations and processes, it is therefore possible to design new plants more efficiently by coordinating a series of *unit actions*, each one of which operates according to certain laws of physics regardless of the other operations being performed along with it. The unit operation of incineration, for example, is used in many different types of industries; many of the critical design parameters for the incinerator, however, are common to all incinerators and independent of the particular industry.

There are usually five conceptual steps to be considered with design of equipment. These are:

1. Identification of the parameters that must be specified.
2. Application of the fundamentals underlying theoretical equations or concepts.
3. Enumeration, explanation, and application of simplifing assumptions.
4. Possible use of correction factors for nonideal behavior.

5. Identification of other factors that must be considered for adequate equipment specification.

For a HWI plant, the major pieces of equipment that must be considered are all of the following:

Incinerator (rotary kiln or liquid injection)
Quench tower
Storage and handling facilities
Venturi scrubber
Absorber (packed tower or dry scrubber)
Baghouse
Fan(s)
Stack
Pumps

Calculation procedures for most of these have been presented in Part III. The design of a complete HWI facility is the topic of Chapter 14. Since design calculations are generally based on the maximum throughput capacity for the proposed process or for each piece of equipment, these calculations are never completely accurate. It is usually necessary to apply reasonable safety factors when setting the final design. Safety factors vary widely and are a strong function of the accuracy of the data involved, calculational procedures, and past experience. Attempting to justify these is a difficult task.

Unlike many of the problems encountered and solved by the engineer, there is no absolutely correct solution to a design problem; however, there is usually a *better* solution. Many alternative designs when properly implemented will function satisfactorily, but one alternative will usually prove to be economically more efficient and/or attractive than the others.

Although all engineers approach design problems somewhat differently, six major steps are generally required. These six steps are discussed here and may also be applied to the design of a HWI facility:

1. The first step is to conceptualize and define the process. A designer must know the bases and assumptions that apply, the plant capacity, and the process time allowed. Some of the answers to a host of questions pertaining to the process operation will be known from past experience.

2. After the problem has been defined, a method of solution must be sought. Although a method is seldom obvious, a good starting point is the construction of a process flow sheet. This effort usually produces several advantages. For example, it may suggest to the designer ways of reducing the complexity of the problem; it can allow for easier execution of material and energy balances which in turn can point up the most important process variables; it is an efficient way to become familiar with the process; and information that is lacking often becomes evident.

3. The third step is the actual design of the process equipment that involves the numerous calculations needed to arrive at specifications of operating conditions, equipment geometry, size, and so on. As part of this step, equipment costs must be

established. Cost estimating precision is dependent on the desired accuracy of the estimate. If the decision based on an estimate is positive, a detailed project control or contractor's estimate will follow.

4. An economic analysis must also be performed in order to determine the process feasibility. The main purpose of this step is to answer the question of whether a process will ultimately be profitable or not. To answer this, raw material, labor, equipment, and other processing costs are estimated to give an accurate economic forecast for the proposed operation.

5. In a case where alternate design possibilities exist, economics and engineering optimization is necessary. (This topic is discussed in the next section.) Since this is often the case, optimization calculations are usually applied several times during most design projects.

6. The final step of this design scheme is the compiling of a design report. A design report may represent the only relevant product of months or even years of effort. This is also discussed in the next section. At a minimum the report should look impressive. An attractive cover with the logo of the company or university is an important factor.

These six activities are prominent steps in the development of all modern chemical processes, including hazardous waste incineration.

The Design Report

As pointed out in Step 6, a comprehensive plant design project report is often required. This material should be written up in a clear and concise fashion. In addition, the project leader might be requested to make informal and/or formal presentations to management on the study. The report and presentation should explain what has been accomplished and how it has been carried out.

There are many different formats for design reports; the format will vary with the organization and the project. One possible outline for a project report is given in the following list:

- Title page
- Table of contents
- Abstract
- Prefatory comments (optional)
- Introduction
- Discussion
- Comprehensive process design including annualized costs
- Calculation and design limitations
- Optimization studies or suggestions
- Illustrative calculations
- Appendix

Great care should be exercised with the preparation of the abstract. It is recommended that the abstract be no longer than one single-spaced typewritten page. It should contain a short introduction, important results, and pertinent recommendations and conclusions. It should *not* refer to the body of the report. In many instances it is only the abstract that upper level management will initially review.

The use of this somewhat crude approach to a plant design presented in this chapter can produce valuable information. Perhaps its greatest value is that it can provide early signals when the proposed process is technically unfeasible or economically prohibitive.

REFERENCES

1. G.D. Ulrich, *A Guide to Chemical Engineering Process Design and Economics*, Wiley, New York, 1984.
2. T. Shen, U. Choi, and L. Theodore, *Hazardous Waste Incineration Manual*, U.S. EPA Air Pollution Training Institute, Research Triangle Park, NC, to be published.
3. R.M. Felder, and R.W. Rousseau, *Elementary Principles of Chemical Processes*, Wiley, New York, 1978.
4. R. Perry, and D. Green, *Perry's Chemical Engineering Handbook*, 6th ed., McGraw-Hill, New York, 1984.
5. W.L. McCabe and J.C. Smith, *Unit Operations of Chemical Engineering*, 3rd ed., McGraw-Hill, New York, 1976.

13

Economic Considerations

Contributing author: Benedetto Conetta

13.1 INTRODUCTION

Incineration has been recognized as one of the highest-cost options for hazardous waste management. Hence, an understanding of the economics involved in incineration is quite important in making decisions at both the engineering and management levels. Every engineer should be able to execute an economic evaluation of a proposed project. If the project is not profitable, it should obviously not be pursued and the earlier such a project can be identified, the fewer are the resources that will be wasted.

Before the cost of a HWI facility can be evaluated, the factors contributing to the cost must be recognized. There are two major contributing factors: capital costs and operating costs; these are discussed in Sections 13.2 and 13.3, respectively. Once the total cost of the facility has been estimated, the engineer must determine whether or not the project will be profitable. This involves converting all cost contributions to an annualized basis, a method that is discussed in Section 13.4, *Project Evaluation*. If more than one project proposal is under study, this method provides a basis for comparing alternate proposals and for choosing the best proposal. *Project Optimization* is the subject of Section 13.5, where a brief description of a perturbation analysis is presented. The chapter concludes with examples and problems that illustrate the material presented.

Detailed cost estimates are beyond the scope of this text. Such procedures are capable of producing accuracies in the neighborhood of ±5%; however, such estimates generally require many months of engineering work. This chapter is designed to give the reader a basis for a *preliminary cost analysis* only, with an expected accuracy of approximately ±20%.

13.2   CAPITAL COSTS

Capital Equipment Costs

Equipment cost is a function of many variables, one of the most significant of which is *capacity*. Other important variables include incinerator type, residence time, operating temperature, and degree of equipment sophistication. Preliminary estimates are often made from simple cost-capacity relationships that are valid when the other variables are confined to narrow ranges of values; these relationships can be represented by approximate linear (on log–log coordinates) cost equations of the form[1]

$$C = \alpha Q^{\beta} \tag{13.2.1}$$

where C = cost
   Q = some measure of equipment capacity
   α, β = empirical "constants" that depend mainly on equipment type

It should be emphasized that this procedure is suitable for rough estimation only; actual estimates from vendors are more preferable. Only major pieces of equipment such as the incinerator itself, waste heat boiler, and scrubber are included in this analysis; smaller peripheral equipment such as pumps and compressors are not discussed.

The most common types of HWI systems in use are the liquid injection, rotary kiln, and to a lesser extent, the multiple hearth (among others). Cost-capacity relationships for two types of liquid injection systems are presented here. Both include a waste heat boiler and a flue gas scrubbing system, in addition to the incinerator.

The first type, designated as *Case A* in Table 13.2.1, is designed to burn relatively clean liquids that contain an insignificant amount of chlorine or salts. Normal operating temperatures seldom exceed 1800°F and residence times are usually short (0.3–0.5 s). The incinerator (described in Chapter 7) consists of a horizontally aligned cylindrical steel shell, combustion air blower, flame safeguard, and combustion control instrumentation. *Case B* is a liquid injection system designed to handle problematic liquid wastes such as highly chlorinated materials, salty aqueous wastes, and liquefied tars. Units in this category are obviously more costly and are designed to accommodate operating temperatures up to 2200 to 2400°F; residence times range from 1.5 to 2.5 s. These units can be either vertical or horizontal cylindrical steel shells and include combustion air blowers, flame safeguards, and combustion control instrumentation. The applicable capacity range for both Case A and Case B systems is from 1 to 100 million Btu/h.

The rotary-kiln system in Table 13.2.1 includes the kiln–afterburner and auxiliaries plus waste heat boiler (optional) and a scrubbing–flue gas handling system. The cost of this system is considerably higher than that of the liquid injection systems. The applicable capacity range for the constants in Table 13.2.1 is 5–100 million Btu/h. The kiln temperatures range from 1600 to 1800°F while the afterburner temperatures range from 1800 to 2200°F with a residence time of approximately 2 s.

Note: A venturi scrubber is usually required for particulate control.

TABLE 13.2.1. Values of α and β for Use with Eq. (13.2.1)[a]

Incinerator[b]	α (dollar/(Btu/h))	β
Liquid injector (Case A)	30.1	0.503
Liquid injector (Case B)	41.6	0.512
Rotary kiln	220	0.478
Multiple hearth (Case A)	517	0.351
Multiple hearth (Case B)	586	0.630

Waste Heat Boiler[c]	α (dollar/(lb/h))	β
Case A	462	0.515
Case B	1944	0.401
Case C	9231	0.278

Scrubber[d]	α (dollar/acfm)	β
Case A	145	0.717
Case B	46	0.816

[a] Costs calculated using these constants are based on the value of the dollar in 1982.
[b] Units of Q are Btu/h.
[c] Units of Q are lb/h.
[d] Units of Q are acfm.

Hearth incineration systems include the multiple chamber combustion unit, a waste heat boiler (optional), and a scrubbing–flue gas handling system. *Case A* is valid for capacities ranging from 1 to 10 million Btu/h and *Case B* for capacities ranging from 10 to 50 million Btu/h.

Waste heat boilers, discussed in Chapter 8, are primarily used in facilities that can accommodate 10 to 50 million Btu/h, where the steam can be used elsewhere. Their use is less likely in smaller facilities and in commerical facilities where there is usually no market for the steam.[1] In Table 13.2.1, waste heat boilers are divided into three categories. Case A is for smaller facilities using a fire-tube boiler with a gas flow rate ranging from 2000 to 100,000 lb/h. Case B is for water-tube boilers of standard design with a gas flow rate range from 15,000 to 150,000 lb/h. Water-tube boilers operating with extremely high particulate loadings and high acid gas environments make up *Case C*, where the applicable range is 15,000 to 150,000 lb/h. The constants α and β for all three boiler types are based on inlet temperatures of 1800 to 2200°F and outlet temperatures of approximately 500°F. For the fire-tube boiler, the steam pressure is in the neighborhood of 150 psig. Water-tube steam pressures range from 150 to 200 psig.

Scrubbing system designs and costs also vary considerably from one system to the next. Typical scrubbing and flue gas handling systems include quenchers, venturi

TABLE 13.2.2. Cost Adjustment Factors for Scrubbing Systems[a]

Adjustment from Base Line Scenario	CAF
Extremely high venturi ΔP(100 in. H_2O)	2.0
Same, but no acid gas absorption	1.2
No venturi required	0.85
No absorption system required	0.6

[a] See ref. (1).

scrubbers, acid gas absorbers, caustic recycle systems, ID fans, stacks, and auxiliaries. System costs calculated from Table 13.2.1 are based on a venturi pressure drop of 30 in. H_2O.

In Table 13.2.1, scrubbers are divided into two categories based on whether or not a waste heat boiler is involved. *Case A* is for systems receiving hot combustion gas directly from the incinerator (no waste heat boiler) at a temperature of 1800 to 2200°F. The applicable range of capacities for the constants listed in Table 13.2.1 for this case is 1000–200,000 acfm. *Case B* involves scrubbing systems receiving gas from waste heat boilers at a temperature of approximately 550°F. Because of the much lower operating temperatures, less costly materials of construction and smaller-sized equipment can be employed. The applicable capacity range is from 1000 to 50,000 acfm. Because scrubbing systems vary so much in type, *cost adjustment factors* (CAF) are used to account for this variation[1]; these are listed in Table 13.2.2. To use the CAFs for any of the four cases listed, the cost resulting from Eq. (13.2.1) is multiplied by the appropriate CAF.

Similar methods for estimating the costs of storage tanks, waste conveyors, pumps, compressors, site development, and so on, are available in the literature.[1] If greater accuracy is needed, however, actual quotes from vendors should be used.

Again, the equipment cost estimation model just described is useful for a very preliminary estimation. If more accurate values are needed and if old price data are available, the use of an indexing method is better, although a bit more time consuming. The method consists of adjusting the earlier cost data to present values using factors that correct for inflation. A number of such indices are available; one of the most commonly used is the *Chemical Engineering Fabricated Equipment Cost Index* (FECI)[2], past values of which are listed in Table 13.2.3. Other indices for construction, labor, buildings, engineering, and so on are also available in the literature.[2] Generally, it is not wise to use past cost data older than 5 to 10 yr, even with the use of the cost indices. Within that time span, the technologies used in the processes may have changed drastically. The use of the indices could cause the estimates to be much greater than the actual costs. Such an error might lead to the choice of an alternative proposal other than the least costly.

Capital Costs

The usual technique for determining the *capital costs* (i.e., *total* capital costs, which include equipment design, purchase, and installation) for the incineration facility is

TABLE 13.2.3. Fabricated Equipment Cost Index

Year	Index
1986	312.5
1985	336.9
1984	334.1
1983	327.4
1982	326.0
1981	321.8
1980	291.6
1979	261.7
1978	238.6
1977	216.6
1976	200.8
1975	192.2

Source: Chemical Engineering, McGraw-Hill, New York.

based on the *factored method* of establishing direct and indirect installation costs as a function of the known equipment costs. This is basically a *modified Lang method*, whereby cost factors are applied to known equipment costs.[3,4]

The first step is to obtain from vendors (or, if less accuracy is acceptable, from one of the estimation techniques previously discussed) the purchase prices of the primary and auxiliary equipment. The total base price, designated by X, which should include instrumentation, control, taxes, freight costs, and so on, serves as the basis for estimating the direct and indirect installation costs. The installation costs are obtained by multiplying X by the cost factors, which are available in the literature.[3-8] For more refined estimates, the cost factors can be adjusted to more closely model the proposed system by using adjustment factors that take into account the complexity and sensitivity of the system.[3,4]

The second step is to estimate the direct installation costs by summing all the cost factors involved in the direct installation costs, which include piping, insulation, foundation and supports, and so on. The sum of these factors is designated as the DCF (*direct installation cost factor*). The direct installation costs are then the product of the DCF and X.

Step 3 consists of estimating the indirect installation costs. The procedure here is the same as that for the direct installation costs; that is, all the cost factors for the indirect installation costs (engineering and supervision, startup, construction fees, etc.) are added; the sum is designated by ICF (indirect installation cost factor). The indirect installation costs are then the product of ICF and X.

Once the direct and indirect installation costs have been calculated, the *total capital cost* (TCC) may be evaluated as

$$\text{TCC} = X + (\text{DCF})(X) + (\text{ICF})(X) \tag{13.2.11}$$

This cost is then converted to *annualized* capital costs with the use of the *Capital Recovery Factor* (CRF), which is described in Section 13.5. The annualized capital cost (ACC) is the product of the CRF and TCC and represents the total installed equipment cost distributed over the lifetime of the facility.

Some guidelines in purchasing equipment are listed here.

- Do not buy or sign any documents unless provided with certified independent test data.
- Previous clients of the vendor company should be contacted and their facilities visited.
- Prior approval from local regulatory officials should be obtained.
- A guarantee from the vendors involved should be required. Start-up assistance is usually needed and assurance of prompt technical assistance should be obtained in writing. A complete and coordinated operating manual should be provided.
- Vendors should provide key replacement parts if necessary.
- Finally, 10 to 15% of the cost should be withheld until the installation is completed.

13.3 OPERATING COSTS

Operating costs can vary from site to site since these costs, in part, reflect local conditions, for example, staffing practices, labor, and utility costs. Operating costs, like capital costs, may be separated into two categories: direct and indirect costs. *Direct* costs are those that cover material and labor and are directly involved in operating the facility. These include labor, materials, maintenance and maintenance supplies, replacement parts, waste (i.e., residues after incineration) disposal fees, utilities, and laboratory costs. *Indirect* costs are those operating costs associated with but not directly involved in operating the facility; costs such as overhead (e.g., building–land leasing and office supplies), administrative fees, local property taxes, and insurance fees fall into this category.

The major direct operating costs are usually those associated with labor and materials. *Materials* costs for incineration systems involve the cost of chemicals needed for the operation of the gas absorbers in the air pollution control systems.[9] *Labor* costs differ greatly, depending on whether or not the facility is located on-site or off-site and the degree of controls and/or instrumentation; on-site facilities are usually owned and operated by the company generating the waste; off-site facilities are usually run by commercial incinerator companies. Off-site facilities require a greater amount of operating labor. Typically, there are three working shifts per day with one supervisor per shift. On the other hand, on-site incinerators at manufacturing plants, particularly the liquid injection type, may be manned by a single operator for only one third or one half of each shift; usually only an operator, supervisor, and site manager are necessary to run the facility.[10] Salary costs vary from state to state and depend significantly on the location of the facility. The cost of *utilities* generally consists of that for electricity, water, fuel, and steam. The annual costs are estimated with the use of material and energy balances. Costs for *waste disposal*, that is, for the incinerator (ash) and scrubber (water) residues, can be estimated on a per ton capita basis. Costs of landfilling the ash can run significantly upwards of $100/ton if the material is hazardous, to $10/ton if it is nonhazardous. The cost of handling the scrubber effluent can vary depending on the method of disposal. For example, if conventional sewer disposal is used, the effluent probably has to be cooled and neutralized before disposal; the cost for this depends on the solids concentration. Annual *maintenance* costs can be estimated as a percentage of the capital equipment cost; the following values have been recommended: 5% for liquid injection incinerators, 7% for hearth types, and 10% for rotary kiln facilities.[9] The annual costs of replacement parts can be computed by dividing the cost of the individual part by its expected lifetime. The life expectancies can be found in the literature.[4] *Laboratory* costs depend on the number of samples tested and the extent of these tests; these costs can be estimated as 10 to 20% of the operating labor costs.

The *indirect* operating costs consist of overhead, local property tax, insurance, administration, less any credits. The *overhead* comprises payroll, fringe benefits, social security, unemployment insurance, and other compensation that is indirectly paid to the plant personnel. This cost can be estimated as 50 to 80% of the operating labor, supervision, and maintenance costs.[8,9] Local *property taxes* and *insurance* can be estimated as 1 to 2% of the total capital cost (TCC) while administration costs can be estimated as 2% of the TCC.

The total operating cost is the sum of the direct operating costs and the indirect operating costs less any credits that may be recovered (e.g., the value of recovered steam). Unlike capital costs, operating costs are always calculated on an annual basis.

13.4 PROJECT EVALUATION

In comparing alternate processes or different options of a particular process from an economic point of view, it is recommended that the total capital cost be converted to an annual basis by distributing it over the projected lifetime of the facility. The sum of both the *annualized capital costs* (ACC) and the *annual operating costs* (AOC) is known as the *total annualized cost* (TAC) for the facility. The economic merit of the proposed facility, process, or scheme can be examined once the total annual cost is available. Alternate facilities or options (e.g., a baghouse versus an electrostatic precipitator for particulate control, or two different processes for accomplishing the same degree of waste destruction) may also be compared. Note, a small flaw in this procedure is the assumption that the operating costs remain constant throughout the lifetime of the facility. However, since the analysis is geared to comparing different alternatives, the changes with time should be somewhat uniform among the various alternatives, resulting in little loss of accuracy.

The conversion of the total capital cost to an annualized basis involves an economic parameter known as the *capital recovery factor* (CRF). These factors can be found in any standard economics text[10,11] or can be calculated directly from Eq. (13.4.1).

$$\text{CRF} = \frac{(i)(1 + i)^n}{(1 + i)^n - 1} \tag{13.4.1}$$

where n = projected liftime of the system (yr)
i = annual interest rate (expressed as a fraction)

The CRF is a positive, fractional number. The ACC is computed by multiplying the TCC by the CRF. The annualized capital cost reflects the cost associated in recovering the initial capital outlay over the depreciable life of the system. Investment and operating costs can be accounted for in other ways, such as a *present worth* analysis. However, the capital recovery method is preferred because of its simplicity and versatility. This is especially true when comparing control systems having different depreciable lives. There are usually other considerations in such decisions besides the economics, but if all the other factors are equal, the proposal with the lowest total annualized cost should be the most viable.

If an on-site (internal) incineration system is under consideration for construction, the total annualized cost should be sufficient to determine whether or not the proposal is economically attractive as compared to other proposals. If, however, a commercial incineration process is being considered, the profitability of the proposed operation becomes an additional factor. The method presented here assumes a facility lifetime of 10 yr and that the land is already available.

One difficulty in this analysis is estimating the revenue generated from the facility, because both technology and costs can change from year to year. Also affecting the revenue generated is the amount of waste handled by the facility. Naturally, the more waste handled, the greater are the revenues. If a reasonable estimate as to the revenue that will be generated from the facility can be made, a rate of return can be calculated.

This method of analysis is known as *the discounted cash flow method using an end-of-year convention*, that is, the cash flows are assumed to be generated at the end of the year, rather than throughout the year (the latter obviously being the real case). An expanded explanation of this method can be found in any engineering economics text.[10] The data required for the analysis are the TCC, the annual after-tax cash flow (A) and the working capital (WC). Generally, for HWI facilities, WC includes the on-site fuel inventory, caustic soda solution, maintenance materials (spare parts, etc.), and wages for approximately 30 days. The WC is expended at the startup of the plant (time = 0 yr) and is assumed to be recoverable after the life of the facility (time = 10 yr). For simplicity, it is assumed that the WC is 10% of the TCC and that the TCC is spread evenly over the number of years used to construct the facility. (For this example, this construction period is assumed to be 2 yr.)

Usually, an after-tax rate of return on the initial investment of at least 30% is desirable. The method used to arrive at a rate of return will be discussed briefly. An annual after-tax cash flow can be computed as the annual revenues (R) less the annual operating costs (AOC) and less income taxes (IT). Income taxes can be estimated at 50% (this number may be lower with the passage of the new tax laws) of taxable income (TI).

$$IT = 0.5(TI) \tag{13.4.2}$$

The taxable income is obtained by subtracting the AOC and the depreciation of the plant (D) from the revenues generated (R) or

$$TI = R - AOC - D \tag{13.4.3}$$

For simplicity, straight-line depreciation is assumed, that is, the plant will depreciate uniformly over the life of the plant. For a 10-yr lifetime, the facility will depreciate 10% each year.

$$D = 0.1(TCC) \tag{13.4.4}$$

The annual after-tax cash flow (A) is then

$$A = R - \text{AOC} - \text{IT} \tag{13.4.5}$$

This procedure involves a trial-and-error solution. There are both positive and negative cash flows. The positive cash flows consist of A and the recoverable working capital in year 10. Both should be discounted *backward* to time $= 0$, the year the facility begins operation. The negative cash flows consist of the TCC and the initial WC. In actuality, the TCC is assumed to be spent evenly over the 2-yr construction period. Therefore, one half of this flow is adjusted *forward* from after the first construction year (time $= -1$ yr) to the year the facility begins operating (time $= 0$). The other half, plus the WC, is assumed to be expended at time $= 0$. Forward adjustment of the 50% TCC is accomplished by multiplying by an economic parameter known as the *single-payment compound amount factor F/P*, given by

$$F/P = (1 + i)^m \tag{13.4.6}$$

where $i =$ rate of return (fraction)
  $m =$ the number of years (in this case, 1 yr)

For the *positive* cash flows, the annual after-tax cash flow (A) is discounted backward by using a parameter known as the *uniform series present worth factor* (P/A). This factor is dependent on both interest rate (rate of return) and the lifetime of the facility and is defined by

$$P/A = \frac{(1 + i)^n - 1}{i(1 + i)^n}$$

where $n =$ lifetime of facility (in this case, 10 yr)

Note: The P/A is the inverse of the CRF (*capital recovery factor*). The recoverable working capital at year 10 is discounted backward by multiplying WC by the *single present worth factor* (P/F) which is given by

$$P/F = \frac{1}{(1 + i)^n} \tag{13.4.7}$$

where $n =$ lifetime of the facility (in this case, 10 yr)

The positive and negative cash flows are now equated and the value of i, the rate of return, may be determined by trial and error from Eq. (13.4.8).

$$\text{Term 1} + \text{Term 2} = \text{Term 3} + \text{Term 4} \tag{13.4.8}$$

where Term 1 $= [((1 + i)^{10} - 1)/i(1 + i)^{10}]A$;   worth at year $= 0$ of annual after tax cash flows
  Term 2 $= (1/(1 + i)^{10})\text{WC}$;   worth at year $= 0$ of recoverable WC after 10 yr
  Term 3 $= (\text{WC} + \frac{1}{2}\text{TCC})$;   assumed expenditures at year $= 0$

Term $4 = \frac{1}{2}(TCC)(1 + i)^1$; worth at year $= 0$ of assumed expenditures at year $= -1$

This method (i.e., *the discounted cash flow method using an end-of-year conversion*) is demonstrated in Example 13.6.5.

13.5 PERTURBATION STUDIES IN OPTIMIZATION

Once a particular process scheme has been selected, it is common practice to optimize the process from a capital cost and O&M (operation and maintenance) standpoint. There are many optimization procedures available, most of them too detailed for meaningful application to a hazardous waste facility. These sophisticated optimization techniques, some of which are routinely used in the design of conventional chemical and petrochemical plants, invariably involve computer calculations. Use of these techniques in HWI analysis is not warranted, however.

One simple optimization procedure that is recommended is the *perturbation study*. This involves a systematic change (or *perturbation*) of variables, one by one, in an attempt to locate the optimum design from a cost and operation viewpoint. To be practical, this often means that the engineer must limit the number of variables by assigning constant values to those process variables that are known beforehand to play an insignificant role. Reasonable guesses and simple or short-cut mathematical methods can further simplify the procedure. Much information can be gathered from this type of study since it usually identifies those variables that significantly impact on the overall performance of the process and also helps identify the major contributors to the total annualized cost.

13.6 ILLUSTRATIVE EXAMPLES

Example 13.6.1. Plans for the construction of a hazardous waste incinerator were initiated in 1977. The HCl emission control device was to be an absorber. The cost for this unit at that time was determined to be $136,000. Estimate the cost of the absorber in terms of both 1981 and 1986 dollars.

SOLUTION. This problem involves the use of the CE (*Chemical Engineering*) Fabricated Equipment Cost Index to obtain the FECI factors for the years 1977, 1981, and 1986. The cost for a particular year is the product of the cost in 1977 and the ratio of the FECI factor for that year to that for 1977. From Table 13.2.3,

$$1977 \text{ FECI} = 216.6$$
$$1981 \text{ FECI} = 321.8$$
$$1986 \text{ FECI} = 312.5$$

Cost in 1981 dollars is therefore

$$\text{cost } (1981) = \text{cost } (1977)(1981 \text{ FECI}/1977 \text{ FECI}) = (\$136,000)(321.8/216.6)$$
$$= \$202,050$$

Similarly, the cost in 1986 is

cost (1986) = cost (1977)(1986 FECI/1977 FECI) = ($136,000)(312.5/216.6)

$$= \$196,000$$

Example 13.6.2. A vendor has provided a total installed capital cost estimate for a small proposed incinerator facility of $25.36/acfm (350°F) of flue gas at the fan inlet. A bahouse is to be installed at the site for particulate control employing Teflon felt bags, each with an area of 12 ft^2 at an air-to-cloth ratio (acfm/ft^2) of 5.81 to 1.0. The cost per bag is $75.00. The total pressure drop across the system is 16.5 in. H_2O, including 3.0 in. H_2O for the baghouse. Determine the installed capital, operating, and maintenance costs for the proposed facility on an annualized basis. The following data are available and are applicable at the time of the purchase:

Estimated flue gas flow rate (at 350°F), $Q_a = 70,000$ acfm
Overall fan efficiency, $E_f = 60\%$ (at 350°F)
Operating time = 6240 h/yr
Electrical power cost = $0.15/kW-h
Fuel requirements for incineration of waste = none
Yearly maintenance cost (entire facility) = $50,000/yr
Replacement parts = 25% of bags each year
Lifetime of facility, $n = 15$ yr
Interest rate, $i = 8\%$
Salvage value = 0

Note: The salvage value is the amount that can be obtained, in dollars, for the equipment at the end of the facility or equipment lifetime.

SOLUTION. The problem involves the determination of tne total capital cost, the operating costs, and the maintenance costs. The CRF is used to convert the total capital cost to an annualized basis. The TCC is

$$\text{TCC} = (\$24.36/\text{acfm})(Q_a) = (\$25.36/\text{acfm})(70,000 \text{ acfm})$$

$$= \$1,775,200$$

To convert this cost to an annualized basis the CRF is calculated from Eq. (13.4.1).

$$\text{CRF} = (i)(1+i)^n/[(1+i)^n - 1] = (0.08)(1+0.08)^{15}/[(1+0.08)^{15} - 1]$$

$$= 0.11683$$

The ACC is the product of TCC and CRF.

$$\text{ACC} = (\text{TCC})(\text{CRF}) = (\$1,775,200)(0.11683)$$

$$= \$207,400/\text{yr}$$

The AOC is

$$AOC = \frac{(Q_a)(\Delta P)(\text{operating time})(\text{cost of electricity})}{E_f}$$

$$= (70{,}000)(16.5)(6240)(0.15)(1.175 \times 10^{-4})/(0.6)$$

$$= \$211{,}720/\text{yr}$$

Note: 1.175×10^{-4} is a conversion factor used to obtain the correct units. It is obtained as:

$$\frac{[5.20(\text{lb}_f/\text{ft}^2)/(\text{in. } H_2O)](60\,\text{min/h})}{2.6552 \times 10^6(\text{ft-lb}_f/\text{kW-h})} = 1.175 \times 10^{-4}$$

The annual maintenance cost can be computed as the sum of the yearly maintenance cost and the cost of replacement parts.

The total number of bags (N) necessary for operating the baghouse at the waste facility is calculated as follows:

$$N = (Q_a)/(\text{ACR}) \, (\text{area of bag})$$

$$= (70{,}000\,\text{acfm})/(5.81\,\text{acfm/ft}^2)(12\,\text{ft}^2)$$

$$= 1004 \text{ bags}$$

$$MC = \$50{,}000 + (0.25)(\text{cost of a bag})(N)$$

$$= \$50{,}000 + (0.25)(\$75.00/\text{bag})(1004 \text{ bags})$$

$$= \$68{,}825/\text{yr}$$

The total annualized cost of the facility is calculated as the sum of the ACC, AOC, and annual maintenance cost (MC).

$$TAC = ACC + AOC + MC = \$207{,}400/\text{yr} + \$211{,}720/\text{yr} + \$68{,}825/\text{yr}$$

$$= \$487{,}945/\text{yr}$$

Example 13.6.3. A stream of 100,000 acfm of particulate-contaminated gas from a waste facility is to be cleaned according to environmental regulations. You have been requested to find the best way to install a pollution control device at the site. A reputable vendor has provided information on the cost of the equipment, as well as installation, operating, and maintenance costs according to the condition of the gas treated. Given are all the data you have collected. Determine what equipment you would select for this work in order to minimize costs on an annualized basis.

	Electrostatic Precipitator	Venturi Scrubber	Baghouse
Equipment cost	$2.5/acfm	$1.5/acfm	$2.0/acfm
Installation cost	$0.65/acfm	$1.1/acfm	$0.8/acfm
Operating cost	$0.05/acfm-yr	$0.05/acfm-yr	$0.075/acfm-yr
Maintenance cost	$11,000/yr	$22,000/yr	$7500/yr
Lifetime of equipment	20 yr	15 yr	20 yr

Costs are based on comparable overall collection efficiencies. Interest rate is 10% and there is zero salvage value.

SOLUTION. The first step is to convert the equipment, installation, and operating costs to total costs by multiplying each by the total gas flow, 100,000 acfm. Hence, for the electrostatic precipitator, the total costs are

$$\text{equipment cost} = 100,000\,\text{acfm}\ (\$2.5/\text{acfm}) = \$250,000$$

$$\text{installation cost} = 100,000\,\text{acfm}\ (\$0.65/\text{acfm}) = \$65,000$$

$$\text{operating cost} = 100,000\,\text{acfm}\ (\$0.05/\text{acfm-yr}) = \$5000/\text{yr}$$

Note that the operating costs are on an annualized basis. The equipment cost and the installation cost must then be converted to an annual basis using the CRF. From Eq. (13.4.1),

$$\text{CRF} = (0.1)(1 + 0.1)^{20}/[(1 + 0.1)^{20} - 1] = 0.11746$$

The annual costs for the equipment and the installation is the product of the CRF and the total costs of each.

$$\text{equipment annual cost} = \$250,000(0.11746)$$

$$= \$29,365/\text{yr}$$

$$\text{installation annual cost} = \$65,000(0.11746)$$

$$= \$7635/\text{yr}$$

The calculations for the scrubber and the baghouse are performed in the same manner. The three devices can be compared after all the annual costs are added. The results are tabulated here:

	Electrostatic Precipitator	Venturi Scrubber	Baghouse
Equipment cost	$250,000	$150,000	$200,000
Installation cost	$ 65,000	$110,000	$ 80,000
CRF	0.11746	0.13147	0.11746
Annual equipment	$ 29,365	$ 19,721	$ 23,492
Annual installation	$ 7,635	$ 14,462	$ 9,397
Annual operating	$ 5,000	$ 5,000	$ 7,500
Annual maintenance	$ 11,000	$ 22,000	$ 7,500
Total annual $	$ 53,000	$ 61,183	$47,889

According to the analysis, the baghouse is the most economically attractive device since the annual cost is the lowest.

Example 13.6.4. Plans are underway to construct and operate a commercial hazardous waste facility in Dumpsville in the state of Egabrag. The company is still undecided as to whether to install a liquid injection or rotary kiln incinerator at the waste site. The liquid injection unit is less expensive to purchase and operate than a comparable rotary kiln system. However, projected waste treatment income from the rotary kiln unit is higher since it will handle a larger quantity of wastes. In addition, the rotary kiln will be treating solids as well as liquid wastes.

Based on the economic and financial data provided here, select the incinerator that will yield the higher annual profit.

Costs/Credits	Liquid Injection	Rotary Kiln
Capital ($)	2,625,000	2,975,000
Installation ($)	1,575,000	1,700,000
Operation ($/yr)	400,000	550,000
Maintenance ($/yr)	650,000	775,000
Income ($/yr)	2,000,000	2,500,000

Calculations should be based on an interest rate of 12% and a process lifetime of 12 yr for both incinerators.

SOLUTION. For both units:

$$CRF = (0.12)(1 + 0.12)^{12}/[(1 + 0.12)^{12} - 1]$$

$$= 0.1614$$

Annual capital and installation costs for the liquid injection (LI) unit are

$$LI\ costs = (2,625,000 + 1,575,000)(0.1614)$$

$$= \$677,880/yr$$

Annual capital and installation costs for the rotary kiln (RK) unit are

$$RK\ costs = (2,975,000 + 1,700,000)(0.1614)$$

$$= \$754,545/yr$$

A comparison of costs and credits for both incinerators is given in the following table.

	Liquid Injection	Rotary Kiln
Total installed ($/yr)	678,000	755,000
Operation ($/yr)	400,000	550,000
Maintenance ($/yr)	650,000	775,000
Total annual cost ($/yr)	1,728,000	2,080,000
Income credit ($/yr)	2,000,000	2,500,000
Profit ($/yr)	272,000	420,000

A rotary kiln incinerator is recommended.

Example 13.6.5. Two small commercial incineration facility designs are under consideration. The first design involves a liquid injection incinerator and the second a rotary kiln incinerator. For the liquid injection system, the total capital cost (TCC) is $2.5 million, the annual operating costs (AOC) are $1.2 million, and the annual revenue generated from the facility (R) is $3.6 million. For the rotary kiln system, TIC, AOC, and R are $3.5, 1.4, and 5.3 million, respectively. Using the discounted cash flow method, which design is more attractive? Assume a 10-yr facility lifetime and a 2-yr construction period.

SOLUTION. The solution involves the calculation of the rate of return for each of the two proposals.

1. *The Liquid injection System.* The working capital (WC), the depreciation of the plant (D), the taxable income (TI), the income tax to be paid (IT), and the annual after tax cash flow (A) must first be calculated.
   The depreciation is

$$D = 0.1(\text{TCC}) = 0.1(\$2,500,000) = \$250,000$$

The WC is sct at 10% of the TCC.

$$\text{WC} = 0.1(\text{TCC}) = 0.1(\$2,500,000) = \$250,000$$

The TI is calculated using Eq. (13.4.3).

$$\text{TI} = R - \text{AOC} - D = \$3,600,000 - \$1,200,000 - \$250,00 = \$2,150,000$$

The IT is calculated from Eq. (13.4.2).

$$\text{IT} = 0.5(\text{TI}) = 0.5(\$2,150,000) = \$1,075,000$$

The after-tax cash flow is calculated using Eq. (13.4.5).

$$A = R - \text{AOC} - \text{IT} = 3,600,000 - \$1,200,000 - \$1,075,000 = \$1,325,000$$

The rate of return can be computed by solving Eq. (13.4.8) by trial and error.

$$\left[\frac{(1+i)^{10} - 1}{i(1+i)^{10}}\right] A + \left(\frac{1}{(1+i)^{10}}\right)\text{WC} = \text{WC} + \frac{1}{2}\text{TCC} + \frac{1}{2}\text{TCC}(1+i)^1$$

or

$$\left[\frac{(1+i)^{10} - 1}{i(1+i)^{10}}\right] 1.325 \times 10^6 + \left[\frac{1}{(1+i)^{10}}\right] 0.250 \times 10^6$$
$$= 0.250 \times 10^6 + 1.250 \times 10^6 + 1.250 \times 10^6 (1+i)^1$$

A rate of return (i) is assumed until both sides of the equation are equal. Initially assuming a rate of 30%, substitution of $i = 0.30$ in this equation yields

$$\$4,115,034 = \$3,125,000$$

The difference between the two sides of the equation is \$990,036. For a second try, assume $i = 0.40$. This equation becomes

$$\$3,207,192 = \$3,250,000$$

with a difference between the two sides of $-\$42,807$. The rate of return is very close to 40%. Linearly interpolating,

$$\frac{40 - i}{40 - 30} = \frac{(-\$42,807) - 0}{(-\$42,807) - \$990,036}$$

$$i = 39.6\%$$

2. *The Rotary Kiln System*. As with the liquid injection system,

$$\text{WC} = D = 0.1(\$3,500,000) = \$350,000$$

$$\text{TI} = \$5,300,000 - \$1,400,000 - \$350,000 = \$3,550,000$$

$$\text{IT} = 0.5(\$3,550,000) = \$1,775,000$$

The annual after tax cash flow is

$$A = \$5,300,000 - \$1,400,000 - \$1,775,000 = \$2,125,000$$

Equation (3.4.8) becomes

$$\left[\frac{(1 + i)^{10} - 1}{i(1 + i)^{10}} \right] 2.125 \times 10^6 + \left[\frac{1}{(1 + i)^{10}} \right] 0.350 \times 10^6$$

$$= 0.350 \times 10^6 + 1.750 \times 10^6 + 1.750 \times 10^6 (1 + i)^1$$

Assuming $i = 0.40$ yields

$$\$5,141,850 = \$4,550,000$$

The difference is \$591,850
Assuming $i = 0.45$ yields

$$\$4,615,519 = \$4,637,500$$

with a difference of $-\$21,981$. Once again using linear interpolation, the interest rate is found to be 44.8%

Hence, by the *discounted cash flow method*, the rate of return on the initial capital investment is $\sim 5\%$ greater for the rotary kiln system than the liquid injection system. From a purely financial standpoint, the rotary kiln system is the more attractive.

PROBLEMS

1. A hazardous waste incinerator cost $985,000 in 1982. You intend to install a similar type incinerator in your facility in 1986. What is the cost of the new incinerator? If the total (direct plus indirect) installation cost is 60% of the cost of the incinerator, what is the annualized capital cost of this unit? The expected life of the incinerator is 10 yr; assume an interest rate of 10%.

2. A 200,000-acfm stream of contaminated air is to be treated using one of three devices, an electrostatic precipitator, a venturi scrubber, or a baghouse. The following data were obtained from a reliable vendor. Which air pollution control device should be selected? Assume an interest rate of 10%.

	Electrostatic Precipitator	Venturi Scrubber	Baghouse
Total capital cost	$3.5/acfm	$2.8/acfm	$3.2/acfm
Operating cost	$0.03/acfm	$0.035/acfm	$0.08/acfm
Maintenance costs	$12,000/yr	$15,000/yr	$13,000/yr
Lifetime	10 yr	10 yr	10 yr

3. There are two options available to an engineer designing a hazardous waste facility. The engineer can use either a "conventional" incinerator or an incinerator with heat recovery. The engineer compiled the data given here. From an economic point of view, which incinerator should the engineer select? The lifetime of each incinerator is 10 yr; the interest rate is 10%.

	Incinerator with Heat Recovery	Incinerator w/o Heat Recovery
Incinerator cost	$1,294,000	$1,081,000
Installation cost	$ 786,000	$ 659,000
Operating labor	$ 39,900/yr	$ 8,500/yr
Maintenance	$ 43,000/yr	$ 17,000/yr
Utilities	$ 958,000/yr	$ 821,000/yr
Overhead	$ 51,300/yr	$ 13,900/yr
Taxes, insurance and administration	$ 86,200/yr	$ 72,600/yr
Credits (from steam recovery)	$ 380,000/yr	0

4. Resolve Example 13.5.4 if the equipment lifetime and interest rate are 20 yr and 10%, respectively.

REFERENCES

1. R.J.McCormick and R.J. DeRosier, *Capital and O&M Cost Relationships for Hazardous Waste Incineration*, Acurex Corp., Cincinnati, OH, EPA Report 600/2–84–175, October, 1984.

2. J. Matley, "CE Cost Indexes Set Slower Pace," *Chem. Eng.*, 75–76, April 29 (1985).

3. R.B. Neveril, *Capital and Operating Costs of Selected Air Pollution Control Systems*, Gard, Inc., Niles, IL, EPA Report 450/5–80–002, December, 1978.

4. W.M. Vatavuk, and R.B. Neveril, "Factors for Estimating Capital and Operating Costs," *Chem. Eng.*, 157–162, November 3, (1980).

5. G.A. Vogel, and E.J. Martin, "Hazardous Waste Incineration Part 1—Equipment Sizes and Integrated-Facility Costs," *Chem. Eng.*, 143–146, September 5, (1983).

6. G.A. Vogel, and E.J. Martin, "Hazardous Waste Incineration Part 2—Estimating Costs of Equipment and Accessories," *Chem. Eng.*, 75–78, October 17, (1983).

7. G.A. Vogel, and E.J. Martin, "Hazardous Waste Incineration Part 3—Estimating Capital Costs of Facility Components," *Chem. Eng.*, 87–90, November 28, (1983).

8. G.D. Ulrich, *A Guide to Chemical Engineering Process Design and Economics*, Wiley, New York, 1984.

9. G.A. Vogel and E.J. Martin, "Hazardous Waste Incineration Part 4—Estimating Operating Costs," *Chem. Eng.*, 97–100, January 9, (1984).

10. E.P. DeGarmo, J.R. Canada, and W.G. Sullivan, *Engineering Economy*, 6th ed., Macmillan, New York, 1979.

11. *C. Hodgman, S. Selby, and R. Weast, editors, CRC Standard Mathematical Tables*, 12th ed., Chemical Rubber Company, Cleveland, OH, 1961 (presently CRC Press, Boca Raton, FL).

14

Design of a Hazardous Waste Incineration Facility

Contributing author: Rocco Grassi

14.1 INTRODUCTION

This chapter presents two problems that illustrate design techniques associated with hazardous waste incineration (HWI). The example problem in Section 14.2 requires the application of some of the design and calculational procedures developed by the authors. The second example problem, Section 14.3, requires the use of several of the EPA worksheets presented in Chapter 6. Although most of the procedures used to design the equipment are approximate, they will provide fairly accurate results. In both examples, incineration calculations are first performed to determine the combustion flue products and the operating temperatures. These are followed by quench and/or waste heat boiler calculations, which in turn are followed by design calculations for the air pollution control equipment. Both examples conclude with horsepower requirements and economic analyses. This chapter will expose the reader to the type of engineering calculations that are involved in the design of modern HWI facility.

14.2 ILLUSTRATIVE DESIGN PROBLEM I

A chemical company is generating appreciable amounts of chlorobenzene (C_6H_5Cl) at one of its plants. Management has decided that it would be economically advantageous to design and construct a HWI facility on-site. The company's corporate engineering staff is in the process of preparing the preliminary process design for this facility. Since the waste is a liquid, a liquid injection incinerator is to be employed. Pilot studies indicate that burning with 50% excess air will satisfy the *four nines* (99.99% DRE) regulatory requirement.

Data and information pertinent to the incinerator are given in the following list.

429

1. The optimum length-to-diameter ratio (L/D) has been determihed to be four.
2. The net heating value (NHV) for C_6H_5Cl is assumed to be high enough so that no auxiliary fuel is required.
3. The C_6H_5Cl contains 1.0 wt% of particulates in the form of fly ash.
4. The maximum capacity of the unit is to be 5000 lb C_6H_5Cl/h.

Economic studies have shown that a waste heat boiler should be used to recover energy from the flue gas. Steam at 80 psia is needed to drive a turbine elsewhere in the plant; water at 100°F is available for this purpose. A boiler of the fire-tube variety is to be designed to cool the flue gases to 550°F and generate the required steam. The tubes are to be 1.5-in. o.d., BWG 18; the optimum mass flow rate of flue gas is 100 lb/h through each.

A venturi scrubber is to be installed to collect the particulates using water as the scrubbing medium. Previous studies on a similar process indicate that the optimum L/G ratio and throat velocity should be 9.5 gal/1000 ft^3 and 250 ft/s, respectively. The correlation coefficient k (which is a function of the system geometry and operating conditions) is 0.200 acf/gal. Assume that the average particle size of the fly ash discharged is 1.2 μm and that the particle density is 46 lb/ft^3. Once the particulates are collected, the wastewater is pumped to an on-site wastewater treatment system.

The HCl in the flue gas is to be removed with water at 100°F using a packed absorption tower. The absorber operating temperature and pressure are 150°F and 1 atm, respectively, and the packing is 3.5-in. ceramic Raschig rings.

System pressure drop data have also been provided from preliminary studies. Estimated pressure drops across the incinerator, waste heat boiler, and absorber are 3, 2, and 1 in. H_2O, respectively. The pressure drops for valves, fittings, elbows, and ductwork (including expansion and contraction losses) totals 5 in. H_2O. Several manufacturers of induced draft fans have suggested an overall fan motor efficiency of 65%.

Cost estimates indicate that the annual operating cost (AOC) will be $800,000/yr. The total capital (TCC) and maintenance costs (MC) have been estimated to be $4,000,000 and $100,000/yr, respectively. For an economic analysis, a 25-yr facility lifetime and an 8% interest rate are assumed.

You and several other engineers have been called on to prepare a preliminary process design for the proposed HWI facility. The analysis should include:

Combustion calculations
Incinerator design
Waste heat boiler design
Venturi scrubber calculations
Packed tower absorber design
Horsepower requirements
Economic evaluation, including annualized costs

SOLUTION

Preliminary Calculations

Basis: 5000 lb of cholorobenzene, C_6H_5Cl. (The 1% ash in the waste will be neglected in the stoichiometric calculations.) The NHV of C_6H_5Cl can be estimated by a modified form of Du Long's equation that includes the chlorine content.

$$NHV = 14,000m_C + 45,000[m_H - (m_O/8)] - 760m_{Cl} + 4500m_S \qquad (5.4.9)$$

where
$$NHV = \text{net heating value (Btu/lb)}$$
$$m_C = \text{weight fraction of C (carbon) in the waste–fuel mixture}$$
$$C, H, O, Cl, S = \text{subcripts indicating carbon, hydrogen, oxygen, and sulfur,}$$
$$\text{respectively}$$

Using the basis of 5000 lb of C_6H_5Cl, the mass fractions are determined.

$$(5000\,lb \text{ of } C_6H_5Cl)\frac{6 \times 12\,lb \text{ of } C}{112\,lb \text{ of } C_6H_5Cl} = 3214\,lb \text{ of } C$$

$$(5000\,lb \text{ of } C_6H_5Cl)\frac{5 \times 1\,lb \text{ of } H}{112\,lb \text{ of } C_6H_5Cl} = 223\,lb \text{ of } H$$

$$(5000\,lb \text{ of } C_6H_5Cl)\frac{1 \times 35\,lb \text{ of } Cl}{112\,lb \text{ of } C_6H_5Cl} = 1563\,lb \text{ of } Cl$$

The weight percentage of each compound is 64.28% C, 31.26% Cl, and 4.46% H. The NHV is calculated directly by Eq. (5.4.9).

$$NHV = 14,000(0.6428) + 45,000(0.0446) - 760(0.3126)$$

$$= 10,769\,Btu/lb$$

This number is slightly lower than that predicted in Example 5.7.9. To be conservative, this (lower) value will be used. Because a value of 10,769 Btu/lb is reasonably high for the NHV, no auxiliary fuel is required.

Combustion Calculations

The chemical equation for the combustion of 1.0 lbmol of chlorobenzene with stoichiometric air is

$$C_6H_5Cl + 7O_2 + 26.3N_2 \longrightarrow CO_2 + 2H_2O + HCl + 26.3N_2 \qquad (14.2.1)$$

From Eq. (14.2.1), the stoichiometric amount of air is 33.3 lbmol ($O_2 + N_2$). Since 50% excess air is to be employed, the total amount of air is 1.5 (33.3) = 50 lbmol, and the amount of excess air is 16.7 lbmol. Equation (14.2.1) is now rewritten for 50% excess air. Also shown are the molecular weight and mass flow rate (w_i) of each component (based on the maximum capacity of 5000 lb of C_6H_5Cl/h).

$$C_6H_5Cl + 10.5O_2 + 39.5N_2 \longrightarrow 6CO_2 + 2H_2O + HCl + 39.5N_2 + 3.5O_2$$

MW	112	32	28	44	18	36	28	32
w_i	5000	14,986	49,328	11,774	1606	1606	49,328	4995	(14.2.2)

Table 14.2.1 shows flue gas quantities and gas partial pressures (assuming a total pressure of 1 atm) of each component in the flue gas.

The temperature of the flue gas may be estimated by Eq. (6.4.20).

TABLE 14.2.1. Flue Gas Composition (Illustrative Problem No. 1)

Component	Mass Flow (lb/h)	Molar Flow (lbmol/h)	Mole Fraction	Partial Pressure (atm)
CO_2	11,774	268	0.1155	0.1155
H_2O	1,606	89.2	0.0384	0.0384
HCl	1,606	44.6	0.0192	0.0192
N_2	49,328	1,762	0.7595	0.7595
O_2	4,995	156	0.0672	0.0672

$$T = 60 + \frac{NHV}{(0.3)[1 + (1 + EA)(7.5 \times 10^{-4})(NHV)]}$$

$$= 60 + \frac{10769}{0.3[1 + (1 + 0.50)(7.54 \times 10^{-4})(10,769)]}$$

$$= 2784°F$$

where T = temperature (°F)
NHV = net heating value of the waste (Btu/lb)
EA = fractional amount of excess air

The reader may refer back to Chapter 6 for additional details on this equation.

Bècause sufficient amounts of HCl and O_2 are present, the following reaction also takes place.

$$2HCl + 0.5O_2 \longleftrightarrow H_2O + Cl_2$$

In this reaction, chlorine, a hazardous pollutant, is formed. To determine the partial pressure of Cl_2 in the flue gas, the equilibrium constant (K_p) for this reaction must first be obtained. Equation (14.2.3) may be used for this purpose[1]:

$$K_P = Ae^{BX} \tag{14.2.3}$$

where $A = 0.229 \times 10^{-3}$
$B = 7340$
$X = 1/T$; T in kelvins (K)

Since $T = 2784°F = 1802$ K

$$K_P = 0.229 \times 10^{-3} e^{(7340/1802)}$$

$$= 0.0134 \text{ (atm)}^{-1}$$

The equilibrium expression may be represented in the following form

$$K_P = \frac{P_{H_2O} P_{Cl_2}}{P_{HCl}^2 P_{O_2}^{0.5}} \tag{14.2.4}$$

TABLE 14.2.2. Corrected Flue Gas Composition (Illustrative Problem No. 1)

Component	Mass Flow (lb/h)	Molar Flow (lbmol/h)	Mole Fraction	Partial Pressure (atm)
CO_2	¹1,792	268	0.1155	0.1155
H_2O	1,607	89.2	0.03843	0.03843
HCl	1,598	44.4	0.01913	0.01913
N_2	49,336	1,762	0.7596	0.7596
O_2	4,989	155.9	0.06718	0.06718
Cl_2	5	0.0766	0.000033	0.000033

The partial pressures in atmospheres of H_2O, HCl, and O_2 are given in Table 14.2.1.

$$P_{H_2O} = 0.0384, \quad P_{HCl} = 0.0192, \quad P_{O_2} = 0.0672$$

These are the initial partial pressures. Let Z represent the partial pressure of Cl_2 at equilibrium. The equilibrium partial pressures that appear in Eq. (14.2.4) are given here.

$$P_{H_2O} = 0.0384 + Z, \quad P_{HCl} = 0.0192 - 2Z, \quad P_{O_2} = 0.0672 - 0.5Z$$

Substituting these values into Eq. (14.2.4) and solving the equation for Z by trial and error gives

$$0.0134 = \frac{(0.0384 + Z)(Z)}{(0.0192 - 2Z)^2(0.0672 - 0.5Z)^{0.5}}$$

with

$$Z = 3.3 \times 10^{-5} \text{ atm} = 33 \text{ ppm}$$

The mole fractions and partial pressures given in Table 14.2.1 were determined without taking into account the formation of Cl_2. Since HCl and O_2 are reacting to form H_2O and Cl_2, their partial pressures will decrease and those of H_2O and Cl_2 will increase. The results of this approximate calculation for the corrected flue gas quantities are given in Table 14.2.2.

Incinerator Design

The amount of heat released by the waste during operation may be expressed as

$$q = w(\text{NHV}) \qquad (14.2.5)$$

where q = rate of heat released (Btu/h)
w = mass flow rate of waste (lb/h)

Since $w = 5000$ lb/h and the NHV = 10,769 Btu/lb, q can be calculated.

$$q = (5000\,\text{lb/h})(10{,}769\,\text{Btu/lb})$$

$$= 5.38 \times 10^7\,\text{Btu/h}$$

Assuming a typical heat release rate (q_H) of 25,000 Btu/h-ft^3 for liquid injection incinerators, the volume of the incinerator may now be calculated.

$$V = q/q_H = (5.38 \times 10^7\,\text{Btu/h})/(25{,}000\,\text{Btu/h-ft}^3)$$

$$= 2152\,\text{ft}^3 \tag{14.2.6}$$

The diameter (D) of the incinerator can be determined using an L/D ratio of four.

$$V = \pi D^2 L/4 = \pi D^2 (4D)/4 = \pi D^3$$

$$D = (V/\pi)^{1/3} = (2152/\pi)^{1/3} = 8.8\,\text{ft} \simeq 9.0\,\text{ft} \tag{14.2.7}$$

The length of the incinerator is

$$L = 4(9) = 36\,\text{ft}$$

The corrected volume is also calculated.

$$V = \pi D^2 L/4 = \pi(9)^2(36)/4 = 2290\,\text{ft}^3$$

Assuming the ideal gas law to apply, the flow rate of gas (acfm) can be determined. From Table 14.2.1, the total molar flow rate $(\dot{n})$ is 2320 lbmol/h.

$$Q_a = \dot{V} = \dot{n}RT/P = (2320)(10.73)(3244)/(14.7) = 5.5 \times 10^6\,\text{ft}^3/\text{h}$$

$$= 91{,}667\,\text{ft}^3/\text{min (acfm on a wet basis)}$$

Using Charles' Law, Eq. (4.4.8), the standard cubic feet per minute of the gas can be calculated. Standard conditions will be taken as 60°F and 1 atm.

$$Q_s = Q_a\left(\frac{T_s}{T_a}\right) = 91{,}667\left(\frac{520}{3244}\right) = 14{,}694\,\text{scfm}$$

The superficial velocity can be calculated by Eq. (14.2.8)

$$v = \frac{Q_a}{A} = \frac{(91{,}667/60)}{(\pi 9^2/4)}$$

$$= 24\,\text{ft/s} \tag{14.2.8}$$

where v = superficial velocity (ft/s)
    A = incinerator cross sectional area (ft^2)

The residence time (θ) is calculated from Eq. (14.2.9).

$$\theta = \frac{V}{Q_a} = \frac{(2290)(60)}{(91{,}667)} = 1.5\,\text{s} \tag{14.2.9}$$

This is a typical residence time for liquid injection incinerators. The gas residence time calculated previously is an *approximate* value because it does not take into consideration flow variations that occur during chemical reaction and temperature changes in the incinerator. In actuality, a residence time distribution exists within the unit.

Since the actual volumetric flow rate is known, the particulate loading of the flue gas leaving the incinerator may now be determined. From the problem statement, the waste contains 1 wt% of particulates in the form of fly ash. The particulate emission rate can be calculated by Eq. (14.2.10).

$$w_p = 0.01w = 0.01\frac{5000}{60} = 0.833 \, \text{lb/min} \tag{14.2.10}$$

where w_p = particulate emission rate (lb/h)
w = waste feed flow rate (lb/h)

The dry particulate loading (W_i) is calculated by Eq. (14.2.11).

$$W_i = \frac{w_p}{Q_{s,d}} \tag{14.2.11}$$

where $Q_{s,d}$ = the scfm of gas on a dry basis (dscfm).

$Q_{s,d}$ is calculated from Eq. (14.2.12).

$$Q_{s,d} = Q_s (1 - \text{moisture content}) \tag{14.2.12}$$

The moisture content is the mole fraction of water in the flue gas. A value of 0.03843 is obtained from Table 14.2.2. Therefore,

$$Q_{s,d} = 14,694(1 - 0.03843) = 14,129 \, \text{dscfm}$$

The particulate loading (gr/dscf) is calculated from Eq. (14.2.11).

$$W_i = \frac{(0.833)(7000)}{(14,129)} = 0.412 \, \text{gr/dscf}$$

Note: 1 lb = 7000 gr.

Waste Heat Boiler Design

A waste heat boiler is to be designed to cool 69,327 lb/h of flue gas from 2784 to 550°F. The unit will be designed using Ganapathy's method,[2] which was discussed in Chapter 8. Using the information given in the problem statement, the total surface area and the quantity of steam generated in the shell can be determined.

The first step is to calculate the temperature difference ratio (ϕ):

$$\phi = \frac{T_{h1} - T_c}{T_{h2} - T_c} = \frac{2784 - 312}{550 - 312} = 10.4$$

where T_{h1} = temperature of the inlet flue gas
T_{h2} = the exit temperature of the flue gas
T_c = temperature of saturated steam in the shell

On Fig. 8.4.1, connect $(w_h/N) = 100$ with $D_i = 1.4$ and extend the line to intersect *cut line No. 1* at A. Calculate T_{ha}, the average flue gas temperature.

$$T_{ha} = 0.5(T_{h1} + T_{h2}) = 0.5(2784 + 550) = 1667°F$$

Connect T_{ha} with ϕ (10.4) and extend the resulting line to intersect *cut line No. 2* at B. Join A and B to intersect the L scale at a tube length of 16 ft. Since the number of tubes (N) is $69,327/100 = 694$, the total surface area can be calculated by Eq. (14.2.13).

Note: The surface area per linear foot of 1.5-in. o.d., BWG 18 tubes is $0.3925\,\text{ft}^2$.

$$A = (N)(L)[\text{surface per linear ft (ft}^2)] = 694(16)(0.3925)$$

$$= 4358\,\text{ft}^2 \tag{14.2.13}$$

The heat transfer rate may now be calculated via Eq. (8.2.9a).

$$q = w_h c_P(T_{h2} - T_{h1}) \tag{8.2.9a}$$

where w_h = mass flow rate of flue gas (lb/h)
c_P = average heat capacity of the flue gas (Btu/lb-°F)

Assuming an average heat capacity of approximately 0.3 Btu/lb-°F,

$$q = 69,327(0.30)(2784 - 550) = 4.64 \times 10^7\,\text{Btu/h}$$

If feed water enters at 100°F, the quantity of steam generated can be calculated from Eq. (14.2.14).

$$w_s = \frac{q}{h_s - h_{cl}} = \frac{4.64 \times 10^7}{1183 - 68} = 4.16 \times 10^4\,\text{lb/h} \tag{14.2.14}$$

where w_s = steam rate (lb/h)
h_s = enthalpy of saturated steam of 80 psia (Btu/lb)
h_{cl} = enthalpy of feed water at 100°F (Btu/lb)

The enthalpy values are obtained from the steam tables.

Venturi Scrubber Calculations

The venturi is designed to capture the fly ash in the gas stream. The unit also quenches the gases, however. This quench operation is assumed to lower the temperature of the gases from 550°F to approximately the adiabatic saturation temperature of the flue gas (which is in the neighborhood of 180°F).

Using the Nukiyama and Tanasawa relationship, the mean droplet diameter (d_1) can be calculated.

$$d_1 = (16,400/v_t) + 1.45R^{1.5}$$

$$= \frac{16,400}{250} + 1.45(9.5)^{1.5}$$

$$= 108\,\mu m = 3.5 \times 10^{-4}\,\text{ft} \qquad (10.3.3)$$

where v_t = throat velocity (ft/s)
    R = liquid-to-gas ratio (gal/1000 ft³)
    d_1 = droplet diameter (μm)

The average particle size (d_p) is

$$d_p = 1.2\,\mu m = 3.94 \times 10^{-6}\,\text{ft}$$

The temperature of the hot flue gas entering the venturi scrubber is 550°F. From Table 14.2.2, N_2 has a mole fraction of 0.7596. Since the flue gas is essentially N_2, the gas viscosity is close to that of nitrogen at 550°F (1.92×10^{-5} lb/ft-s).

The inertial impaction number (ψ) can be calculated using Eq. (10.3.2). The Cunningham correction factor has been neglected.

$$\psi = \frac{d_p^2 \rho_p v_t}{9\,\mu d_1} = \frac{(3.94 \times 10^{-6})^2(46)(250)}{(9)(1.92 \times 10^{-5})(3.54 \times 10^{-4})}$$

$$= 2.92$$

where d_p = average particle size
    ρ_p = particle density

The Johnstone equation, Eq. (10.3.1), is used to calculate the overall efficiency of the venturi scrubber.

$$E_i = 1 - e^{-KR\,\psi^{0.5}} = 1 - e^{(-0.200)(9.5)(2.92)^{0.5}} = 0.9611 = 96.11\%$$

The outlet particulate concentration does not have to be corrected to 50% EA because the inlet particulate concentration was determined on a dry basis using 50% EA in the combustion calculations. Therefore, the correction to 50% EA has been "built into" the calculation.

$$W_e = W_i - E_i W_i = 0.412 - (0.9611)0.412 = 0.0160\,\text{gr/dscf}$$

where W_e = the outlet particulate loading (gr/dscf)
    W_i = the inlet particulate loading (gr/dscf)
    E_i = overall venturi scrubber efficiency

This value is in compliance with the regulations since the outlet particulate loading is < 0.080 gr/dscf corrected to 50% EA.

The pressure drop across the venturi scrubber is calculated using Eq. (10.3.6)[3]:

$$\Delta P = \frac{v_t^2 \rho_G A_t^{0.133} R^{0.78}}{1270} \tag{10.3.6}$$

In Eq. (10.3.6), A_t represents the cross-sectional area of the throat. The value of A_t is calculated as

$$A_t = \frac{Q_a}{v_t} = \frac{91,667}{(60)(250)} = 6.0 \, \text{ft}^2$$

The gas density is calculated by the ideal gas law.

$$\rho_G = \frac{MP}{RT} = \frac{(28)(14.7)}{(10.73)(180 + 460)} = 0.060 \, \text{lb/ft}^3$$

From Eq. (10.3.6),

$$\Delta P = \frac{250^2(0.060)6^{0.133}9.5^{0.78}}{1270} = 22 \, \text{in. H}_2\text{O}$$

Packed Tower Absorber Design

Table 14.2.2 gives the component flow rates entering the absorber. One of the components, Cl_2, would normally be absorbed with a caustic solution since it is not readily soluble in water. However, for purposes of preliminary process design calculations, the low concentration of Cl_2 (33 ppm) will be neglected in the absorber calculation. Therefore, only the HCl will be absorbed by the water. The regulations state that HCl emissions $> 4 \, \text{lb/h}$ must be controlled at an efficiency of 99%. Since equilibrium data are not available, it is assumed that m (slope of equilibrium curve) approaches zero. This is not an unreasonable assumption for most solvents that preferentially absorb the pollutant. Water and HCl fall into this category. For this condition, the number of overall gas transfer units (N_{OG}) can be calculated from Eq. (10.5.2a).

$$N_{OG} = \ln(w_1/w_2) \tag{10.5.2a}$$

where w_1 = inlet mass flow rate of HCl
w_2 = outlet mass flow rate of HCl

If the absorber operates at a 99% removal efficiency, w_2 may be determined as follows:

$$E = (w_1 - w_2)/w_1$$
$$w_2 = 1598 - (0.99)(1598)$$
$$= 15.98 \, \text{lb/h}$$

The number of gas transfer units is calculated from Eq. (10.5.2a).

$$N_{OG} = \ln(1598/15.98) = 4.605$$

Information on the height of a gas transfer unit is well documented for water systems. For the 3.5-in. diameter ceramic packing used in this absorber, the approximate H_{OG} is 5.5 ft. The height of packing (Z) is obtained from Eq. (10.5.1) with a safety factor included:

$$Z = N_{OG}H_{OG} \text{ (safety factor)} \qquad (10.5.1)$$

Experience has shown that safety factors can range from 1.25 to 1.5. Using a safety factor of 1.3, Z becomes

$$Z = (4.605)(5.5)(1.3)$$
$$= 33 \, \text{ft}$$

Using a typical superficial throughout velocity of 8 ft/s, the cross-sectional area of the tower can be estimated.

$$A = \frac{Q_a}{v} = \frac{1528}{8} = 191 \, \text{ft}^2$$

The diameter is then

$$D = \left(\frac{4A}{\pi}\right)^{0.5} = \left(\frac{4(191)}{\pi}\right)^{0.5} = 15.6 \, \text{ft} \simeq 16 \, \text{ft}$$

We want to determine the actual process gas flow rate at the absorber operating temperature of 150°F. With an HCl removal efficiency of 99%, the number of moles in the gas stream decreases from 2320 to 2276. The gas stream leaving the absorber will be essentially saturated with water vapor. Since the gas entering the absorber is already nearly saturated, this slight increase in flow rate is neglected in the following calculation. Assuming ideality, Q_a is determined.

$$Q_a = (2276)(10.73)[(460 + 150)/14.7)]$$
$$= 1.013 \times 10^6 \, \text{acfh} = 16{,}890 \, \text{acfm}$$

Horsepower Requirements

The total pressure drop (ΔP) between the incinerator and the induced draft fan is calculated as follows:

$$\Delta P = \Delta P_e + \Delta P_f = (3 + 2 + 1 + 22) + 5$$
$$= 33 \, \text{in. H}_2\text{O}$$

where ΔP_e = pressure drop across equipment (in. H_2O)
ΔP_f = pressure drop for valves, fittings, elbows, ductwork, and expansion–contraction losses (in. H_2O)

The brake horsepower (BHP) may be calculated from Eq. (14.2.16).

$$
\begin{aligned}
\text{BHP} &= Q_a \Delta P (1.575 \times 10^{-4})/(\text{fan efficiency}) \\
&= (16{,}890\,\text{acfm})(33\,\text{in. } H_2O)(1.575 \times 10^{-4})/(0.65) \\
&= 135
\end{aligned}
\tag{14.2.16}
$$

The BHP represents an energy consumption rate, which is converted to kilowatts by multiplying by 0.746. The product of power in kilowatts and the number of hours of usage gives the energy consumed in kilowatt hours (kW-h). Utility companies usually charge on the basis of kilowatt-hours.

Economic Considerations

The last chapter showed that the total capital investment includes *direct costs* (e.g., cost for process equipment) and *indirect costs* (e.g., cost for contingencies). The total capital investment is recovered over the operating lifetime of the facility. The amount recovered is determined from a *present worth analysis*. However, the process of analyzing direct and indirect costs and after-tax cash flows is complicated. This was also shown in the last chapter. A present worth analysis will not be performed here. The TCC, AOC and MC will be used to calculate the TAC through the use of the CRF.

The CRF is given by Eq. (13.4.1)

$$
\text{CRF} = \frac{(1 + i)^n i}{(1 + i)^n - 1}
\tag{13.4.1}
$$

where i = interest rate (fraction)
n = equipment life (years)

The annualized capital cost is then

$$
\begin{aligned}
\text{ACC} &= (\text{TCC})(\text{CRF}) = 4{,}000{,}000(1 + 0.08)^{25}(0.08)/[(1 + 0.08)^{25} - 1] \\
&= \$374{,}715/\text{yr}
\end{aligned}
$$

The TAC is given by

$$
\begin{aligned}
\text{TAC} &= \text{ACC} + \text{AOC} + \text{MC} = 374{,}715 + 800{,}000 + 100{,}000 \\
&= \$1.275 \times 10^6/\text{yr}
\end{aligned}
$$

14.3 ILLUSTRATIVE DESIGN PROBLEM II

Several chemical plants located in an industrial area are generating chlorinated wastes in solid and liquid form. The design and construction of a HWI facility has been proposed. An engineering construction firm is in the process of preparing a preliminary design of a multipurpose incinerator. Data on a somewhat similar system indicate that an operating temperature in the 2000 to 2500°F range with 75–125% excess air (to help insure adequate mixing and sufficient oxygen) would insure a 99.99% DRE for this waste. Consultants recommend that a rotary kiln ($L/D = 4$) with an afterburner ($L/D = 3$) be installed in order to treat both solid and liquid wastes. Preheated combustion air at 550°F is to be employed. The minimum excess air requirement calculation for both the kiln and the afterburner should be based on an operating temperature of 2500°F.

The combustion gases are to be quenched in a spray tower from 2500 to 200°F, which is approximately the adiabatic saturation temperature of the flue gas produced. During the quenching operation, it is assumed that no absorption of HCl occurs.

Water, entering a packed absorption tower, is used to absorb HCl from the gas stream. The operating temperature and pressure are 120°F and 1 atm, respectively. The tower is packed with 2-in. ceramic Raschig rings.

The proposed preliminary design does not call for any particulate control equipment at this time. Only trace amounts of particulates and ash are anticipated, and it is assumed that some of these will be captured in the quencher and, to a lesser degree, in the absorber.

The pressure drop across the system including valves, elbows, fittings, and so on, has been estimated to be 11 in. H_2O. According to the fan manufacturer, the overall fan-motor efficiency is 0.65.

In a preliminary cost analysis, the AOC, TCC, and MC were estimated to be $600,000/yr, $3,100,000, and $80,000/yr, respectively.

Waste characteristic data and additional system information are given here. (The notation used is that of the original EPA worksheet. See Table 6.3.2.)

Waste feed flow rate to the kiln, $m_1 + m_2 = 4000\,lb/h$

Fraction of the solid waste to the kiln, $n_1 = 0.7\,lb$ solid/lb feed

Fraction of the liquid waste to the kiln, $n_2 = 0.3\,lb$ liquid/lb feed

Waste feed flow rate to the afterburner, $m_3 = 1500\,lb/h$

Mass fractions of the waste feed stream to the kiln:

	Solids (1)	Liquid (2)	
Carbon (C)	0.70	0.30	lb/lb waste
Hydrogen (H)	0.10	0.11	lb/lb waste
Moisture (H_2O)	0.03	0.42	lb/lb waste
Oxygen (O)	0.12	0.10	lb/lb waste
Chlorine (Cl)	0.05	0.07	lb/lb waste
NHV	14,262	9,067	Btu/lb

Mass fractions of the liquid waste feed to the afterburner:

Carbon (C_a)	0.28	lb/lb waste
Hydrogen (H_a)	0.15	lb/lb waste
Moisture (H_2O_a)	0.40	lb/lb waste
Oxygen (O_a)	0.08	lb/lb waste
Chlorine (Cl_a)	0.09	lb/lb waste
NHV_a	10,602	Btu/lb

The HWI facility is to be designed using preliminary process design principles. Calculations should include the following:

Air requirements.

Combustion calculations.

Incinerator design.

Coolant flow rate for quench operation.

Quench tower design.

Packed tower absorber design.

Horsepower requirements.

Total annualized costs (TAC) based on a 30-yr lifetime at a 10% interest rate.

SOLUTION. As indicated in the introduction, this example will be solved in part by using the EPA worksheets presented in tabular form in Chapter 6.

Note: The net heating values of these wastes are high enough so that no auxiliary fuel is required.

Stoichiometric Air and Combustion Calculations*

Calculate the composition of each component of the waste feed to the kiln. The mass fraction of each component is given here.

$$n_C = (n_1 C_1 + n_2 C_2) = (0.7)(0.70) + (0.3)(0.30) = 0.58 \, \text{lb/lb feed}$$

$$n_H = (0.7)(0.10) + (0.3)(0.103) = 0.103 \, \text{lb/lb feed}$$

$$n_{H_2O} = (0.7)(0.03) + (0.3)(0.42) = 0.147 \, \text{lb/lb feed}$$

$$n_O = (0.7)(0.12) + (0.3)(0.10) = 0.114 \, \text{lb/lb feed}$$

$$n_{Cl} = (0.7)(0.05) + (0.3)(0.07) = 0.056 \, \text{lb/lb feed}$$

Subscripts 1 and 2 refer to the solid and liquid, respectively.

The stoichiometric oxygen requirement for the kiln can now be calculated.

$$(O_2)_{\text{stoich}} = n_C(2.67 \, \text{lb} \, O_2/\text{lb C}) + [n_H - (n_{Cl})/35.5](8 \, \text{lb} \, O_2/\text{lb H}) - n_O$$

$$= (0.58)(2.67) + [0.103 - (0.056)/35.5](8) - 0.114$$

$$= 2.246 \, \text{lb} \, O_2/\text{lb feed}$$

*See Table 6.3.2.

The next step is to determine the mass fraction of each component in the combustion gas per pound of kiln feed.

$$N_{CO_2} = n_C(3.67 \text{ lb } CO_2/\text{lb C}) = 0.58(3.67)$$
$$= 2.13 \text{ lb } CO_2/\text{lb feed}$$

$$N_{H_2O} = n_H(9.0 \text{ lb } H_2O/\text{lb H}) + n_{H_2O} = (0.103)(9.0) + (0.147)$$
$$= 1.07 \text{ lb } H_2O/\text{lb feed}$$

$$N_{N_2} = (O_2)_{stoich}(3.31 \text{ lb } N_2/\text{lb } O_2) = (2.246)(3.31)$$
$$= 7.434 \text{ lb } N_2/\text{lb feed}$$

$$N_{HCl} = n_{Cl}(1.03 \text{ lb } HCl/\text{lb Cl}) = (0.056)(1.03)$$
$$= 0.0577 \text{ lb } HCl/\text{lb feed}$$

The combustion products per pound of kiln feed (CP) are obtained by the following simple relationship:

$$CP = N_{CO_2} + N_{H_2O} + N_{N_2} + N_{HCl} = 10.7 \text{ lb/lb feed}$$

Calculate the composition of each component of the waste feed to the afterburner. The mass fraction of each component is given here.

$$n_{C,a} = (C)_3 = 0.28 \text{ lb/lb feed}$$
$$n_{H,a} = (H)_3 = 0.15 \text{ lb/lb feed}$$
$$n_{H_2O,a} = (H_2O)_3 = 0.40 \text{ lb/lb feed}$$
$$n_{O,a} = (O)_3 = 0.08 \text{ lb/lb feed}$$
$$n_{Cl,a} = (Cl)_3 = 0.09 \text{ lb/lb feed}$$

Subscript 3 refers to the components in the liquid feed to the afterburner.

The stoichiometric oxygen requirement for the afterburner can now be determined.

$$(O_2)_{stoich,a} = n_{C,a}(2.67 \text{ lb } O_2/\text{lb C}) + [n_{H,a} - (n_{Cl,a})/35.5](8 \text{ lb } O_2/\text{lb H}) - n_{O,a}$$
$$= 0.28(2.67) + [(0.15) - (0.09)/35.5](8) - 0.080$$
$$= 1.85 \text{ lb } O_2/\text{lb feed}$$

It is now possible to calculate the mass fraction of each component in the gas per pound of afterburner feed.

$$N_{CO_2,a} = n_{C,a}(3.67 \text{ lb } CO_2/\text{lb C}) = (0.28)(3.67)$$
$$= 1.03 \text{ lb } CO_2/\text{lb feed}$$

$$N_{H_2O,a} = n_{H,a}(9 \text{ lb } H_2O/\text{lb H}) + n_{H_2O,a} = (0.15)(9) + (0.40)$$
$$= 1.75 \text{ lb } H_2O/\text{lb feed}$$

$$N_{N_2, a} = (O_2)_{stoich, a}(3.31 \text{ lb } N_2/\text{lb } O_2) = (1.85)(3.31)$$
$$= 6.12 \text{ lb } N_2/\text{lb feed}$$

$$N_{HCl, a} = n_{Cl, a}(1.03 \text{ lb HCl/lb Cl}) = (0.09)(1.03)$$
$$= 0.093 \text{ lb HCl/lb feed}$$

The combustion products per pound of afterburner feed (CP_a) is determined here.

$$CP_a = N_{CO_2, a} + N_{H_2O, a} + N_{N_2, a} + N_{HCl, a} = 8.99 \text{ lb/lb feed}$$

The ratio of total afterburner feed to total kiln feed (n_{ak}) is next determined.

Solid waste to kiln: $m_1 = (0.7)(4000) = 2800 \text{ lb/h}$
Liquid waste to kiln: $m_2 = (0.3)(4000) = 1200 \text{ lb/h}$
Liquid waste to afterburner: $m_3 = 1500 \text{ lb/h}$

$$n_{ak} = m_3/(m_1 + m_2) = 1500/(2800 + 1200)$$
$$= 0.375 \text{ lb afterburner feed/lb kiln feed}$$

Calculate the total combustion gas flow based on the stoichiometric oxygen requirements.

$$N_{CO_2, t} = (N_{CO_2} + N_{CO_2, a}n_{ak})/(1 + n_{ak})$$
$$= 1.83 \text{ lb/lb feed}$$
$$N_{H_2O, t} = (N_{H_2O} + N_{H_2O, a}n_{ak})/(1 + n_{ak})$$
$$= 1.26 \text{ lb/lb feed}$$
$$N_{N_2, t} = (N_{N_2} + N_{N_2, a}n_{ak})/(1 + n_{ak})$$
$$= 7.07 \text{ lb/lb feed}$$
$$N_{HCl, t} = (N_{HCl} + N_{HCl, a}n_{ak})/(1 + n_{ak})$$
$$= 0.067 \text{ lb/lb feed}$$

The total combustion products are the sum of these:

$$CP_t = 10.23 \text{ lb/lb feed}$$

Rotary Kiln Excess Air Calculation*

Calculate the enthalpy input to the kiln using an air preheat treatment temperature of 550°F.

$$\Delta H_1' = 1.12(T_{air} - 77)(O_2)_{stoich} = 1.12(550 - 77)(2.246)$$
$$= 1190 \text{ Btu/lb feed}$$

*See Table 6.4.6.

The maximum heat generated in the kiln from the combustion of the waste is

$$\Delta H_2 = n_1(\text{NHV}_1) + n_2(\text{NHV}_2)$$
$$= 0.7(14{,}262) + 0.3(9067) = 12{,}704 \text{ Btu/lb feed}$$

Assuming a 5% loss, the heat loss through the walls of the kiln is estimated as

$$q = 0.05(\Delta H_2) = 0.05(12{,}704) = 635 \text{ Btu/lb feed}$$

The enthalpy of the combustion products leaving the kiln is

$$\Delta H_3 = 0.26(N_{\text{CO}_2} + N_{\text{N}_2})(T - 77) + 0.49(N_{\text{H}_2\text{O}})(T - 77)$$
$$= 0.26(2.13 + 7.434)(2500 - 77) + 0.49(1.07)(2500 - 77)$$
$$= 7296 \text{ Btu/lb feed}$$

The enthalpy of excess air leaving the kiln is given by

$$\Delta H_4' = 1.12(T - 77)(\text{O}_2)_{\text{stoich}} = 1.12(2500 - 77)(2.246)$$
$$= 6095 \text{ Btu/lb feed}$$

The percentage of excess air is

$$\text{EA} = 100(\Delta H_1' + \Delta H_2 - Q - \Delta H_3)/(\Delta H_4' - \Delta H_1')$$
$$= 100(1190 + 12{,}704 - 635 - 7296)/(6095 - 1190)$$
$$= 122\%$$

Afterburner Excess Air Calculation*

Calculate the enthalpy input to the kiln and afterburner using an air preheat treatment temperature of 550°F.

$$\Delta H_1' = 1.12(T_{\text{air}} - 77)[(\text{O}_2)_{\text{stoich, k}} + (\text{O}_2)_{\text{stoich, a}}] = 1.12(550 - 77)(2.246 + 1.85)$$
$$= 2170 \text{ Btu/lb feed}$$

The heat generated in the kiln and afterburner by combustion of the total waste feed is given by

$$\Delta H_2 = [\Delta H_k + n_{ak}\Delta H_{2,a}]/(1 + n_{ak})$$

For this equation,

*See Table 6.4.6.

$$\Delta H_k = 0.7(\text{NHV}_1) + 0.3(\text{NHV}_2) = 12,704 \, \text{Btu/lb}$$

$$\Delta H_{2,a} = \text{NHV}_3 = 10,602 \, \text{Btu/lb}$$

$$\Delta H_2 = [12,704 + (0.375)(10,602)]/(1 + 0.375)$$

$$= 12,131 \, \text{Btu/lb feed}$$

Assuming a 5% loss, the heat loss through the walls of the kiln and the after-burner is

$$q = 0.05(\Delta H_2) = 606 \, \text{Btu/lb feed}$$

The enthalpy of the combustion products leaving the afterburner is given by

$$\Delta H_3 = 0.26(N_{\text{CO}_2} + N_{\text{N}_2})(T - 77) + 0.49 N_{\text{H}_2\text{O}}(T - 77)$$

$$= 6582 \, \text{Btu/lb feed}$$

The enthalpy of the excess air leaving the afterburner is

$$\Delta H_4' = 1.12(T - 77)[(O_2)_{\text{stoich}} + (O_2)_{\text{stoich, a}}] = 1.12(2500 - 77)(2.246 + 1.85)$$

$$= 11,116 \, \text{Btu/lb feed}$$

The excess air percentage in the afterburner is

$$\text{EA} = 100(2170 + 12,131 - 606 - 6582)/(11,116 - 2170)$$

$$= 79.5\%$$

This represents the average excess air feed to the kiln–afterburner per unit mass of feed. The calculations to follow will be based on 80% excess air. It is now possible to calculate the additional N_2 and O_2 present in the combustion gases due to excess air feed.

$$(O_2)_{\text{EA}} = \text{EA}[(O_2)_{\text{stoich}} + n_{\text{ak}}(O_2)_{\text{stoich, a}}]/(1 + n_{\text{ak}})$$

$$= 0.8[(0.375)(1.85)]/(1.375)$$

$$= 1.71 \, \text{lb} \, O_2/\text{lb feed}$$

$$(N_2)_{\text{EA}} = 1.71(3.31 \, \text{lb} \, N_2/\text{lb} \, O_2)$$

$$= 5.66 \, \text{lb} \, N_2/\text{lb feed}$$

The total combustion gas flow is

$$\text{CG}_t = \text{CP}_t + (O_2)_{\text{EA}} + (N_2)_{\text{EA}} = 17.6 \, \text{lb/lb feed}$$

The mass fraction of each combustion gas component is given here.

TABLE 14.3.1. Flue Gas Composition[a] **(Illustrative Problem No. 2)**

Component	Mass Flow (lb/h)	Molar Flow (lbmol/h)	Mole Fraction	Partial Pressure (atm)
CO_2	10,067	229	0.067	0.067
H_2O	6,970	387	0.113	0.113
HCl	387	10.6	0.003	0.003
N_2	136,227	2,500	0.731	0.731
O_2	9,390	293	0.086	0.086

[a] This neglects any chlorine (Cl_2) formation.

$$CO_2 \quad N_{CO2}/CG_t = 0.104$$

$$H_2O \quad N_{H_2O}/CG_t = 0.072$$

$$N_2 \quad [N_{N_2,t} + (N_2)_{EA}]/CG_t = 0.723$$

$$O_2 \quad (O_2)_{EA}/CG_t = 0.097$$

$$HCl \quad N_{HCl,t}/CG_t = 0.004$$

The total amount of combustion gas per hour is

$$5500 \text{ lb/h}(17.6 \text{ lb gas/lb feed}) = 96,800 \text{ lb gas/h}$$

Table 14.3.1 gives combustion gas quantities assuming a pressure of 1 atm.

The volumes of the combustion products per pound of feed at standard conditions of 68°F and 1 atm are

$$CO_2 \quad (N_{CO_2,t})/0.114 \text{ lb/scf} = 16.0 \text{ scf/lb feed}$$

$$H_2O \quad (N_{H_2O,t})/0.0467 \text{ lb/scf} = 27.0 \text{ scf/lb feed}$$

$$N_2 \quad [N_{N_2,t} + (N_2)_{EA}]/0.0727 \text{ lb/scf} = 175 \text{ scf/lb feed}$$

$$O_2 \quad (O_2)_{EA}/0.083 \text{ lb/scf} = 20.6 \text{ scf/lb feed}$$

$$HCl \quad (N_{HCl,t})/0.0945 \text{ lb/scf} = 0.71 \text{ scf/lb feed}$$

$$\text{Total volume per lb feed} = 239 \text{ scf/lb feed}$$

$$\text{Total volumetric flow rate } (Q_s) = (5500 \text{ lb/h})(239 \text{ scf/lb feed})$$

$$= 1.31 \times 10^6 \text{ scfh}$$

The air requirement for the kiln feed (AR_k) is

$$AR_k = (O_2)_{stoich}(4.31 \text{ lb air/lb } O_2)(\text{EA ratio } + 1.0) = 2.246(4.31)(2.22)$$

$$= 21.5 \text{ lb air/lb kiln feed}$$

$$AR_{k,t} = 21.5(4000 \text{ lb kiln feed/h})$$

$$= 86,000 \text{ lb air/h}$$

The total air requirement for both the kiln and afterburner (AR) is

$$AR = (O_2)_{stoich}(4.31\,lb\ air/lb\ O_2)(EA\ ratio) = (1.71/0.8)(4.31)(1.8)$$
$$= 16.6\,lb\ air/lb\ afterburner\ feed$$
$$AR_t = 16.6(5500\,lb/h) = 91,300\,lb\ air/h$$

Approximately 5000 lb air/h is added to the afterburner.

Incinerator Design

The rate of heat released by the waste in the kiln is given here.

$$q_k = 4000[0.7(NHV_1) + 0.3(NHV_2)]$$
$$= 5.1 \times 10^7\,Btu/h \tag{14.3.1}$$

Assuming a typical heat release rate (q_H) of 25,000 Btu/h-ft^3, the volume of the kiln is calculated.

$$V_k = q/q_H = 5.1 \times 10^7/25,000 = 2040\,ft^3 \tag{14.3.2}$$

The diameter (D_k) of the kiln can be determined using an L/D ratio of four.

$$V_k = \pi D_k^2 L/4 = \pi D_k^3$$
$$D_k = (V_k/\pi)^{1/3} = (2040/\pi)^{1/3} = 8.6\,ft \approx 9.0\,ft \tag{14.3.3}$$

The length of the kiln is $L_k = 4(9) = 36$ ft. The corrected volume is

$$V_k = \pi 9^2(36)/4 = 2290\,ft^3 \tag{14.3.4}$$

The heat released from the afterburner feed is

$$q_a = (1500\,lb/h)(10,602\,Btu/lb) = 1.59 \times 10^7\,Btu/h$$

The minimum afterburner volume, using a somewhat high afterburner heating rate of 50,000 Btu/h-ft^3, is given by

$$V_a = 1.59 \times 10^7/50,000 = 318\,ft^3$$

Using the L/D ratio of three, the length and diameter of the afterburner are

$$D_a = [4V_a/3\pi]^{1/3} = [4(318)/3\pi]^{1/3} = 5.13\,ft \approx 5.5\,ft \tag{14.3.5}$$

Using 5.5 ft as the *design* diameter,

$$L_a = 3(5.5) = 16.5\,ft$$
$$new\ V_a = 392\,ft^3$$

Retention Time Calculation

The actual combustion gas flow rate is calculated using Charles' law:

$$Q_a = Q_s(T_a/T_s) = 1.31 \times 10^6(2500 + 460)/(68 + 460)$$

$$= 7.34 \times 10^6 \, \text{acfh} \tag{4.4.8a}$$

The residence time (θ) is based on the total volume and the actual cubic feet per hour.

$$\theta = (V_k + V_a)/Q_a = (2290 + 392)/(7.34 \times 10^6)$$

$$= 3.65 \times 10^{-4}\,\text{h} = 1.3\,\text{s} \tag{14.3.6}$$

Quench Spray Tower Design

The quench water requirement is first calculated. For combustion gases being cooled, the quench water requirement may be approximated by[1]

$$w_{H_2O} = 0.5w_{CG} \tag{14.3.7}$$

where w_{H_2O} and w_{CG} are the water and flue gas flow rates. The flue gas rate has been determined to be 96,800 lb/h. Therefore w_{H_2O} can be obtained.

$$w_{H_2O} = (0.5)96,800 = 48,400\,\text{lb/h}$$

A simplifed, reasonably accurate calculational procedure for designing a water quench spray tower is available.[1] The area can be calculated as follows:

$$A = 0.15w_{in}/\rho \tag{14.3.8}$$

where w_{in} = total mass flow of hot gases entering quench (lb/s)
ρ = hot gas density (lb/ft^3)
A = tower cross-sectional area (ft^2)

The diameter is therefore

$$D = (4A/\pi)^{0.5}$$

and the tower height is

$$Z = 2.5D \quad \text{or} \quad V = 2D^3$$

The hot gas density will be taken as that of N_2 because the flue gas is approximately 75% N_2. At 1 atm and 2500°F,

$$\rho = 28(14.7)/[(2500 + 460)(10.73)] = 0.013\,\text{lb/ft}^3$$

The quench tower area, diameter, and height are calculated as follows:

$$A = 0.15 \left(\frac{96,800\,\text{lb/h}}{0.013\,\text{lb/ft}^3} \right) \left(\frac{1\,\text{h}}{3600\,\text{s}} \right) = 310\,\text{ft}^2$$

$$D = [(4)(310)/\pi]^{0.5} = 20\,\text{ft}$$

$$Z = 2.5(20) = 50\,\text{ft}$$

Absorber Design

Since equilibrium data are not available, it is assumed that the slope of the equilibrium curve (m) approaches zero and that Eq. (10.5.2a) is applicable.

$$N_{OG} = \ln(w_1/w_2) \qquad\qquad (10.5.2a)$$

where w_1 = inlet mass flow rate of HCl
   w_2 = outlet mass flow rate of HCl

From Table 14.3.1, the inlet mass flow of HCl is 387 lb/h. First determine the outlet mass flow of HCl assuming that the removal efficiency is 99%.

$$E = (w_1 - w_2)/w_1$$

$$w_2 = 387 - 0.99(387) = 3.87\,\text{lb/h of HCl}$$

The number of gas transfer units via Eq. (10.5.2a) is

$$N_{OG} = \ln(387/3.87) = 4.605$$

From data supplied by the manufacturer, the H_{OG} value for water systems is approximately 3.0 ft for 2-in. ceramic packing. Assuming a safety factor of 1.35, the packed height is calculated from Eq. (10.5.1)

$$Z = (N_{OG})(H_{OG})(\text{safety factor}) = 4.605(3.0)(1.35)$$

$$= 18.6\,\text{ft} \qquad\qquad (10.5.1)$$

Using a typical superficial throughput velocity of 6 ft/s, the cross-sectional area of the tower can be estimated. It is desired to determine the actual process gas flow rate at the operating temperature of the absorber, 120°F. With an HCl removal efficiency of 99%, the number of moles in the gas stream decreases from 3420 to 3037. However, there is also a slight increase in the water content and flow rate for saturated conditions at 120°F. The flue gas water content is approximately 0.0775 lb of H_2O/lb of dry air (see Table 14.3.1) compared to the saturated humidity at 120°F of 0.081 lb of H_2O/lb of dry air. Therefore, assume the molar flow rate of flue gas at the absorber outlet to be 3420 lbmol/h. Assuming ideal gas behavior, the actual cubic feet per minute is then

$$Q_a = \frac{(3420)(10.73)(460 + 120)}{(14.7)(60)}$$

$$= 24,132\,\text{acfm}$$

Horsepower Requirements

The BHP may be calculated from Eq. (14.3.9).

$$\text{BHP} = Q_a(\text{acfm})\Delta P(\text{in. } H_2O)(1.575 \times 10^{-4})/(\text{fan efficiency})$$
$$= 24{,}132(11)(1.57 \times 10^{-4})/0.65 = 64 \, \text{hp} \tag{14.3.9}$$

Economic Considerations

The ACC is given by

$$\text{ACC} = (\text{TCC})(\text{CRF})$$

$$\text{CRF} = \frac{(1+i)^n(i)}{(1+i)^n - 1}$$

$$\text{ACC} = 3{,}100{,}000[(1+0.1)^{30}0.1]/[(1+0.1)^{30} - 1]$$
$$= \$328{,}846/\text{yr} \tag{13.4.1}$$

The TAC is now determined.

$$\text{TAC} = \text{ACC} + \text{AOC} + \text{MC}$$
$$= 328{,}846 + 600{,}000 + 80{,}000 = \$1.01 \times 10^6/\text{yr} \tag{14.3.10}$$

PROBLEMS

1. Resolve Illustrative Design Problem I (Section 14.2) if the chlorobenzene is combusted with 100% excess air.

2. The construction of a hazardous waste incineration facility has been proposed. Using the approaches employed in the EPA worksheets and simple plant design principles, prepare a preliminary design of a multipurpose incineration facility fueled with natural gas that is to be operated at 2050°F with 100% excess air. Since the incinerator must be able to handle solid waste and/or liquid waste, a rotary kiln with an afterburner has been recommended. Calculations should include:

   Air requirements
   Fuel requirements
   Incinerator design
   Retention time
   Coolant flow requirement for combustion gas quench–absorber operation
   Auxiliary equipment
   (a) Absorption tower
   (b) Horsepower requirements
   Annualized costs (25-yr lifetime at 8% interest rate)

Waste characteristic data and system information are given here.

Waste feed flow rate to the kiln, $m_1 + m_2 = 4000\,\text{lb/h}$
Fraction of the solid waste to the kiln, $n_1 = 0.7\,\text{lb solid/lb feed}$
Fraction of the liquid waste to the kiln, $n_2 = 0.3\,\text{lb liquid/lb feed}$
Waste feed flow rate to the afterburner, $m_3 = 1500\,\text{lb/h}$

Mass fractions of the waste feed stream to the kiln:

	Solids (1)	Liquid (2)	
Carbon (C)	0.70	0.30	lb/lb waste
Hydrogen (H)	0.10	0.11	lb/lb waste
Moisture (H_2O)	0.03	0.42	lb/lb waste
Oxygen (O)	0.12	0.10	lb/lb waste
Chlorine (Cl)	0.05	0.07	lb/lb waste

Mass fractions of the liquid waste feed to the afterburner:

Carbon (C_a)	0.28 lb/lb waste
Hydrogen (H_a)	0.15 lb/lb waste
Moisture (H_2O_a)	0.40 lb/lb waste
Oxygen (O_a)	0.08 lb/lb waste
Chlorine (Cl_a)	0.09 lb/lb waste

Kiln solid waste heating value, $NHV_1 = 14{,}300\,\text{Btu/lb}$
Kiln liquid waste heating value, $NHV_2 = 9150\,\text{Btu/lb}$
Afterburner waste heating value, $NHV_3 = 10{,}600\,\text{Btu/lb}$
Fuel-to-waste ratio for the kiln and afterburner, $n_f = 0.0\,\text{lb/lb}$
Excess air for the kiln and afterburner = 100%
Length-to-diameter ratio of the kiln = 4.0
Length-to-diameter ratio of the afterburner = 3.0
Recommended operating temperature during the incineration = 2050°F
Minimum retention time across both incineration units = 0.75 s
Required removal efficiency of chlorine = 99.99%
Estimated pressure drop across the system = 11.0 in. H_2O
Pressure requirement of combustion air for kiln = 6.0 in. H_2O
Pressure requirement of combustion air for afterburner = 5.0 in. H_2O
Absorber description: packed, 2-in.-diameter ceramic Raschig rings; operating temperature = 128°F; enthalpy of combustion gas entering and leaving quench–absorber unit = 625 and 130 Btu/lb, respectively; coolant water temperature entering and leaving quench–absorber unit = 106 and 177°F
Overall fan-motor efficiency = 65%
Estimated AOC = $\$6 \times 10^5/\text{yr}$
Estimated TCC = $\$3.1 \times 10^6$
Estimated MC = $\$8 \times 10^4/\text{yr}$

REFERENCES

1. L. Theodore, personal notes, 1987.
2. V. Ganapathy, "Size or Check Waste Heat Boilers Quickly," *Hydrocarbon Processing*, 169–170, September (1984).
3. H.E. Hesketh, "Atomization and Cloud Behavior in Wet Scrubbers," US–USSR Symposium on Control of Fine Particulate Emissions, January, 15–18, 1974.

APPENDIX

A.1 THE METRIC SYSTEM

The need for a single worldwide coordinated measurement system was recognized over 300 years ago. Gabriel Mouton, Vicar of St. Paul in Lyons, proposed in 1670 a comprehensive decimal measurement system based on the length of 1 minute of arc of a great circle of the earth. In 1671, Jean Picard, a French astronomer, proposed the length of a pendulum beating seconds as the unit of length. (Such a pendulum would have been fairly easily reproducible, thus facilitating the widespread distribution of uniform standards.) Other proposals were made, but over a century elapsed before any action was taken.

In 1790, in the midst of the French Revolution, the National Assembly of France requested the French Academy of Sciences to "deduce an invariable standard for all the measures and all the weights." The commission appointed by the academy created a system that was, at once, simple and scientific. The unit of length was to be a portion of the earth's circumference. Measures for capacity (volume) and mass (weight) were to be derived from the unit of length, thus relating the basic units of the system to each other and to nature. Furthermore, the larger and smaller versions of each unit were to be created by multiplying or dividing the basic units by 10 and its multiples. This feature provided a great convenience to users of the system by eliminating the need for such calculations as dividing by 16 (to convert ounces to pounds) or by 12 (to convert inches to feet). Similar calculations in the metric system could be performed simply by shifting the decimal point. Thus, the metric system is a "base 10" or "decimal" system.

The commission assigned the name "metre" (which which we now spell *meter*) to the unit of length. This name was derived from the Greek word "metron," meaning "a measure." The physical standard representing the meter was to be constructed so that it would equal one ten-millionth of the distance from the North Pole to the . equator along the meridian of the earth running near Dunkirk in France and Barcelona in Spain.

454

The metric unit of mass, called the "gram," was defined as the mass of one cubic centimeter (a cube that is 1/100 of a meter on each side) of water at its temperature of maximum density. The cubic decimeter (a cube 1/10 of a meter on each side) was chosen as the unit of fluid capacity. This measure was given the name "liter."

Although the metric system was not accepted with enthusiasm at first, adoption by other nations occurred steadily after France made its use compulsory in 1840. The standardized character and decimal features of the metric system made it well suited to scientific and engineering work. Consequently, it is not surprising that the rapi spread of the system coincided with an age of rapid technological development. In the United States, by act of Congress in 1866, it was made "lawful throughout the United States of America to employ the weights and measures of the metric system in all contracts, dealings, or court proceedings."

By the late 1860s, even better metric standards were needed to keep pace with scientific advances. In 1875, an international treaty, the "Treaty of the Meter," set up well-defined metric standards for length and mass, and established permanent machinery to recommend and adopt further refinements in the metric system. This treaty, known as the Metric Convention, was signed by 17 countries, including the United States.

As a result of the treaty, metric standards were constructed and distributed to each nation that ratified the Convention. Since 1893, the internationally agreed-to metric standards have served as the fundamental weights and measures standards of the United States.

By 1900 a total of 35 nations, including the major nations of continental Europe and most of South America, had officially accepted the metric system. Today, with the exception of the United States and a few small countries, the entire world is using the metric system predominantly or is committed to such use. In 1971, the Secretary of Commerce, in transmitting to Congress the results of a three-year study authorized by the Metric Study Act of 1968, recommended that the United States change to predominant use of the metric system through a coordinated national program.

The International Bureau of Weights and Measures located at Sèvres, France, serves as a permanent secretariat for the Metric Convention, coordinating the exchange of information about the use and refinement of the metric system. As measurement science develops more precise and easily reproducible ways of defining the measurement units, the General Conference of Weights and Measures—the diplomatic organization made up of adherents to the Convention—meets periodically to ratify improvements in the system and the standards.

A.2 THE SI SYSTEM

In 1960, the General Conference adopted an extensive revision and simplification of the system. The name "*Le Système Internationale d'Unités*" (International System of Units), with the international abbreviation SI, was adopted for this modernized metric system. Further improvements in and additions to SI were made by the General Conference in 1964, 1968, and 1971.

The basic units in the SI system are the kilogram (mass), meter (length), second (time), kelvin (temperature), ampere (electric current), candela (the unit of luminous intensity), and radian (angular measure). All are commonly used by the engineer. The Celsius scale of temperature ($0°C = 273.15$ K) is commonly used with the absolute Kelvin scale. The important derived units are the newton (SI unit of force), the joule (SI unit of energy), the watt (SI unit of power), the pascal (SI unit of pressure), and the hertz (unit of frequency). There are a number of electrical units: coulomb (charge), farad (capacitance), henry (inductance), volt (potential), and weber (magnetic flux). One of the major advantages of the metric system is that larger and smaller units are given in powers of 10. In the SI system a further simplification is introduced by recommending only those units with multipliers of 10^3. Thus, for lengths in engineering, the micrometer (previously micron), millimeter, and kilometer are recommended and the centimeter is generally avoided. A further simplification is that the decimal point may be replaced by a comma (as in France, Germany, and South Africa), while the other numbers, before and after the comma, will be separated by spaces between groups of three; for example, one million dollars will be "$1 000 000,00." More details are provided in the following section.

Seven Base Units

1. *Length–meter (m).* The meter (common international spelling, *metre*) is defined as 1,650,763.73 wavelengths in vacuum of the orange-red line of the spectrum of krypton-86. The SI unit of area is the square meter (m^2). The SI unit of volume is the cubic meter (m^3). The liter (0.001 cubic meter), although not an SI unit, is commonly used to measure fluid volume.

2. *Mass—kilogram (kg).* The standard for the unit of mass, the kilogram, is a cylinder of platinum–iridium alloy kept by the International Bureau of Weights and Measures at Sèvres. A duplicate in the custody of the National Bureau of Standards serves as the mass standard for the United States. This is the only base unit still defined by an artifact. The SI unit of force is the newton (N). One newton is the force which, when applied to a 1-k mass, will give the kilogram mass an acceleration of 1 meter per second per second. $1 N = 1$ kg·m/s^2. The SI unit for pressure is the pascal (Pa). 1 Pa $= 1$ N/m^2. The SI unit for work and energy of any kind is the joule (J). $1 J = 1$ N·m. The SI unit for power of any kind is the watt (W). $1 W = 1$ J/s.

3. *Time—second (s).* The second is defined as the duration of 9,192,631,770 cycles of radiation associated with a specified transition of the cesium-133 atom. It is realized by tuning an oscillator to the resonance frequency of cesium-133 atoms as they pass through a system of magnets and a number of periods or cycles per second is called frequency. The SI unit for frequency is the hertz (Hz). One hertz equals one cycle per second (1 Hz $= 1$ cps). The SI unit for speed is the meter per second (m/s). The SI unit for acceleration is the (meter per second) per second (m/s^2).

4. *Electric current—ampere (A).* The ampere is defined as that current which, if maintained in each of two long parallel wires separated by one meter in free space, would produce a force between the two wires (due to their magnetic fields) of 2×10^{-7} Newton for each meter of length. The SI unit of voltage is the volt (V). $1 V = 1$ W/A. The SI unit of electrical resistance is the ohm (Ω). $1 \Omega = 1$ V/A.

5. *Temperature—kelvin (K).* The kelvin is defined as the fraction 1/273.16 of the thermodynamic temperature of the triple point of water. The temperature 0 K is called "absolute zero." On the commonly used Celsius temperature scale, water freezes at about 0°C and boils at about 100°C. The °C is defined as an interval of 1 K, and the Celsius temperature 0°C is defined as 273.15 K. 1.8 Fahrenheit degrees are equal to 1.0°C or 1.0 K; the Fahrenheit scale uses 32°F as a temperature corresponding to 0°C.

6. *Amount of substance—mole (mol).* The mole is the amount of substance of a system that contains as many elementary entities as there are atoms in 0.012 kilogram of carbon-12. When the mole is used, the elementary entities must be specified and may be atoms, molecules, ions, electrons, other particles, or specified groups of such particles. The SI unit of concentration (of amount of substance) is the mole per cubic meter (mol/m^3).

7. *Luminous intensity—candela (cd).* The candela is defined as the luminous intensity of 1/600 000 of a square meter of a blackbody at the temperature of freezing platinum (2045 K). The SI unit of light flux is the lumen (lm). A source having an intensity of 1 candela in all directions radiates a light flux of 4π lumens.

Two Supplementary Units

1. *Plane angle—radian (rad).* The radian is the plane angle with its vertex at the center of a circle that is subtended by an arc equal in length to the radius.

2. *Solid angle—steradian (sr).* The steradian is the solid angle with its vertex at the center of a sphere that is subtended by an area of the spherical surface equal to that of a square with sides equal in length to the radius.

Multiples and prefixes assigned to SI units are given in Table 4.2.4. Common abbreviations for SI units are given in Table 4.2.3.

Index